The IMA Volumes
in Mathematics
and Its Applications

Volume 2

Series Editors
George R. Sell   Hans Weinberger

Institute for Mathematics and Its Applications
IMA

The **Institute for Mathematics and Its Applications** was established by a grant from the National Science Foundation to the University of Minnesota in 1982. The IMA seeks to encourage the development and study of fresh mathematical concepts and questions of concern to the other sciences by bringing together mathematicians and scientists from diverse fields in an atmosphere that will stimulate discussion and collaboration.

The IMA Volumes are intended to involve the broader scientific community in this process.

Hans Weinberger, Director
George R. Sell, Associate Director

IMA Programs

1982–1983 Statistical and Continuum Approaches to Phase Transition

1983–1984 Mathematical Models for the Economics of Decentralized Resource Allocation

1984–1985 Continuum Physics and Partial Differential Equations

1985–1986 Stochastic Differential Equations and Their Applications

1986–1987 Scientific Computation

1987–1988 Applied Combinatorics

1988–1989 Nonlinear Waves

Springer Lecture Notes from the IMA

The Mathematics and Physics of Disordered Media
 Editors: Barry Hughes and Barry Ninham
 (Lecture Notes in Mathematics, Volume 1035, 1983)

Orienting Polymers
 Editor: J. L. Ericksen
 (Lecture Notes in Mathematics, Volume 1063, 1984)

New Perspectives in Thermodynamics
 Editor: James Serrin
 (in press)

Models of Economic Dynamics
 Editor: Hugo Sonnenschein
 (Lecture Notes in Economics, Volume 264, 1986)

Oscillation Theory, Computation, and Methods of Compensated Compactness

Edited by
Constantine Dafermos, J.L. Ericksen,
David Kinderlehrer, and Marshall Slemrod

With 67 Illustrations

Springer-Verlag
New York Berlin Heidelberg
London Paris Tokyo

Constantine Dafermos
Division of Applied Mathematics, Brown University, Providence, RI 02912, U.S.A.

J.L. Ericksen
School of Mathematics and Department of Aerospace Engineering and Mechanics, University of Minnesota, Minneapolis, MN 55455, U.S.A.

David Kinderlehrer
School of Mathematics, University of Minnesota, Minneapolis, MN 55455, U.S.A.

Marshall Slemrod
Department of Mathematical Sciences, Rensselaer Polytechnic Institute, Troy, NY 12180, U.S.A.

AMS Classification: 76-XX, 35A35, 35A40, 35B10, 35JXX, 35LXX, 25NXX, 35Q20

Library of Congress Cataloging in Publication Data
Oscillation theory, computation, and methods of
  compensated compactness.

(The IMA volumes in mathematics and its
applications ; v. 2)
  "Represents the proceedings of a workshop which
was an integral part of the 1984-85 IMA program on
Continuum physics and partial differential equations"—P.
  Includes bibliographies.
  1. Conservation laws (Physics)—Mathematics—
Congresses. 2. Oscillations—Mathematics—Congresses.
3. Differential equations, Nonlinear—Numerical
solutions—Congresses. 4. Differential equations,
Hyperbolic—Numerical solutions—Congresses.
I. Dafermos, C.M. (Constantine M.) II. Series.
QC73.8.C6083 1986   530   86-17781

© 1986 by Springer-Verlag New York Inc.
All rights reserved. No part of this book may be translated or reproduced in any form without written permission from Springer-Verlag, 175 Fifth Avenue, New York, New York 10010, U.S.A.
Permission to photocopy for internal or personal use, or the internal or personal use of specific clients, is granted by Springer-Verlag New York Inc. for libraries registered with the Copyright Clearance Center (CCC), provided that the base fee of $0.00 per copy, plus $0.20 per page is paid directly to CCC, 21 Congress Street, Salem, MA 01970, U.S.A. Special requests should be addressed directly to Springer-Verlag New York, 175 Fifth Avenue, New York, New York 10010, U.S.A.
96401-0/1986 $0.00 + .20

Printed and bound by R.R. Donnelley & Sons, Harrisonburg, Virginia.
Printed in the United States of America.

9 8 7 6 5 4 3 2 1

ISBN 0-387-96401-0 Springer-Verlag New York Berlin Heidelberg
ISBN 3-540-96401-0 Springer-Verlag Berlin Heidelberg New York

The IMA Volumes in Mathematics and Its Applications

Current Volumes:

Volume 1: Homogenization and Effective Moduli of Materials and Media
  Editors: J.L. Ericksen, David Kinderleherer, Robert Kohn, and J.-L. Lions
Volume 2: Oscillation Theory, Computation, and Methods of Compensated Compactness
  Editors: Constantine Dafermos, J.L. Ericksen, David Kinderlehrer, and Marshall Slemrod

Forthcoming Volumes:

1984–1985: Continuum Physics and Partial Differential Equations
  Theory and Applications of Liquid Crystals
  Amorphous Polymers and Non-Newtonian Fluids
  Metastability and Incompletely Posed Problems
  Dynamical Problems in Continuum Physics

1985–1986: Stochastic Differential Equations and Their Applications
  Random Media
  Percolation Theory and Ergodic Theory of Infinite Particle Systems
  Hydrodynamic Behavior and Interacting Particle Systems and Applications
  Stochastic Differential Systems, Stochastic Control Theory and Applications

CONTENTS

Foreword viii
Preface ix

Convection of Microstructures by Incompressible and Slightly Compressible
Flows 1
    T. Chacon and O. Pironneau

Oscillations in Solutions to Nonlinear Differential Equations 23
    Ronald J. DiPerna

Geometry and Modulation Theory for the Periodic Nonlinear Schrodinger Equation 35
    M. Gregory Forest and Jong-Eao Lee

On High-Order Accurate Interpolation for Non-Oscillatory Shock Capturing . . . 71
Schemes
    Ami Harten

On the Weak Convergence of Dispersive Difference Schemes 107
    Peter D. Lax

Nonlinear Geometric Optics for Hyperbolic Systems of Conservation Laws 115
    Andrew Majda

On the Construction of a Modulating Multiphase Wavetrain for a
Perturbed KdV Equation 167
    David W. McLaughlin

Evidence of Nonuniqueness and Oscillatory Solutions in Computational Fluid
Mechanics 197
    J.W. Nunziato, D.K. Gartling, and M.E. Kipp

Very High Order Accurate TVD Schemes 229
    Stanley Osher and Sukumar Chakravarthy

Convergence of Approximate Solutions to Some Systems of Conservative Laws:
A Conjecture on the Product of the Riemann Invariants 275
    Michel Rascle

Applications of the Theory of Compensated Compactness 289
    M.E. Schonbek

A General Study of a Commutation Relation given by L. Tartar 295
    Denis Serre

Interrelationships among Mechanics, Numerical Analysis, Compensated Compactness,
and Oscillation Theory 309
    M. Slemrod

The Solution of Completely Integrable Systems in the Continuum Limit of the
Spectral Data 337
    Stephanos Venakides

Stability of Finite-Difference Approximations for Hyperbolic
Initial-Boundary-Value Problems 357
    Robert F. Warming and Richard M. Beam

Construction of a Class of Symmetric TVD Schemes 381
    Helen Yee

Information About Other Volumes in this Program 397

FOREWORD

This IMA Volume in Mathematics and its Applications

Oscillation Theory, Computation, and Methods of Compensated Compactness

represents the proceedings of a workshop which was an integral part of the 1984-85 IMA program on CONTINUUM PHYSICS AND PARTIAL DIFFERENTIAL EQUATIONS. We are grateful to the Scientific Committee:

> J.L. Ericksen
> D. Kinderlehrer
> H. Brezis
> C. Dafermos

for their dedication and hard work in developing an imaginative, stimulating, and productive year-long program.

> George R. Sell
> Hans Weinberger

PREFACE

Historically, one of the most important problems in continuum mechanics has been the treatment of nonlinear hyperbolic systems of conservation laws. The importance of these systems lies in the fact that the underlying equations of mass, momentum, and energy are described by conservation laws. Their nonlinearity and hyperbolicity are consequences of some common constitutive relations, for example, in an ideal gas.

The I.M.A. Workshop on "Oscillation theory, computation, and methods of compensated compactness" brought together scientists from both the analytical and numerical sides of conservation law research. The goal was to examine recent trends in the investigation of systems of conservation laws and in particular to focus on the roles of dispersive and diffusive limits for singularily perturbed conservation laws. Special attention was devoted to the new ideas of compensated compactness and oscillation theory. Here the hope is that these new methods may lead to new existence theorems for systems of conservation laws and perhaps also provide a greater understanding of convergence of finite difference schemes. As this last issue is intimately related to the study of the higher order difference schemes a portion of the program was devoted to this topic as well.

The conference committee would like to express its appreciation especially to Mitchell Luskin and Luc Tartar for their help in assuring the success of this workshop.

We also take this opportunity to thank the staff of the I.M.A., Professors Weinberger and Sell, Mrs. Susan Anderson, Mrs. Pat Kurth, and Mr. Robert Copeland, for their assistance in arranging the workshop. Special thanks are due to Mrs. Debbie Bradley and Mrs. Kaye Smith for their preparation of the manuscripts. We gratefully acknowledge the support of the National Science Foundation and the Army Research Office.

                                              C. Dafermos
                                              J.L. Ericksen
                                              D. Kinderlehrer
                                              M. Slemrod
                                                        conference committee

CONVECTION OF MICROSTRUCTURES BY INCOMPRESSIBLE AND SLIGHTLY COMPRESSIBLE FLOWS

T. Chacon
INRIA 78153 Le Chesnay France

O. Pironneau
INRIA and University of Paris-13

Abstract

In this paper we wish to extend the work of McLaughlin-Papanicolaou-Pironneau [11] to compressible flows. Thus we shall first summarize the results for imcompressible fluids then present the current state of numerical simulation of these problems and finally make some preliminary statements on the extension to compressible flows and the possible applications to turbulence and acoustics.

1. Introduction

Although turbulent flows are solutions of mathematically well defined partial differential equations it is a well known fact that these solutions are too complicated to be represented in all their details with a number of parameters consistent with computer memories. The aim of turbulence modeling is to establish equations for mean quantities like the mean flow (u), the mean kinetic energy (κ) or the mean rate of energy dissipated by viscous effect (ϵ).

For incompressible fluids there are several such models but most of them are derived by empirical methods, dimensional analysis and physical considerations. Among these the best known are the subgrid model of Smagorinsky [20] (see also Reynolds [17], Schumann [19]), the length scale, kinetic energy and/or κ-ϵ models initiated by Launder [9] (see also Saffman [18]) and the spectral models of Kraichnan and Orszag [13] (see also Mathieu [10]).

From a theoretical point of view the domain validity of these models is not well known in general. On the experimental side the situation is not so bad for homogeneous turbulence (the $\kappa^{-5/3}$ law of Kolmogorov is reasonably correct) and for boundary layer turbulence (see the log law) but for detached flows with large eddies where the turbulence is not established or stable none of these models are able to predict all cases.

In Perrier-Pironneau [16] and Papanicolaou-Pironneau [14] an attempt was made to use homogenization techniques (see Bensoussan-Lions-Papanicolaou [2]) to derive equations of the mean flow u by splitting the full flow into a mean part and an oscillatory part (function of a fast variable) with statistics given by the universal laws of homogeneous turbulence. However the general case case could not be handled: some terms in the Navier-Stokes equations had to be neglected. In McLaughlin-Papanicolaou-Pironneau [11] the same problem was studied from another angle: only the initial conditions of the PDE's where assumed to be made of a mean part and a fast oscillatory part. Then it was shown that a suitable asymptotic expansion could lead to a $k-\varepsilon$ type of model for the mean flow, the kinetic energy and the helicity. However an additional term was found which produced oscillations instead of viscosity. Since then Frisch et al [6] also found an "elastic" term in the Kuramoto-Sivashinsky equation by homogenization techniques and Chacon [3] reproduced numerically the oscillations predicted by the theory. In the first part of the paper we give a summary of these results.

Naturally for compressible fluids even less is known theoretically (see Monin-Yaglom [12]): turbulence modeling seems to be very difficult. The main advantage of homogenization techniques for turbulence modeling is that one does not need any closure hypothesis; all the hypotheses are made before hand in writing the Ansatz. But since the equations for compressible fluids allow acoustic waves it is much more difficult to propose a reasonable Ansatz. So in the last part of the paper we study some special cases which illustrate the difficulty.

2. The basic model for incompressible flows

Consider the following initial value problem

(1) $\qquad \mathbf{u}^{\varepsilon}{}_{,t} + \mathbf{u}^{\varepsilon}\nabla\mathbf{u}^{\varepsilon} + \nabla p^{\varepsilon} = 0 \qquad \nabla \cdot \mathbf{u}^{\varepsilon} = 0 \quad \text{in} \quad \mathbb{R}^3 \times]0,T[$

(2) $\qquad \mathbf{u}^{\varepsilon}(x,0) = \mathbf{u}^{0}(x) + \mathbf{w}^{0}(x, x/\varepsilon)$

Where $w(x,y)$ is periodic in y on $Y =]0, 2\pi[^3$ (or random) and has zero y-mean

(3) $$\langle w \rangle = \frac{1}{(2\pi)^3} \int_Y w(x,y) dy = 0$$

It was shown in [11] that if one assumes

(4) $$u^\varepsilon(x,t) = u(x,t) + (w(x,t,y,\tau) + \varepsilon u^1(x,t,y,\tau) + \varepsilon^2 u^2 \ldots \Big|_{\substack{y = a(x,t)/\varepsilon \\ \tau = t/\varepsilon}}$$

(5) $$p^\varepsilon(x,t) = p(x,t) + (\pi(x,t,y,\tau) + \varepsilon p^1(x,t,y,\tau) + \varepsilon^2 p^2 \ldots \Big|_{\substack{y = a(x,t)/\varepsilon \\ \tau = t/\varepsilon}}$$

then a meaningful set of equations for all the functions in (4) (5) could be found from (1) and (2):

(6) $$a_{,t} + u \nabla a = 0, \quad a(0) = x$$

(7) $$u_{,t} + u \nabla u + \nabla p + \nabla \cdot q\, R(r/q, \nabla a) = 0$$
$$\nabla \cdot u = 0, \quad u(0) = u^0.$$

(8) $$q_t + u \cdot \nabla q + q \nabla u : R(r/q, \nabla a) + \nabla \cdot [q^{3/2} b(r/q, \nabla a)] = 0$$

(9) $$r_t + u \cdot \nabla r + q \nabla u : S(r/q, \nabla a) + \nabla \cdot [q^{3/2} c(r/q, \nabla a)] = 0$$

Where a is a Lagrangian variable of the problem (see(6)), q and r are the kinetic energy and helicity of w (the leading oscillating part of u^ε):

(10) $$q = \frac{1}{2} \langle w^2 \rangle$$

(11) $$r = \langle w \cdot \nabla_z \times w \rangle \quad (z = \nabla a^{-T} y)$$

and R_{ij}, b_i, S_{ij}, c_i are functions of r/q and ∇a which could, in principle, be tabulated from the τ-y periodic solutions of the equations of $w^\bullet = \nabla a^t w$ and $u^{1\bullet} = \nabla a^t u^1$, on Y:

$$w'_\tau + w^\bullet \nabla_y w^\bullet + (\nabla a^T \nabla a) \nabla_y \pi = 0, \qquad \nabla_y w^\bullet = 0;$$

(12)
$$<w^\bullet> = 0,$$
$$<w^\bullet (\nabla a^T \nabla a)^{-1} w^\bullet> = 2q$$
$$<\nabla a^{-T} w^\bullet \nabla_z x (\nabla a^{-T} w^\bullet)> = r$$

(13)
$$u^{1\bullet}_\tau + u^{1\bullet} \nabla_y w + w \cdot \nabla_y u^{1\bullet} + (\nabla a^t \nabla a) \nabla_y p^1 = f,$$
$$\nabla_{y\bullet} u^{1\bullet} = 0$$

$$<u^1> = 0, \quad <u^1 \bullet w> = 0, \quad <u^1 \nabla_z x w> = 0$$

Where f is a complicated function of $u, \nabla u, w, q, r, p, \pi$. For instance

(14)
$$R_{ij} = \lim_{\tau \to \infty} \tau^{-1} \int_0^\tau <w_i w_j> d\tau$$

It was shown in [11] that a similar result could be obtained with an asymptotic expansion of of the form

$$u^\varepsilon(x,t) = u(x,t) + (\varepsilon^{1/3} w(x,t,y,\tau) + \varepsilon^{2/3} u^1(x,t,y,\tau) + \varepsilon u^2 \ldots |$$
$$y = a(x,t)/\varepsilon$$
$$\tau = t/(\varepsilon^{2/3})$$

the results are:

(6') $a_t + u \nabla a = 0, \quad a(0) = x$

(7') $u_t + u \nabla u + \nabla p + \varepsilon^{2/3} \nabla q R(r/q, \nabla a) = 0$
$\nabla \cdot u = 0, \quad u(0) = u^0$

(8') $q_t + u \cdot \nabla q + q \nabla u : R(r/q, \nabla a) = 0$

(9') $r_t + u \cdot \nabla r + q \nabla u : S(r/q, \nabla a) = 0$

For this simpler form one sees that if $r(0,x) = 0$ then r is zero for all time because $S(0, \nabla a) = 0$.

For one dimensional mean flows in the direction x_1 with small $q(0,x)$ the system (6')-(9') reduces to a wave equation

(15) $$V_{tt} - \beta q^0 V_{zz} = 0 \qquad (z = x_3)$$

where

(16) $$V(z,t) = \int_0^t u_1(x,\sigma)d\sigma$$

(17) $$\beta = -\partial R_{13}(0)/\partial a_{1,3}$$

In [4], it was shown that for two dimensional mean flows with small q^0 and zero r^0, frame invariance of (6)-(9) implies:

(18) $$R(r/q, \nabla a) = \beta(|\nabla a|^2) \nabla a \nabla a^T$$

The function β was tabulated by Begue [1] and Chacon [3] by solving (12) for stationary flows (in τ).

3. Numerical simulations

There are, unfortunately, still a number of loose ends in the model because existence, uniqueness, and differentiability of the solution of (12)-(13) with respect to the parameters are not known. So we turned to numerical simulation to see whether the new set of equations (6)-(9) is relevant to turbulence modeling.

As shown by (15) the leading order term in the Reynolds stress tensor is elastic. It is zero if w is isotropic; this may be why it has been missed so far in the observation. But even if it is small it has a destabilizing transient effect in (7) and thus it may be interesting to keep it.

We took Poiseuille flow and perturbed it by periodic small structure $w^0(x,y)$ as in (2). A similar problem was studied by Patera [15] and the code was made available to us. It solves the Navier-Stokes equation between two flat parallel planes of equations $z = -1$ and $z = +1$ (see figure 1). The flow is assumed periodic in x and y and a no-slip boundary condition is applied on the planes. The code uses Fourier modes in x and y and Chebyshev modes in z to represent the solution u^ϵ; this is particularly convenient to impose (2): we took

(19) $\quad u_1^0(x) = (1-z^2)$ (Poiseuille flow)

(20) $\quad w^0(x) = q^0(z) \sum_{j \in J} u^{jN} \exp 2\pi i \, (j_1 N x_1 + j_2 N x_2)$

$\qquad J = \{+1, 0, -1\}^2 - \{0,0\}$

with $N = 10$ while the total number of modes in each direction is 32 (See figure 2).

Then an approximation of the (x_1, x_2)-mean of $|u|^2$ was plotted as a function of time (see figure 3). The oscillatory behavior is clearly seen and the period is strikingly proportional to $\sqrt{q^0}$ as predicted by (15).

q^0	measured period	predicted period
.2	.105	.082
.1	.148	.116
.04	.220	.184
.01	.368	----

Finally the (x_1, x_2)-mean of u^ε is compared with the time derivative of the solution of

(21) $\quad V_{tt} + [q^0 \exp(\int_0^{-V}, z \, R_{13}(s) \, ds) \, R_{13}(-V, z)]_{,z} = 0$

(of which (15) is a linearization)

See figure 4.

The comparison is not so good but the general phenomenon is qualitatively described properly.

A simulation of 2-D mean flows by (6)-(8) with (18) was done by Begue [1] for the wake of a cylinder but comparisons with direct simulations cannot be made and the results as compared with experiments are so far inconclusive, but they seem to point to the right direction. Thus even though we are still very far away from a clean mathematical derivation of (6)-(9), we have some confidence in the Ansatz (4), (5) and we proceed to see whether it can be applied to compressible fluids.

4. Compressible flows

Consider the case of adiabatic flows governed by the following system:

(22) $$u_t + u \, \nabla u + \frac{1}{\rho} \nabla p = 0$$

(23) $$\rho_{,t} + \nabla \cdot (\rho u) = 0$$

(24) $$p = K \rho^\gamma$$

We shall assume that the solution exists and is smooth (no shocks). Initial data will be of the form

$$u^\varepsilon(x,0) = u^0(x) + \varepsilon^\alpha w^0(x,x/\varepsilon)$$

$$\rho^\varepsilon(x,0) = \rho^0(x) + \varepsilon^\beta \sigma_0(x,x/\varepsilon)$$

The case $\alpha = \beta = 1$ has been investigated by Hunter-Keller [7] and DiPerna-Majda [5]. Here we shall show first that the case $\alpha = 1/3$, $\beta = 2/3$ can be investigated by the same method as above provided that the initial data are compatible; then we shall study the case $\alpha = \beta = 0$ for a special w^0 without vorticity.

4.1 No fast acoustic waves

It is possible to use the techniques of homogenization if the initial conditions are of the form

(25) $$u^\varepsilon(x,0) = u^0(x) + \varepsilon^{1/3} w^0(x,x/\varepsilon)$$

(26) $$\rho^\varepsilon(x,0) = \rho^0(x) + \varepsilon^{2/3} \sigma_0(x,x/\varepsilon)$$

with the following Ansatz

(27) $$u^\epsilon(x,t) = u(x,t) + \epsilon^{1/3} w(x,t,y,\tau) + \epsilon^{2/3} u^1 + \ldots$$

(28) $$\rho^\epsilon(x,t) = \rho(x,t) + \epsilon^{2/3} \sigma(x,t,y,\tau) + \epsilon \rho^1 + \ldots$$

at $y = a(x,t)/\epsilon$, $\tau = t/\epsilon^{2/3}$. As before a is the Lagrangian coordinate given by (6).

Inserting (27) and (28) in (22)--(24) yields a cascade of equations. The first one is an equation for w and $\pi = K\gamma\sigma\rho^{\gamma-1}$ which is the same for incompressible flows:

(29) $$\dot{w}^*_{,\tau} + w^* \nabla_y w^* + (\nabla a^T \nabla a) \nabla_y \pi = 0$$
$$\nabla_y \cdot w = 0$$

(30) $$\dot{w}^* = \nabla a^T w$$

The second equation is an equation for u^1 which is also identical to (13) except for f.

When we apply the compatibility conditions for existence of u^1 we find that f must satisfy 3 equations and these give:

(31) $$u_t + u \nabla u + \frac{1}{\rho} \nabla p + \frac{1}{\rho} \epsilon^{2/3} \nabla \cdot (\rho q R) = 0$$

(32) $$\rho_{,t} + \nabla \cdot (\rho u) = 0$$

(33) $$q_{,t} + u \nabla q + q R : \nabla u = 0$$

(34) $$r_{,t} + u \nabla r + q S : \nabla u = 0$$

(35) $$p = K \rho^\gamma$$

(36) $$R_{ij} = \langle w_i w_j \rangle \quad S_{ij} = \langle w_i (\nabla_z \times w)_j \rangle$$

(See Chacon [3] for more details)

However this analysis requires some compatibility condition between w^0 and σ^0 in (25)-(26). Indeed in principle one could prescribe w^0 and σ^0 independently because (22)-(24) is well posed for u and ρ given at $t = 0$. But from (29) we see that we can specify w^0 but not σ^0. From (24) it is found that the first fluctuating term in the pressure is

$$\Pi = K \gamma \rho^{\gamma-1} \sigma \tag{37}$$

So (29) fixes σ^0 as a function of w^0.

4.2 Fast acoustic waves.

To treat the general initial value problem let us simplify system (22)-(24).

Any vector field u can be written as

$$u = \nabla\phi + v \quad \text{with} \quad \nabla\cdot v = 0 \tag{38}$$

and since

$$u \cdot \nabla u = -u \times \nabla \times u + \nabla\left(\frac{u^2}{2}\right) \tag{39}$$

we find that (22) is also

$$v_{,t} - (v + \nabla\phi) \times \nabla \times u + (K'\nabla\rho^{\gamma-1} + \tfrac{1}{2}|v + \nabla\phi|^2 + \phi_{,t}) = 0 . \tag{40}$$

Where $K' = K_\gamma/(\gamma-1)$. Let

$$q = K'\rho^{\gamma-1} + \tfrac{1}{2}|v + \nabla\phi|^2 + \phi_t \tag{41}$$

Then (22)-(24) can be viewed as a nonlinear hyperbolic system in ϕ and ρ loosely coupled with a Euler type of equation for v:

$$v_{,t} - (v + \nabla\phi) \times \nabla \times v + \nabla q = 0, \quad \nabla\cdot v = 0 \tag{42}$$

(43)
$$\phi_{,t} + \tfrac{1}{2}|v + \nabla\phi|^2 + K' \rho\gamma^{-1} = q$$
$$\rho_{,t} + \nabla\cdot[\rho(v + \nabla\phi)] = 0$$

Let us study this system with the following initial conditions:

(44) $$u^\varepsilon(x,0) = v^0(x) + \varepsilon\,\nabla\phi^0(x, x/\varepsilon)$$

(45) $$\rho^\varepsilon(x,0) = \rho^0(x) + \sigma^0(x, x/\varepsilon)$$

where ϕ^0 and σ^0 are periodic in $y = x/\varepsilon$ and have zero means. Naturally we ask that v^0 be divergence free.

While (25)-(26) corresponds to the study of the transport of microstructures by a velocity field (the mean flow) as for incompressible fluids, (44),(45) is an initial value problem where the acoustic waves will develop in the small scale. Problem (42)-(45) contains as a special case the initial value problem with oscillatory data for the wave equation (see [2] for example), because once linearized, (43) is a wave equation.

Let us assume that

(46) $$\phi^\varepsilon(x,t) = \phi^0(x,t) + \varepsilon\phi^1(x,t,y,\tau) + \phi^2(\ )\ldots|_{y = a(x,t)/\varepsilon,\ t/\varepsilon}$$

where a is given by (6) with $u = v^0 + \nabla\phi^0$ and let us first study the effect of ϕ^ε on v from (42).

Thus, let us take

(47) $$v^\varepsilon(x,t) = v^0(x,t) + \varepsilon v^1(x,t,y,\tau) + \varepsilon^2 v^2 \ldots$$

One can check that

(48) $$v^\varepsilon_{,t} = v^0_{,t} + v^1_{,\tau} + a_{,t}\cdot \boldsymbol{\nabla}v^1 + \varepsilon(v^1_{,t} + v^2_{,\tau} + a_{,t}\cdot \boldsymbol{\nabla}v^2) + \ldots$$

(49) $$\nabla_x v^\varepsilon = \nabla_x v^0 + \nabla a^T\,\boldsymbol{\nabla}v^1 + \varepsilon(\nabla_x v^1 + \nabla a^T\,\boldsymbol{\nabla}v^2) + \ldots$$

When ∇ has no subscript, it is with respect to $y = a(x,t)/\varepsilon$. It is convenient to use the notation $z = \nabla a^{-t} y$ because

(50) $$\nabla a \nabla_y = \nabla_z$$

When these expressions are used in (42) it becomes a polynomial in ε and the ε^0 - term is:

(51) $$v^0{}_{,t} + v^1{}_{,\tau} + a_{,t} \nabla v^1 - (v^0 + \nabla_x \phi^0 + \nabla_z \phi^1) \times (\nabla_z \times v^1 + \nabla_x \times v^0) +$$
$$\nabla_x q^0 + \nabla_z q^1 = 0$$

(52) $$\nabla_x \cdot v^0 + \nabla_z \cdot v^1 = 0$$

Taking the mean in y of (52) yields:

(53) $$\nabla_x \cdot v^0 = 0$$

(54) $$\nabla_z \cdot v^1 = 0$$

Notice also that

$$(v^0 + \nabla_x \phi^0) \times \nabla_z \times v^1 = -(v^0 + \nabla_x \phi^0) \nabla_z v^1 + \nabla_z (v^1 \cdot (v^0 + \nabla \phi^0))$$

so (51) can be rewritten as

(55) $$v^1{}_{,\tau} - \nabla_z \phi^1 \times \nabla_z \times v^1 + \nabla_z q^{1*} = f(x,t) + \nabla_z \phi^1 \times (\nabla_x \times v^0)$$

where f is independent of z and τ.

Let us take the curl of (55):

(56) $$\omega_{,\tau} - \nabla_z \times (\nabla_z \phi^1 \times \omega) = \nabla_z \times [\nabla_z \phi^1 \times \nabla_x \times v^0]$$

Equation (56) with zero initial conditions (imposed by (44)) defines

(57) $$\omega^1 = \nabla_z \times v^1$$

as a linear function of $\nabla \times v^0$. There may be existence and uniqueness problems but the important fact is that $\nabla \times v^0$ acts like a passive parameter on the right hand-side and no other mean flow quantities appear. Now (57) and (54) define v^1 and

q^1; so we have:

(58) $$v^1(x,t,y,\tau) = A(x,t,\nabla a, \phi^1,y,\tau)\nabla_x x v^0$$

(59) $$q^1(x,t,y,\tau) = b(x,t,\nabla a, \phi^1,y,\tau)\nabla_x x v^0$$

Now let us take the y-mean of (51):

(60) $$v^0{,}_t - (v^0 + \nabla_x\phi^0) x \nabla_x x v^0 - \langle \nabla_z\phi^1 x \nabla_z x [A\nabla \ x v^0]\rangle + \nabla q^0 = 0$$

A similar treatment of equation (43) gives

(61) $$\phi^1{,}_\tau + \tfrac{1}{2}|\nabla_z\phi^1|^2 + K'(\sigma + \rho^0)^{\gamma-1} = q^0 - \phi^0{,}_t - \tfrac{1}{2}|v^0 + \nabla\phi^0|^2$$

(62) $$\sigma{,}_\tau + \nabla_z\cdot((\sigma + \rho^0)\nabla_z\phi^1) = 0\ .$$

Here the situation is more complex because ρ^0 appears on the left in (61) and (62). So ϕ^1 and σ are functions of ρ^0, ∇a and of their initial conditions.

The equations for ϕ^0 and ρ^0 are obtained by averaging (61) and the ε^0 term in (43):

(63) $$\phi^0{,}_t + \tfrac{1}{2}|v^0 + \nabla_x\phi^0|^2 + \tfrac{1}{2}\langle|\nabla_z\phi^1|^2\rangle + K\langle(\rho^0 + \sigma^0)^{\gamma-1}\rangle = q^0$$

(64) $$\rho^0{,}_t + \nabla_x(\rho^0(v^0 + \nabla_x\phi^0)) + \langle\nabla_x\cdot(\sigma^0\nabla_z\phi^1)\rangle = 0$$

4.3 Slightly compressible fluids

If ρ^0 is almost constant then one may work with a simplified Euler system (Landau-Lifchitz [8]):

(65) $$u{,}_t + u\nabla u + C\nabla\rho = 0,$$

(66) $$\rho{,}_t + u\nabla\rho + r\nabla\cdot u = 0.$$

In that case system (56)-(60) is unchanged but (61)-(62) become:

(67) $$\phi^1{,}_\tau + \tfrac{1}{2}|\nabla_z\phi^1|^2 + C\sigma = q^0 - \phi^0{,}_t - \tfrac{1}{2}|v^0 + \nabla\phi^0|^2 - C\rho^0$$

(68) $$\sigma,_\tau + \nabla_z \sigma \cdot \nabla_z \phi^1 + r\nabla \cdot \nabla\phi^1 = 0$$

and (63)-(64) become:

(69) $$\phi^0,_t + \frac{1}{2}|v^0 + \nabla_x \phi^0|^2 + \frac{1}{2}<|\nabla_z \phi^1|^2> + C\rho^0 = q^0$$

(70) $$\rho^0,_t + (v^0 + \nabla_x \phi^0)\nabla\rho^0 + r\nabla \cdot \nabla\phi^0 + <\nabla_z\phi^1 \cdot \nabla_x\sigma^0> = 0$$

These can be recombined with (60) into

(71) $$u^0,_t + u^0 \nabla u^0 + C\nabla\rho^0 + A^1 \nabla x u^0 + \nabla b = 0$$

(72) $$\rho^0,_t + u^0 \cdot \nabla\rho^0 + c = 0$$

Recall that A^1 comes from (60), $b = \frac{1}{2}<|\nabla\phi^1|^2>$ and $c = <\nabla_x\sigma \cdot \nabla_z\phi^1>$. They are functions of ∇a, of the initial conditions ϕ^0 and σ^0 in (44)-(45) and of C and r.

This is of course, more a work-plan than a result and to proceed further we would need some better understanding of (67)-(68) and some information on the dependency of A^1, b and c upon their arguments.

Conclusion

To use homogenization techniques to derive a closed set of an equation to understand the interactions between microstructures (turbulence?) and acoustic waves does not seem altogether impossible, but it raises a number of difficult mathematical problems and there seems to be 3 different cases:

i^0) Incompressible microstructures convected by a compressible mean flow (see (31)-(36)).

ii^0) Micro-soundwaves (63) in a compressible flow (see (66)-(67)).

iii^0) Microstructures and micro-soundwaves.

We have sketched arguments which shows that the first two cases may be tackled by the asymptotic expansions that worked for the incompressible case.

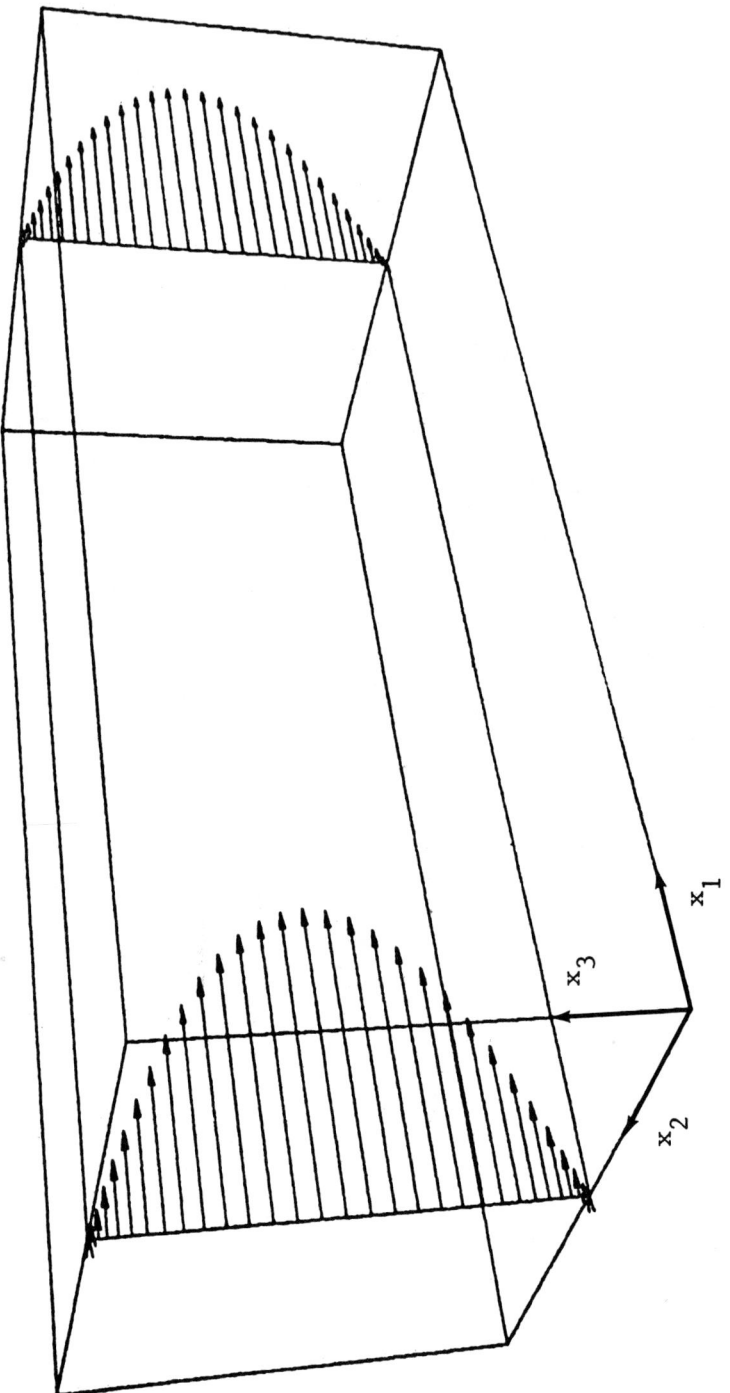

Initial mean velocity

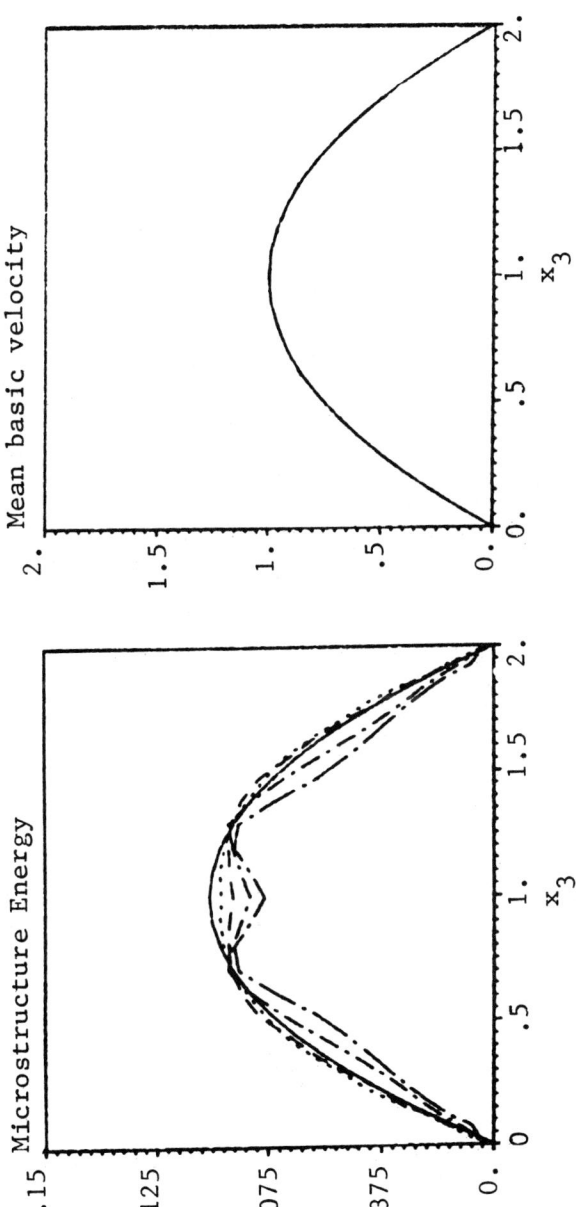

Figure 1: (x_1, x_2) -Mean velocity profiles for Poiseuille flow at two different times: $t = 0$ (top), $t = 2$" (bottom). These are the results of a direct simulation with 32x32x32 modes. It simulates the flow between two horizontal planes with periodic conditions in x_1 and x_2, and initial conditions containing the highly oscillating structures shown in figure 2.

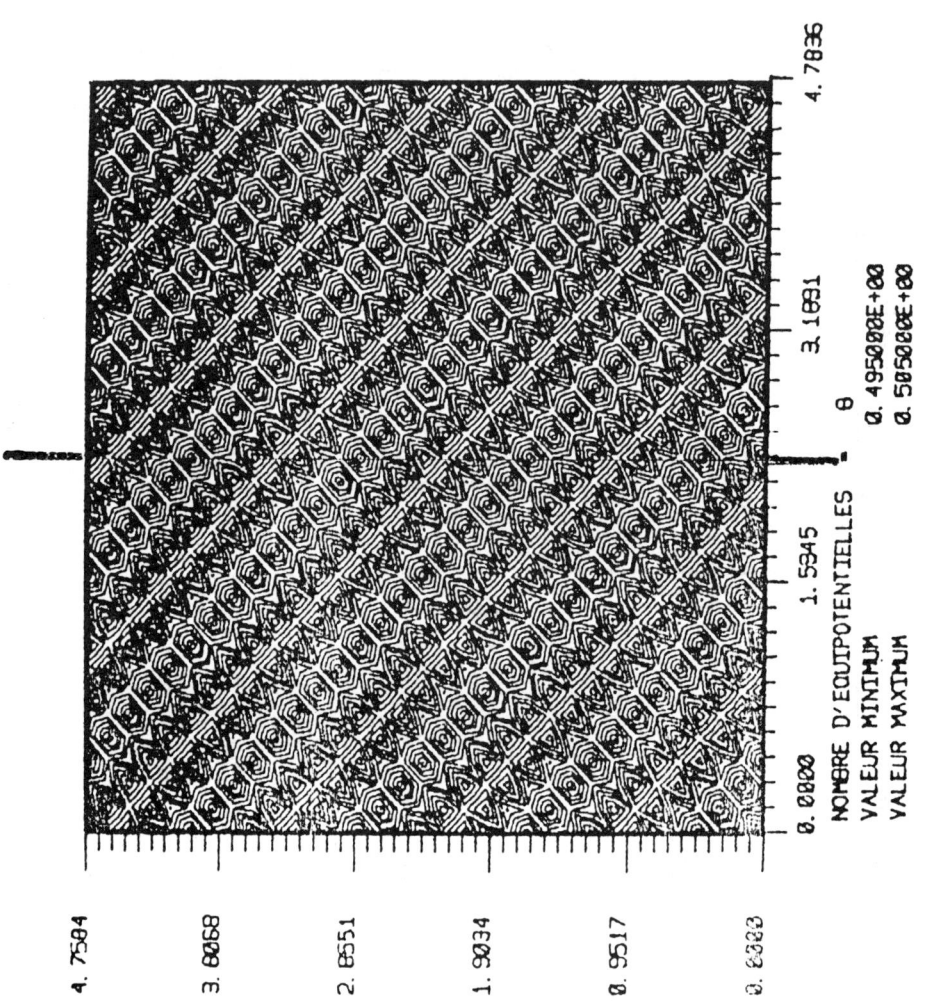

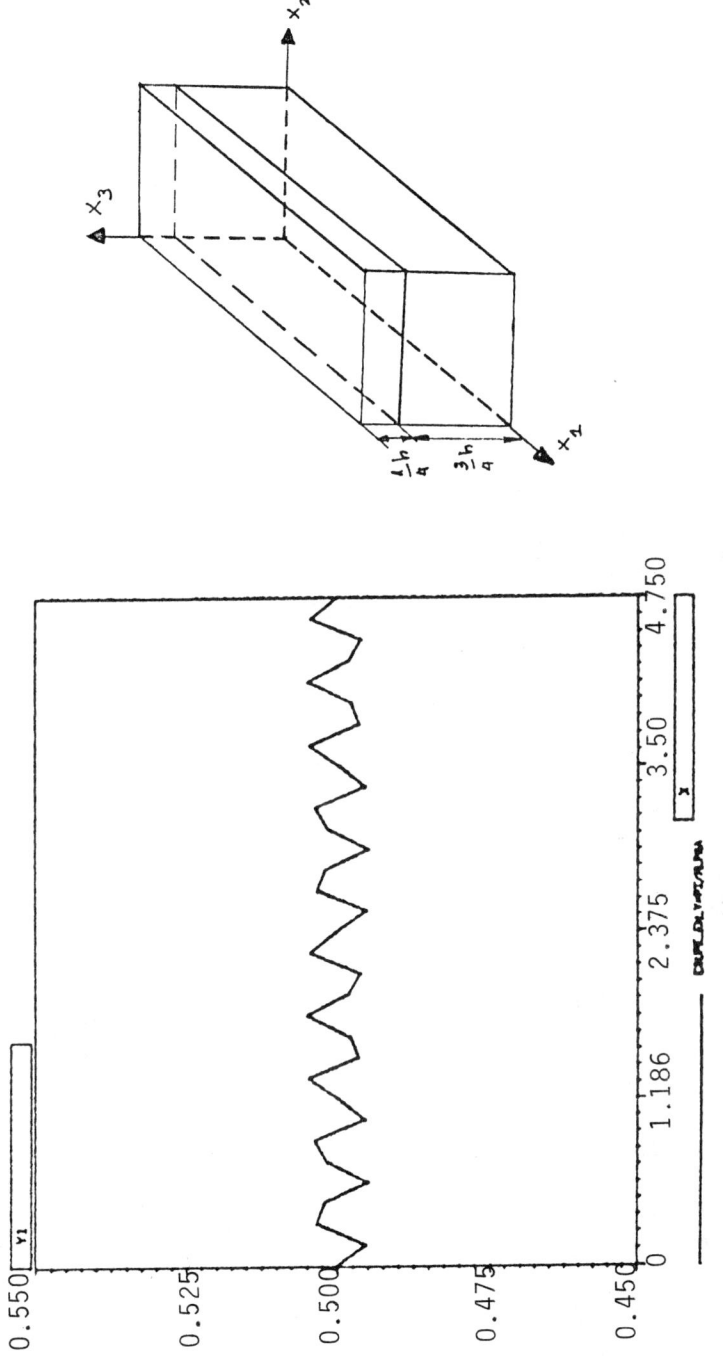

SIMULATION DIRECTE

Figure 2: Top: Level lines of $|u(0)|$ on the plane cross section of the domain in bottom right figure. Bottom left: $|u(0)|$ versus x_1 on the line shown in the top figure.

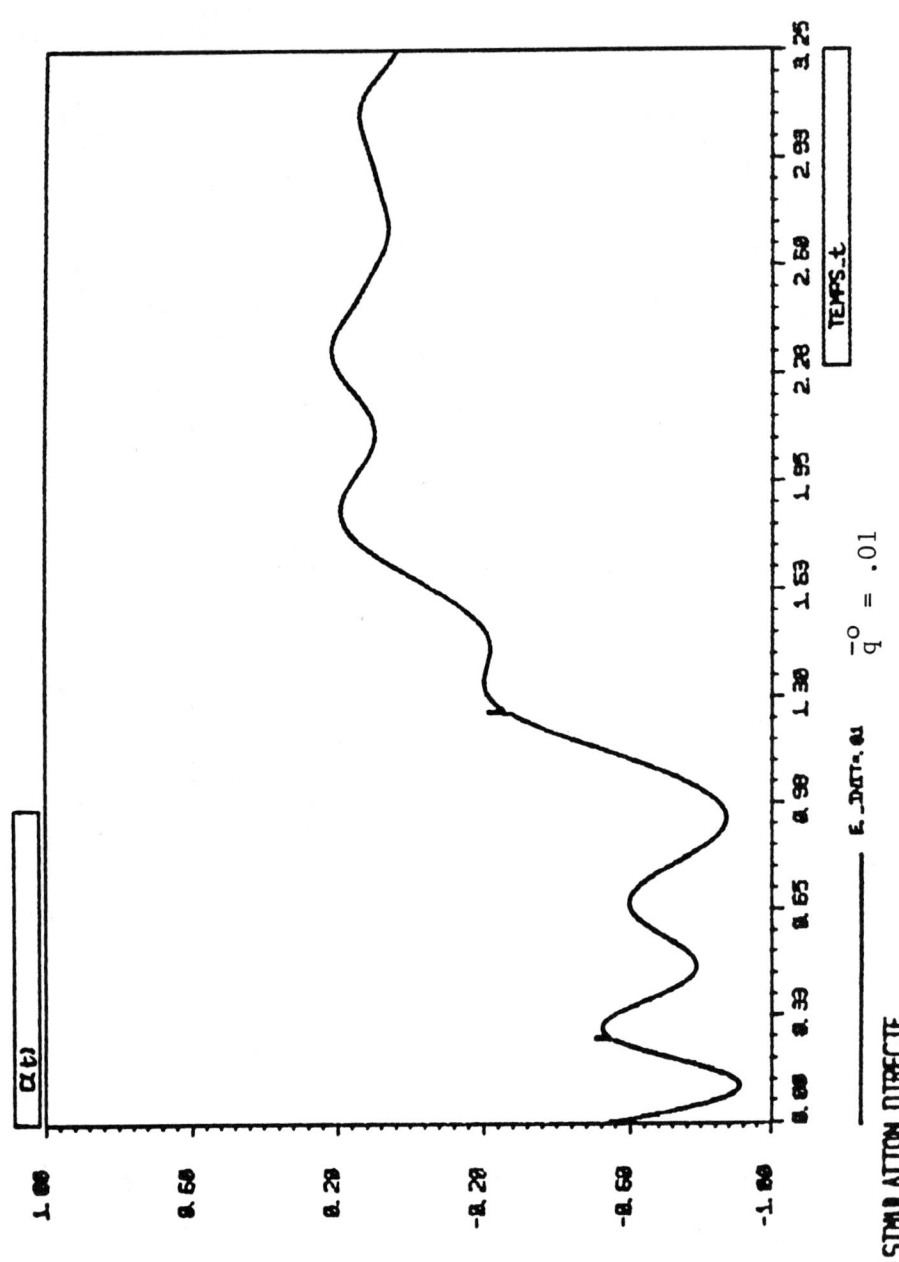

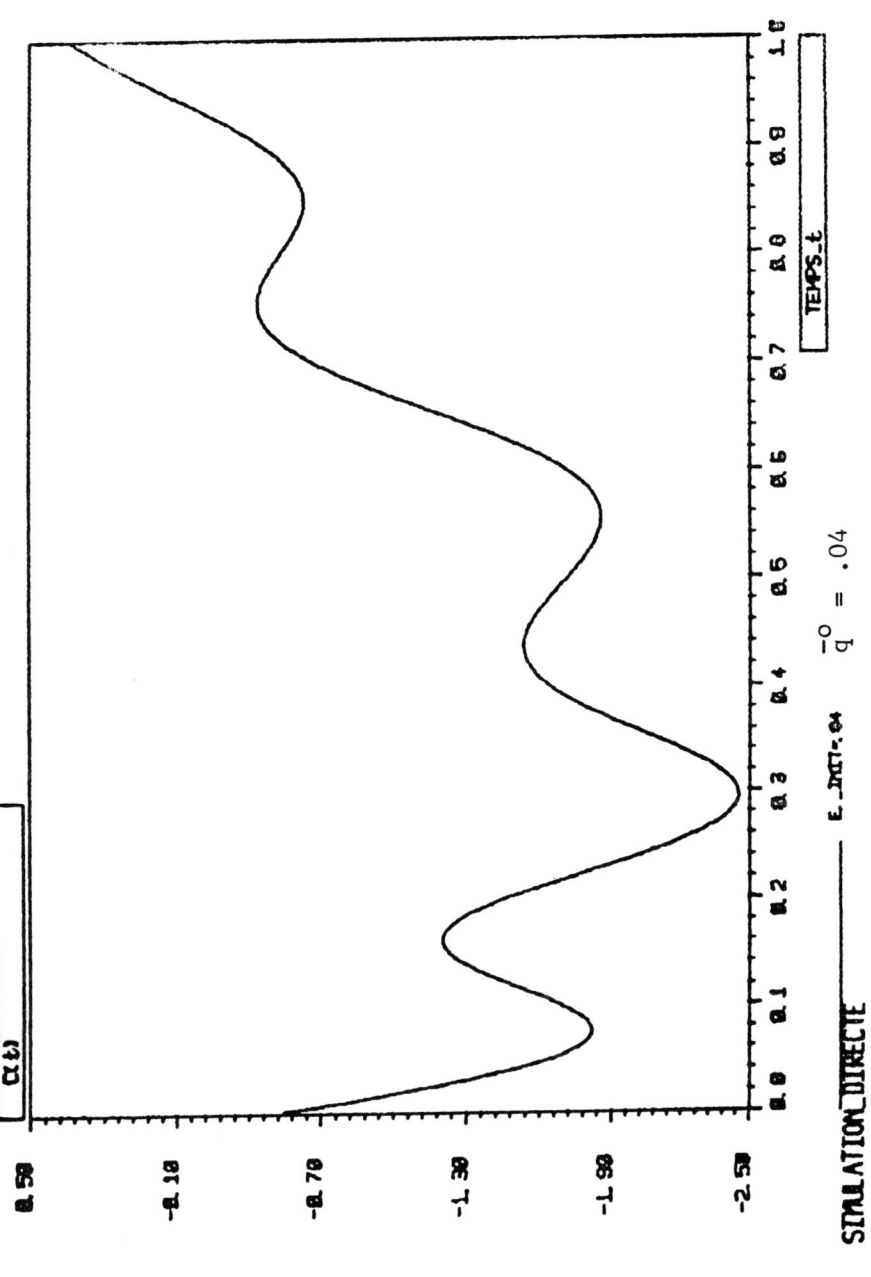

Figure 3: Oscillations of the module of the energy $e(t)$ in the N-th mode of u on a log scale; more precisely $\log(e^{n+1}/e^n)$ is plotted versus n, the time iteration number.

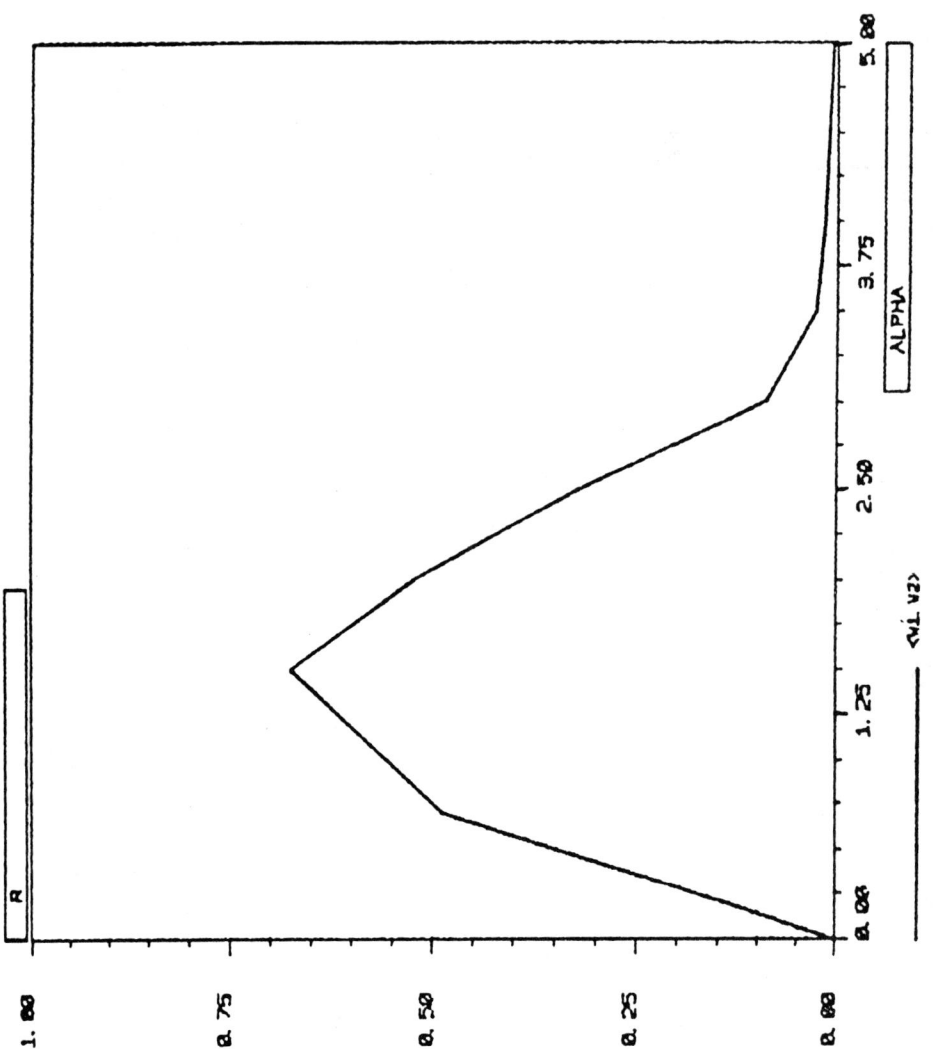

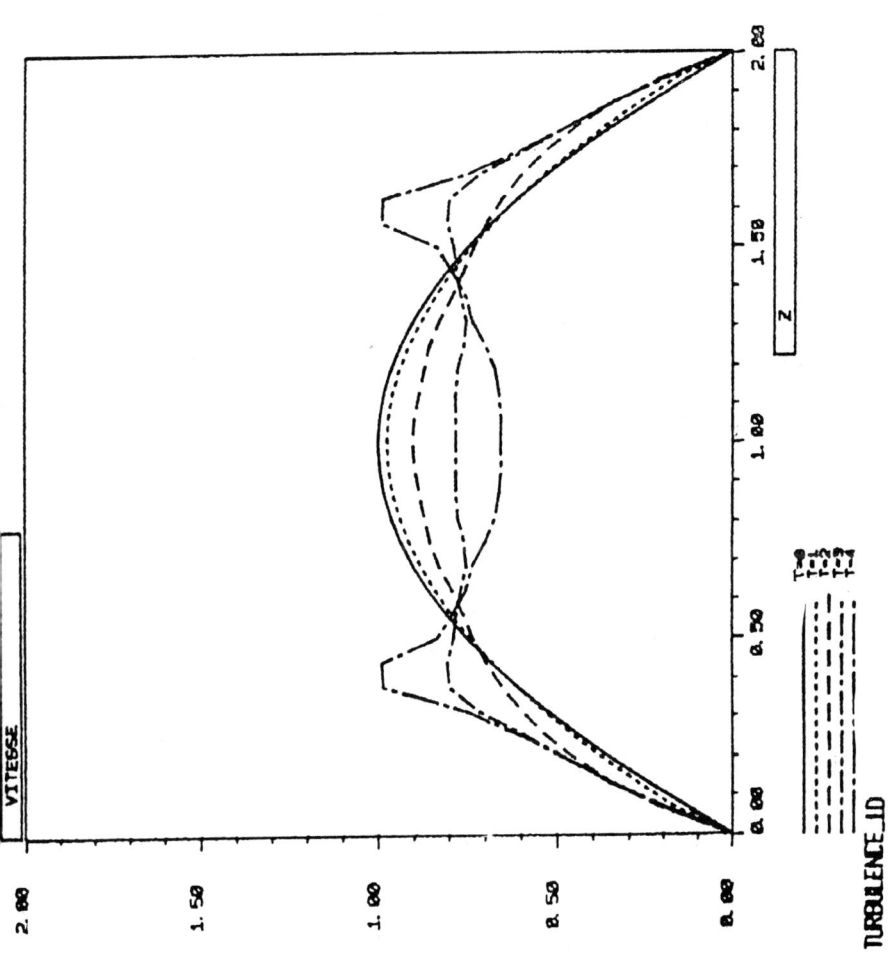

Figure 4: Top: $R_{1,3}$ as a function of α. This curve is an approximation of the one obtained by Begue and Chacon by numerical simulation of (12). Bottom: (x_1, x_2) -Mean of u, for several times, computed by (21) with $R_{1,3}$ shown above. This is to be compared with the direct simulation in Figure 1. At later times strong gradients develop and the numerical scheme is unstable.

Acknowledgement

We would like to thank A. Patera for letting us run his code on our problem.

References

[1] C. Begue: These de 3eme cycle. Universite Paris 6 (1984).

[2] A. Benssoussan, J.L. Lions, G. Papanicolaou: Asymptotic methods for periodic structures. North Holland (1978).

[3] T. Chacon: These de 3eme cycle. Universite Paris 6 (1985).

[4] T. Chacon, O. Pironneau: On the mathematical foundations of the k-ε turbulence model. (To appear)

[5] R. DiPerna, A. Majda: to appear (see also Majda-Hunter-Rosales: Resonantly interacting weakly nonlinear hyperbolic waves. Studies in Applied Mathematics. Elsevier (1984)).

[6] U. Frisch, O. Thual, Z.S. She: Homogenization and visco-elasticity of turbulence Proc. workshop on Turbulence (Nice 1984) To appear in Springer Lecture notes in Physics (Frisch-Keller-Papanicolaou-Pironneau ed).

[7] J. Hunter, J.B. Keller: Weakly nonlinear high frequency waves. Comm Pure Appl. Math. 36(5): 547-569, 1983.

[8] L. Landau, E. Lifschitz: Mecanique des fluides. Moscou (1962) MIR.

[9] B. Launder, D. Spalding: Mathematical models of Turbulence. Academic press (1972

[10] J. Mathieu: Cours de Turbulence Ecole d'ete EDF-CEA 1982.

[11] D. McLaughlin, kG. Papanicoulaou, O. Pironneau: Convection of microstructures. Siam Numer Anal (to appear).

[12] A. Monin, A. Yaglom: Statistical Fluid Mechanics of Turbulence. MIT (1975).

[13] S. Orszag: Lecture on statistical theory of turbulence. (Ecole d'ete des Houches, 1973). Gordon Breach, London 1977.

[14] G. Papanicolaou, O. Pironneau: On the asymptotic behavior of motion in random flow. In Stochastic Nonlinear Systems, Arnold-Lefever eds. Springer (1981).

[15] A. Patera, S. Orszag: 3-D instability in plane channel flow at subcritical Reynolds number. In Numerical and Physical Aspect of aerodynamic flows. (Long Beach 1981) 69-86 Springer (1982).

[16] P. Perrier, O. Pironneau. Subgrid turbulence modeling by homogenization. Math Modeling, Vol. 2, 295-317 (1981).

[17] W.C. Reynolds: Computation of turbulent flows. Annual Rev. Fluid Mech. 8, 183-208 (1976).

[18] P. Saffman: A model for inhomogeneous turbulent flow. Proc. Roy. Soc. London A317 (1970) 417-433.

[19] U. Shumann: Subgrid scale model for finite difference simulations of turbulent flows in plane channel and annuli. J. Comp. Phys. 18, 376-404 (1975).

[20] J. Smagorinsky: Mon. Weather Rev. 91, 99-164.

OSCILLATIONS IN SOLUTIONS TO NONLINEAR DIFFERENTIAL EQUATIONS

Ronald J. DiPerna

Mathematics Department
University of California
Berkeley, California 94720

We shall describe several aspects of a general program dealing with oscillations in solutions to nonlinear partial differential equations. The main problem is to describe the relationship between microscopic oscillations and their macroscopic averages, in terms of both the static structure and the dynamic behavior. One general framework is provided by the following setting.

Consider a sequence of vector fields from physical space to state space that converges weakly:

$$z^\epsilon : R^m \to R^N$$

(1)
$$\lim_{\epsilon \to 0} z^\epsilon = z.$$

It is convenient to regard z^ϵ as the microscopic value of the field containing small scale fluctuations and z, the weak limit, as the macroscopic value of the field obtained by taking local averages. We recall that z^ϵ converges weakly to z if and only if

(2)
$$\lim_{\epsilon \to 0} \int_\Omega z^\epsilon(y)\,dy = \int_\Omega z(y)\,dy$$

for all bounded domains Ω in the physical space R^m.

The fields z^ϵ are subject to linear constant coefficient differential constraints of the form

(3)
$$\sum_{j=0}^{m-1} A_j \partial_j z^\epsilon = 0$$

which represent the conservation laws of physics. Here A_j denotes a constant $r \times N$ matrix with r fixed and $\partial_j \equiv \partial/\partial y_j$. Second, the fields z^ϵ are subject to algebraic constraints which embody the constitutive laws and ask that the values of z^ϵ lie in a specified proper subset M of the state space R^N:

(4)
$$z^\epsilon(y) \in M \subset R^N,$$

for almost all y in R^m. The set M is typically a manifold and will be referred to as the constitutive manifold.

The general problem is to describe the type of oscillations which a weakly convergent sequence can execute subject to differential and algebraic constraints. Specifically, the problem is to describe the admissible oscillations at a typical point y in the physical domain and to describe the propagation of such oscillations in terms of creation, annihilation, correlation, etc. The analytical problem is to characterize the class of Young measures associated with weakly convergence sequences (1) subject to differential and algebraic constraints (3) and (4).

We shall first discuss th problem of determining the impact of the differential constraints (3) alone. The influence of the algebraic constraints will be mentioned subsequently. One of the problems of interest here is to classify the state variables

$$g: R^N \to R,$$

which are weakly continuous when acting on functions that satisfy the differential constraints (3). For which continuous real-valued functions g on the state space R^N do we have

(5) $$\lim g(z^\varepsilon) = g(\lim z^\varepsilon)$$

if (3) holds? Unless otherwise specified, the symbol lim will denote a weak limit. In other words, for which state variables g can we determine the macroscopic value of g, i.e. $\lim g(z^\varepsilon)$, from the macroscopic value of the field, i.e. $z \equiv \lim z^\varepsilon$. This is one of the questions addressed by the Tartar-Murat theory of compensated compactness [8,9,11,12,14,15]. In terms of the Young measure the problem is to classify the state variables g which commute with the Young measure: the equation (5) may be rewritten as

$$\langle \nu_y, g(\lambda) \rangle = g(\langle \nu_y, \lambda \rangle)$$

where ν_y denotes the Young measure at y and λ denotes the generic variable in the state space R^N. For convenience we shall denote the expected value of a state variable g with respect to the Young measure ν_y by the following pairing:

$$\langle \nu_y, g(\lambda) \rangle = \int_{R^N} g(\lambda) d\nu_y.$$

We shall recall three examples before turning to the general theory. The first is the basic div-curl lemma of Tartar and Murat which is relevant to the homogenization theory of elliptic operators [10,15] and the viscosity method for hyperbolic conservation laws [3,4,5] and applies to electrostatics.

Example 1. Consider two sequences of fields D^ε and E^ε converging weakly in L^2 such that the divergence of D^ε and the curl of E^ε are uniformly controlled:

$$D = \lim D^\varepsilon \qquad \text{div } D^\varepsilon \cong 0$$
$$E = \lim E^\varepsilon \qquad \text{curl } E^\varepsilon \cong 0.$$

Then there is precisely one state variable g which is continuous in the weak topology, namely the electrostatic energy density given by the inner product (modulo an affine function):

$$g = \sum_{j=1}^{n} E_j D_j + \text{affine}.$$

Although each of the individual terms forming the inner product is not weakly continuous in general, there exists compensation between the elements of the sum to produce the overall continuity. Thus, from the macroscopic values of the electric induction field D and the electric field E, one may deduce the macroscopic value of just one nonlinear state variable, the energy density.

It is only for the energy density that one need not introduce internal structure or internal variables in order to describe the maroscopic value. It is a general problem to understand the role which internal variables play in the description of propagating oscillations. We shall return to this question below in the context of semilinear hyperbolic systems. As a final technical comment we note that it is sufficient to require that the distributions div D^ε and curl E^ε lie in a compact subset of the negative Sobolev space $W^{-1,2}_{loc}$.

Example 2. This example is a variant of the first and follows a remark of L. Tartar [15] that the density of action is weakly continuous when acting on oscillating solutions of the wave equation in H^1:

(6)
$$w = \lim w^\varepsilon$$
$$w^\varepsilon_{tt} = \Delta w^\varepsilon.$$

The equation (6) may be rewritten as a first order system of two equations with a div-curl structure by introducing the space-time gradient

$$u = \text{grad } w = (w_t, w_{x_1}, \ldots, w_{x_n})$$

and its image $v = Mu$ under the constant diagonal matrix M with entries -1 and 1:

$$v = (-w_t, w_{x_1}, \ldots, w_{x_n}).$$

We have

(7)
$$\text{div } v^\varepsilon = 0$$
$$\text{curl } u^\varepsilon = 0,$$

which express respectively the laws of the conservation of momentum and mass and provides an equivalent formulation of (6). The inner product is the density of action

$$A^\varepsilon = (u^\varepsilon, v^\varepsilon) = \tfrac{1}{2}(w^\varepsilon)^2 - \tfrac{1}{2}|\triangledown w^\varepsilon|^2$$

and is weakly continuous. It is interesting to compare this quadratic form with density of energy

$$E^\varepsilon = \tfrac{1}{2}(w^\varepsilon)^2 + \tfrac{1}{2}|\triangledown w^\varepsilon|^2,$$

which is not weakly continuous in general. As a convex quantity, energy is weakly lower semicontinuous: in general, the energy associated with the macroscopic value of the energy. The presence of microscopic oscillations may trap a finite amount of energy at the microscopic level, but it does not effect the action.

This example is presented in connection with the general problem of determining how small scale oscillations and concentrations in coherent structures may enhance the effective dissipation and reduce the observed energy, which otherwise would have remained constant being a formal invariant of the motion. Example 2 generalizes to the full nonlinear equations of elasticity.

Example 3. This classical example of J. Ball [2] which involves the curl constraint that is significant in the variational case. The associated problem is to determine the fluctuations that are induced by the process of minimizing a given stored energy function. Do they raise or lower the total energy? Consider a sequence of matrices u^ε such that the curl of each of the rows (or columns) is controlled:

$$u = \lim u^\varepsilon \qquad u^\varepsilon: R^n \to R^n$$
$$\text{curl } u^\varepsilon \cong 0.$$

then the determinant of u^ε is weakly continuous, as well as all subdeterminants. As remarked by L. Tartar [15], this example hints at various connections between invariant topological quantities, typically defined by the integration of polynomial forms, and oscillations. With respect to oscillations, the determinant form is particularly robust.

We refer the reader to the work of J. Ball [1,2] for further discussion of example 3 in the variational structure of nonlinear elasticity. We shall next discuss the use of these examples in conjunction with the Young measure in studying weak continuity properties of nonlinear operators. As a point of focus, let us consider the general problem of determining if a given semigroup is continuous in the weak topology. Let

$$S(t): B \to B$$
$$S(t)u_0 = \text{solution at time } t \text{ with initial data } u_0.$$

We shall compare quasilinear and semilinear hyperbolic p.d.e. in the context of this problem: if $u_0 = \lim u_0^\varepsilon$, is

(8) $$S(t)u_0 = \lim S(t)u_0^\varepsilon .$$

Is the macroscopic value of the field at time t, i.e. $\lim S(t)u_0^\varepsilon$, obtained by propagating the macroscopic value of the data and thus equal to $S(t)u_0$. Or, do we need more than just the mean value of the data to determine the macroscopic value of the solution at time t?

Of course, continuity of a nonlinear semigroup as expressed by (8) in the weak topology is a substantial property which is not anticipated without some form of dissipation. This can be illustrated by the following example.

Example 4. Consider the Ricatti p.d.e.,

$$\partial_t u(x,t) = u^2(x,t),$$

which propagates the field at each station x by forming a rational fraction:

(9) $$S(t)u_0 = \frac{u_0(x)}{1 - tu_0(x)} .$$

The lack of continuity of $S(t)$ in the weak topology can be put into evidence by appealing to the Young measure of a sequence of data u_0^ε. The macroscopic value of the solution at time t is given by

$$\lim S(t)u_0 = \int_R \frac{\lambda}{1 - t\lambda} d\nu_x(\lambda) \equiv \langle \nu_x, \frac{\lambda}{1 - t\lambda} \rangle ,$$

and is not equal to the solution at time t corresponding to the macroscopic value of the data, namely

$$S(t)u_0 \equiv \frac{\langle \nu_x, \lambda \rangle}{1 - t\langle \nu_x, \lambda \rangle} .$$

Knowledge of the center of mass of the Young measure of the data does not suffice to characterize oscillations at later times. Indeed, a much stronger statement can be made: if the solution operator is weakly continuous along a particular sequence of initial data then necessarily that sequence is not oscillatory, i.e. it is converging strongly. Analytically, this fact can be seen through a reversal of Jensen's inequality. Observe that, for t_0 fixed, the state variable

$$g(\lambda) = \frac{\lambda}{1 - t_0 \lambda}$$

is a strictly convex function of λ. Weak continuity of the Ricatti semigroup (9) means that

(10) $$\langle \nu_x, g(\lambda) \rangle = g(\langle \nu_x, \lambda \rangle),$$

where ν_x is the Young measure of the sequence of data u_0^ε. Now, Jensen's inequality asserts that

(11) $$\langle \mu, \phi(\lambda) \rangle \geq \phi(\langle \mu, \lambda \rangle)$$

for any convex function ϕ and any probability measure μ. If equality is assumed in Jensen's inequality (11) by a strictly convex function as in (10) then the associated measure must reduce to a Dirac mass. We conclude that weak continuity of the Ricatti semigroup (9) implies that the Young measure of the data reduces to a Dirac mass and consequently that the sequence of data u_0^ε is converging strongly. Hence the solution at time t is converging strongly.

This example is presented in order to direct attention to one of the basic ideas in studying the suppression of oscillations in nonlinear equations which is to examine the relationship between weak limits of distinguished state variables and the corresponding value of these state variables at the center of mass of the Young measure. Approximate inversions of Jensen's inequality have lead to compactness theorems for general elliptic systems. We refer the reader to [5,12] for further discussion envolving the interplay between differential and algebraic constraints.

Let us next return to the question of weak continuity of semigroups in the context of systems of semilinear hyperbolic equations where the modes are coupled.

<u>Example 5.</u> We have in mind discrete velocity models of the Boltzman equation

$$\partial_t U + A \partial_x U = Q(U), \quad U(x,0) = U_0(x)$$

involving a quadratic nonlinearity, perhaps the simplest of which is the two-speed Carleman model:

(12)
$$\partial_t u + \partial_x u = -u^2 + v^2$$
$$\partial_t v - \partial_x v = u^2 - v^2.$$

It is known for (12) that, in order to characterize the weak limit of an oscillating solution sequence it is necessary to have fairly complete knowledge of the Young measure of the data. Consider a sequence of nonnegative data $(\phi_\varepsilon, \psi_\varepsilon)$, that is uniformly bounded in L^∞. Consider the collection of moments of the initial data ϕ_ε for the component u and ψ_ε for the component v:

$$\phi_m \equiv \lim \phi_\varepsilon^m = \int_R \lambda^m d\nu_x$$

$$\psi_m \equiv \lim \phi_\varepsilon^m = \int_R \mu^m d\mu_x.$$

Then, according to a Theorem of L. Tartar [13] the moments of the Young measure of each of the solution components u and v satisfy an explicit tridiagonal system which is derived with the aid of the div-curl lemma and which expresses the coupling of the modes. Furthermore, in the special case of the slow modulation of a periodic function, i.e.

$$\phi_\varepsilon = p(x, x/\varepsilon)$$
$$\psi_\varepsilon = q(x, x/\varepsilon)$$

where p and q are periodic in the second variable, the solution to the infinite tridiagonal system may be derived in two simple steps by solving a nonlinear integrodifferential system of two equations which involves a supplementary variable corresponding to the fast scale and then averaging with respect to the supplementary variable.

Thus, for the Carleman model the process of weak convergence may be described by introducing an auxillary or interval space variable. In addition, no correlations of the initial data are required in order to characterize the oscillations at time t. To what extent is this true for more complicated systems such as the Broadwell model?

At this point it is appropriate to remark that there are two main classes of equations under consideration: those with little or no dissipation for which oscillations in the initial data survive, propagation and interact, and those with substantial dissipation for which oscillations in the initial data are immediately damped out. Examples 4 and 5 belong to the former category while the next example belongs to the latter.

Example 6. Consider the Cauchy problem for a scalar conservation law in one space dimension,

(12) $$\partial_t u + \partial_x f(u) = 0, \quad -\infty < x < \infty, \quad u(x,0) = u_0(x).$$

A theorem of P. Lax [6] asserts that, in the presence of the standard entropy condition [7], the solution operator of (12) is Lipschitz continuous in the weak norm of L^∞:

(13) $$|S(t)u_0 - S(t)v_0|_* \leq K|u_0 - v_0|_*.$$

Furthermore, for the associated diffusion problem

$$\partial_t u + \partial_x f(u) = \epsilon \partial_x^2 u$$

a similar statement holds with a Lipschitz constant K which is independent of the diffusion coefficient ϵ:

(14) $$|S_\epsilon(t)u_0 - S_\epsilon(t)v_0|_* \leq K|u_0 - v_0|_*.$$

In fact (13) is derived from (14) in [6] by passing to the limit as the diffusion parameter ϵ vanishes. We note that, on bounded sets, the weak-star topology of L^∞ is metrizable with a metric induced by a norm. A convenient choice of weak norm here is given by the following formula

$$|w|_* = \sup_{(a,b)} \left| \int_a^b w(y)dy \right|.$$

It is an interesting open problem to determine whether or not analogous inequalities hold for scalar conservation laws in several space dimensions.

It follows from the stability estimate (13) that the entropy-satisfying solution operator is continuous in the weak topology. No matter how widely the initial data oscillate, the loss of information associated with dissipation of entropy allows one to determine the macroscopic value at time t using only the mean value of the data.

In the special case of a strictly convex flux f, the following compactness theorem was established by P. Lax [6].

<u>Theorem</u> Suppose that u^ε is a sequence of admissible solutions to a scalar strictly convex conservation law in one space dimension. Then u^2 contains a subsequence which converges in the strong topology of L^1_{loc}.

The point of the theorem is that no a priori derivative control is required on the solution sequence u^ε in order to extract a strongly convergent subsequence. More recent work on compactness of solutions to hyperbolic equations using the Young measure and the theory of compensated compactness has lead to new results for the scalar conservation law [11] and for systems of two conservation laws [3,4]. The analysis takes into account both the influence of the differential constraints (3) and the algebraic constraints (4) and makes use of state variables which are weakly continuous and weakly lower semicontinuous. We refer the reader to [5,11,12] for a general introduction to the functional analytic theory of conservation laws and the theory of compensated compactness.

References

1. Ball, J.M., On the calculus of variations and sequentially weakly continuous maps, in Lecture Notes in Mathematics, Vol. 564, Springer-Verlag, 1976.
2. Ball, J.M., Convexity conditions and existence theorems in nonlinear elasticity, Arch. Rat. Mech. Anal. 63 (1977), 337-407.
3. DiPerna, R.J., Convergence of approximate solutions to conservation laws, Arch. Rat. Mech. Anal. 82 (1983), 27-70.
4. DiPerna, R.J., Convergence of the viscosity method for isentropic gas dynamics, Comm. in Math. Phys. 91 (1983), 1-30.
5. DiPerna, R.J., Compensated compactness and general systems of conservation laws, to appear in Trans. Amer. Math. Soc. (1986).
6. Lax, P.D., Weak solutions of nonlinear hyperbolic equations and their numerical computation, Comm. Pure Appl. Math. 7 (1954), 159-193.
7. Lax, P.D., Shock waves and entropy, in Contributions to Nonlinear Functional Analysis, ed. E.A. Zarantonello, Academic Press (1971).

8. Murat, F., Compacite par compensation, Ann. Scuola Norm. Sup. Pisa (1978), 489-507.

9. Murat, F., Compacite par compensation: condition necessaire et suffisante de continuite faible sous une hypotheses de rang constant, Ann. Scuola Norm. Sup. Pisa 8 (1981), 69-102.

10. Murat, F. and L. Tartar, Cacul des variations et homogeneisation, preprint.

11. Tartar, L., Compensated compactness and applications to partial differential equations, in Research Notes in Mathematics, Nonlinear Analysis and Mechanics: Heriot-Watt Symposium, Vol. 4, ed. R.J. Knops, Pitman Press, 1979.

12. Tartar, L., The compensated compactness method applied to systems of conservation laws, in Systems of Nonlinear Partial Differential Equations, ed. J.M. Ball, NATO ASI Series, Reidel Pub. Co. (1983).

13. Tartar, L., Solutions oscillantes des equations de Carleman, Seminaire Goulaouic-Meyer-Schwarz, Jan. 1983.

14. Tartar, L., Etude des oscillations dans les equations aux derivees partielles nonlineares, in Trends and Applications of Pure Mathematics to Mechanics, Proceedings of Symposium at Ecole Polytechnique, in Lecture Notes in Physics Vol. 195, Springer-Verlag.

15. Tartar, L., Oscillations in nonlinear partial differential equations: compensated compactness and homogenization, preprint.

GEOMETRY AND MODULATION THEORY FOR THE
PERIODIC NONLINEAR SCHRODINGER EQUATION

by

M. Gregory Forest* and Jong-Eao Lee

Department of Mathematics
Ohio State University

Abstract

We describe the integrable structure of solutions of the nonlinear Schrodinger (NLS) equation under periodic and quasiperiodic boundary conditions. We focus on those aspects of the exact theory which reveal the behavior of these solutions under perturbations of initial conditions (i.e. linearized instabilities), and the effects of slow modulations in space and time, perhaps in the presence of external perturbations. These results and methods continue the investigations of Ercolani, Flaschka, Forest and McLaughlin [1-7] on Korteweg-deVries (KdV), sine-Gordon (sG) and sinh-Gordon wavetrains. Our purpose here is to document the corresponding features of NLS solutions; the rigorous analysis that underlies this paper derives from [1-7] and will appear in the thesis of Lee [8].

I. Inverse Spectral Theory of NLS

We use the first two sections to develop the aspects of the NLS inverse theory which are relevant for the stability questions we address in Sections III, IV. Exact representations of finite degree of freedom periodic and quasiperiodic solutions are described; the remaining infinite number of closed degrees of freedom are also characterized.

The nonlinear Schrodinger equation, (+ focusing, - defocusing)

$$iq_t + q_{xx} \pm 2|q|^2 q = 0, \qquad (I.1\pm)$$

arises as the compatibility condition for the Zakharov-Shabat [9] or AKNS [10] linear system:

$$\begin{pmatrix} \psi_1 \\ \psi_2 \end{pmatrix}_x = \begin{pmatrix} -iE & q \\ -r & iE \end{pmatrix} \begin{pmatrix} \psi_1 \\ \psi_2 \end{pmatrix} \qquad (I.2a)$$

$$\begin{pmatrix} \psi_1 \\ \psi_2 \end{pmatrix}_t = \begin{pmatrix} i(qr - 2E^2) & iq_x + 2Eq \\ ir_x - 2Er & i(2E^2 - qr) \end{pmatrix} \begin{pmatrix} \psi_1 \\ \psi_2 \end{pmatrix}, \qquad (I.2b)$$

$$r = \pm q^*, \text{ where } ()^* \text{ denotes complex conjugation.} \quad (1.2c\pm)$$

That is, $\vec{\psi}_{xt} = \vec{\psi}_{tx}$ iff the potential q satisfies the NLS equation (I.1). The <u>inverse scattering</u> solution of NLS [9,10] applies for vanishing boundary conditions as $|x| \to \infty$, and includes the special case of exact, N-soliton solutions of NLS.

The <u>inverse spectral</u> solution, under periodic and quasiperiodic boundary conditions in x, is our interest here. Only partial, and quite limited, results have appeared. The obstruction is mainly that the linear operator, (I.2a,c+), is <u>non-selfadjoint in the focusing case</u>. The methods of the KdV-Hill inverse theory [11, 12, 13] rely heavily on selfadjoint oscillation theorems, and do not apply to focusing NLS. <u>The defocusing case, however, proceeds just like KdV and sinh-Gordon since</u> (I.2a,c-) <u>is self-adjoint,</u> we omit the description. Formal theta function solutions were found by Its and Kotlyarov [14]. Ablowitz and Ma obtained some results for one and two phase solutions. Tracy [16] rederived the Its-Kotlyarov formula, using the methods in [17,2], and analyzed solutions in the neighborhood of plane waves to exhibit modulational instabilities, using the methods in [17,2].

Daté [17], McKean [23] and Ercolani, Forest and McLaughlin [2-6] have developed techniques from algebraic geometry and Riemann surface theory to characterize all real, C^∞, quasiperiodic solutions of sine-Gordon (also called N-phase solutions). In [2-6], dynamical coordinates for real N-phase solutions are developed for the inverse spectral solution of (sG) and to use in stability considerations. Also, Previato [18] has constructed a Neumann system model, analogous to that of KdV [19], for periodic NLS.

In this paper and [8] we describe the NLS results parallel to references [1-6]. In addition to the exact oscillatory solutions, we study their linearized stability and the stability of slow modulations in these wave trains. Moreover, Venakides' analysis of the weak dispersion limit for periodic KdV [20] utilizes this inverse spectral structure. Any analogous theory for (sG) or NLS will require the ingredients we now describe.

Floquet Theory for NLS Potentials Periodic in x of Period L

The fundamental object of the periodic spectral theory of (I.2a) is the Floquet discriminant, $\Delta(E;q)$. The function space F for $q(x)$ is C_L^∞: C^∞ periodic functions of period L. First compute a fundamental matrix solution, $M_{2\times 2}(x;x_0;E;q)$, of (I.2a), normalized by $M(x=x_0;x_0;E;q) = \text{Id}$. The <u>transfer matrix</u>, $M(x_0+L;x_0;E;q)$, maps solutions of (I.2a) across one period L of the potential q. The <u>Floquet discriminant</u> is

$$\Delta(E;q) = \text{trace } [M(x_0+L;x_0E;q)]. \qquad (I.3a)$$

The roots, $\{E_j\}$, of $\Delta = \pm 2$, are the <u>periodic eigenvalues</u>, and determine the spectrum of (I.2a):

$$\text{SPECTRUM}(I.2a) = \{E \in \mathbb{C} / \Delta(E,q) \in \mathbb{R}, \Delta^2 \leq 4\}. \qquad (I.3b)$$

The NLS equation is an <u>isospectral flow</u> for (I.2a); in the periodic problem this is captured by the invariance of $\Delta(E;q)$ as $q(x,t)$ evolves by (I.1): $\frac{d}{dt} \Delta(E;q(x,t)) = 0$ if q satisfies NLS. Thus, <u>the periodic spectrum of (I.2a) is invariant for solutions</u> $q(x,t)$ <u>of NLS.</u> In the Hamiltonian formalism for periodic solutions, <u>the periodic spectrum</u>, $\Sigma(q) = \{E_k / \Delta(E_k;q) = \pm 2\}$, provides the actions in an action-angle description ([21,22,6,8]). The isospectral classes, M_q = all potentials of (I.2a) with the same periodic spectrum as q, arise as level sets of Δ:

$$M_q = \{r \in F / \Delta(E;q) = \Delta(E;r) \; \forall \; E \in \mathbb{C}\}.$$

<u>The following properties obtain for the focusing NLS periodic spectrum,</u> $\Sigma(q)$, <u>of</u>

(I.2a,c+): (The defocusing spectrum is real by selfadjointness of (I.2a,C-), and is similar to KdV, sinh-Gordon with a band-gap structure on the real axis.)

(I.i) If $\Delta(E_k;q) = \pm 2$, then $\Delta(E_k^*;q) = \pm 2$. Periodic eigenvalues are either real or occur in complex conjugate pairs.

(I.ii) $-2 < \Delta(E;q) < 2$ for all $E \in \mathbb{R}$. The entire real axis is continuous spect

(I.iii) The <u>simple periodic spectrum</u>, $\Sigma^s = \{E_j/\Delta(E_j) = \pm 2, \Delta'(E_j) \neq 0\}$, occurs in complex conjugate pairs off the real axis.

(I.iv) If $q(x,t)$ is a periodic, N-phase NLS solution, then $\Sigma^s(q)$ has precisely 2N elements. An N-phase solution $q(x,t) = q(\theta_1,\ldots,\theta_N)$, is 2π-periodic in each phase, $\theta_j = \kappa_j x + \omega_j t$: $q(\theta_1,\ldots,\theta_j + 2\pi,\ldots,\theta_N) = q(\theta_1,\ldots,\theta_j,\ldots\theta_N)$, $j = 1,\ldots,N$.

(I.v) Decompose $\Sigma = \Sigma^s \cup \Sigma^d$, where $E_d \in \Sigma^d$ are called <u>double points</u>, defined by $\Delta(E_d) = \pm 2$, $\Delta'(E_d) = 0$. (Technically, if higher derivatives vanish, $\Delta^{(k)}(E_d) = 0$, $k=2,\ldots$, then E_d is triple, etc.) Then, there are at most finitely many double points off the real axis. (This fact, proved in [7], places a bound on on the number of linearized instabilities for given NLS solutions; refer to Section IV.)

(I.vi) There is a 1:1 correspondence between degrees of freedom (d.f.) in q and critical points of $\Delta(E;q)$. Each real double point $E_d \in \Sigma^d$ labels a closed d.f., while each pair $(E_{2j}, E_{2j+1}) \in \Sigma^s$ corresponds to an open d.f. For non-real double points $E_d \in \Sigma^d$, the associated degree of freedom <u>may be chosen closed, by ansatz</u>, but there are actually homoclinic orbits associated to non-real E_d. Refer also to Remark (I.xiii); the main reference is [6].

(I.vii) For generic q, $\Sigma^d(q) = \phi$. I.e., in any open neighborhood of an N-phase potential q, all degrees of freedom are excited.

(I.viii) The function theory of $q \in F$ takes place on the Riemann surface $\mathcal{R}$ o $\sqrt{\Delta^2(E,q)-4}$, with <u>branch points</u> E_j precisely at the 2N, N<∞, elements of $\Sigma^s(q)$. If N is finite, q is an N-phase potential, and the appropriate curve is (refer to Sec. II):

$$R^2(E) = \sum_{1}^{2N} (E-E_j), \quad E^*_{2j-1} = E_{2j} \,. \tag{I.4}$$

(I.ix) The spectrum for a typical N-phase potential, N<∞, is depicted in Figure 1.

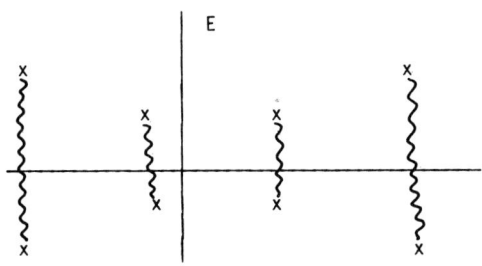

N=4 phase periodic spectrum, focusing NLS

Figure 1

(I.x) <u>Special example</u>: <u>The NLS periodic spectrum for x-independent q</u>.

Consider the x-independent, plane wave solution of focusing NLS,

$$q = a \exp(2ia^2 t), \quad a = \text{constant} \in \mathbb{R} \,. \tag{I.5}$$

The constant coefficient system (I.2a) is easily integrated, and we compute

$$\Delta(E; q = a \exp(2ia^2 t)) = 2 \cos(L\sqrt{E^2 + a^2}). \tag{I.6a}$$

There is one curve of spectrum emanating from $E = 0$ which terminates at $\pm ia$. (Fig.2).

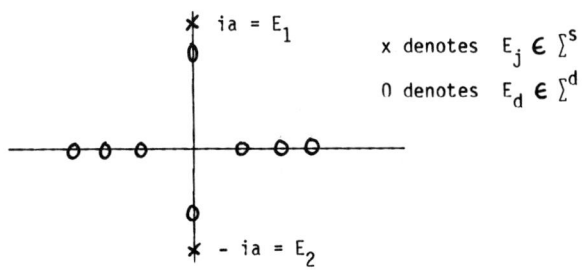

Σ for the NLS plane wave

Figure 2

Some analysis reveals that if L is small enough, all $E_d \in \Sigma^d \mathbb{CR}$.

If $L^2 < \pi^2/a^2$, all double points E_d are real. (I.6b)

As L increases for fixed a, double points coalesce at 0 and migrate into the band between $\pm ia$ on the imaginary axis. It is easy to calculate:

If $\frac{n^2\pi^2}{a^2} < L^2 < \frac{(n+1)^2\pi^2}{a^2}$, there are 2n purely imaginary $E_d \in \Sigma^d$. (I.6c)

We will return to this example in the analysis of linearized instabilities, Sec. IV.

(I.xi) <u>Dynamical spectral coordinates: the Dirichlet spectrum</u> $\{\mu_j\}$. Whereas the trace of the transfer matrix yields constants of the motion, the zeroes of the M_{12} entry provide the angle variables [21,22, 6]:

$$M_{12}(x_0+L;x_0;E = \mu_j;q) = 0.$$ (I.7)

These μ_j are Dirichlet eigenvalues of (I.2a) and are <u>not</u> isospectral. The NLS inversion formula (look ahead to II.9a) for N-phase solutions is:

$$iq_x/q = \sum_1^{2N} E_j - 2 \sum_1^{N-1} \mu_j$$ (I.8)

The dynamical behavior of the $\vec{\mu}$ spectra <u>in the focusing case</u> is quite complicated, just as in the (sG) theory. $\vec{\mu}(x,t)$ cannot be controlled pointwise, but <u>a basis of all closed cycles of</u> $\vec{\mu}(x,t)$ <u>on the Riemann surface of</u> (I.4) <u>is depicted below</u>:

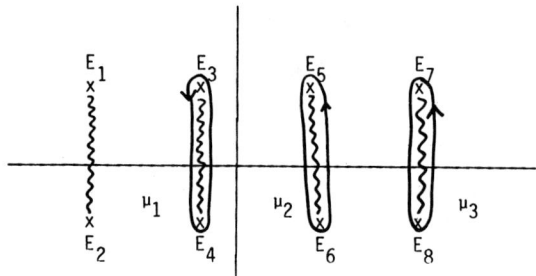

Basis of μ cycles for focusing NLS flows, N = 4.

Figure 3.

This $\vec{\mu}$ behavior follows by the results of Ercolani and Forest [4]. As shown in [6] and Sec. IV. below, these geometric constraints on focusing NLS flows explain the presence of linearized (modulational) instabilities. Moreover, the fact that the μ cycles do not change homology class [4] is crucial to the multiphase averaging of (S-G) and NLS wavetrains [5]; see Sec. III.

(I.xii) For N-phase (quasiperiodic or periodic in X) solutions of NLS, only (N-1) $\vec{\mu}$ variables are free to move, and they coordinate the real tori contained in the isospectral class of q, M_q = the set of all potentials q(x) with the same spectrum $\sum$ of (I.2a) [4]. (The disparity (one less μ_j than phases) is accounted for by the plane wave factor that appears in all NLS wavetrains. Thus, we call the plane wave $q = a \exp(2ia^2 t)$ a single-phase potential, but there is no associated Dirichlet eigenvalue $\mu(t)$ that parameterizes this degree of freedom. This happens because the NLS inversion formulae, (II.9a,b), represent $(\ln q)_x$ and $(\ln q)_t$ in terms of $\{\mu_j\}$, rather than q itself.)

(I.xiii) For a degenerate quasiperiodic solution (N-finite, for example), the remaining $\vec{\mu}$ spectra are tied at the double points $E_d^j \in \sum^d : \mu_j^{tied} = E_d^j$. The exceptions to the closed μ_j tied at double points E_d^j can occur only for non-real double points, of which there are at most a finite number (I.v). This apparent "technicality" is fundamental and is shown in [6] to go hand in hand with: (a) the linearized instabilities of q, and (b) separatrix-type components (homoclinic orbits) in M_q which have previously been undiscovered. We will not concern ourselves in this paper with the separatrices that exist in the periodic focusing NLS phase space F, so we <u>assume</u> from here on

$$\{\mu_j^{tied} \equiv E_d^j \in \sum^d\} = \text{all closed degrees of freedom.} \qquad (I.9)$$

II. Quasiperiodic NLS Solutions and Quadratic Eigenfunctions

For quasiperiodic solutions of soliton equations, there is a straightforward construction, due to Daté, which we now outline for NLS. Details may be found in

[2-4] for (sG). The key step is to replace the linear system (I.2) by an equivalent system for <u>quadratic eigenfunctions</u>. Given $(\psi_1, \psi_2)^T$, $(\phi_1, \phi_2)^T$, two independent solutions of (I.2), define

$$f(x;t;E) = \frac{1}{2}(\phi_1\psi_2 + \phi_2\psi_1) \; ; \; g(x;t;E) = \phi_1\psi_1 \; ; \; h(x;t;E) = \phi_2\psi_2. \quad (II.1)$$

Remarkably, the linear system (I.2a,b) also closes as an equivalent linear system for quadratic eigenfunctions. A straightforward calculation yields:

$$\begin{pmatrix} f \\ g \\ h \end{pmatrix}_x = \begin{pmatrix} 0 & -r & q \\ 2q & -2iE & 0 \\ -2r & 0 & 2iE \end{pmatrix} \begin{pmatrix} f \\ g \\ h \end{pmatrix} \quad (II.2a)$$

$$\begin{pmatrix} f \\ g \\ h \end{pmatrix}_t = \begin{pmatrix} 0 & i(r_x - 2Er) & i(q_x + 2Eq) \\ 2(iq_x + 2Eq) & 2i(qr - 2E^2) & 0 \\ 2(ir_x - 2Er) & 0 & 2i(qr - 2E^2) \end{pmatrix} \begin{pmatrix} f \\ g \\ h \end{pmatrix} \quad (II.2b)$$

where <u>the focusing nonlinear Schrodinger constraint</u> is

$$q = r^*. \quad (II.2c)$$

We now list properties of this system (II.2) which follow by essentially the same arguments in [2-4].

(II.i) The NLS equation (I.1) is the compatibility condition for the nonlinear system (II.2). That is,

$$iq_t + q_{xx} + 2|q|^2 q = 0 \quad \text{iff} \quad (f,g,h)_{xt} = (f,g,h)_{tx}.$$

(II.ii) $P(E) \equiv f^2 - gh$ is a first integral, independent of x and t.
$$\frac{d}{dx} P = \frac{d}{dt} P \equiv 0. \quad (II.3)$$

(II.iii) <u>g,h satisfy the linearized NLS equation</u>,

$$ig_t + g_{xx} = 2q^2 h - 4qrg$$
$$ih_t - h_{xx} = 4qrh - 2hr^2 q, \quad (II.4)$$

and therefore will be fundamental in the study of tangent and normal spaces to the isospectral class for a solution q. We analyze linearized stability of NLS solutions in Section IV.

(II.iv) These quadratic eigenfunctions provide the fundamental conservation law associated to NLS:

$$[f]_t - [4Ef + i(qh + rg)]_x = 0. \tag{II.5}$$

In fact, as we now describe, (II.5) is a quadratic function for the entire infinite family of NLS conservation laws. (Moreover, this representation is central to the modulation theory of NLS wavetrains; refer to Section III.)

(II.v) Normalize this linear system by the condition

$$f^2 - gh = 1. \tag{II.6a}$$

Then, the asympotic behavior near $E = \infty$ is

$$f \sim 1 + \sum_2^\infty f_j(x,t) E^{-j},$$

$$g \sim \sum_1^\infty g_j(x,t) E^{-j}, \quad \text{as } E \to \infty \tag{II.6a}$$

$$h \sim \sum_1^\infty h_j(x,t) E^{-j}.$$

Insert (II.6b) into the ODE's (II.2), and one easily derives a recursive set of ODE's for the coefficients f_j, g_j, h_j with coefficients given by q, q^* and their derivatives. (We omit the general results, but see (II.vi) for the N-phase formulae.) Moreover, one now finds that (II.5) is an infinite series in E, with conservation laws as coefficients:

$$\sum_2^\infty E^{-j} [(T_j)_t + (X_j)_x] = 0, \text{ or more explicitly,}$$

$$\sum_2^\infty E^{-j} [(f_j)_t - (4f_{j+1} + i(qh_j + rg_j))_x] = 0$$

For example, the first few coefficients are:

$$f_2 = -\frac{1}{2}|q|^2, \quad g_1 = -iq, \quad h_1 = -q^*, \quad f_3 = i/4 \ (q^*q_x - qq_x^*), \tag{II.7b}$$

$$h_2 = -\frac{1}{2} q_x^* , \quad q_2 = \frac{1}{2} q_x .$$

(II.vi) N phase quasiperiodic solutions of NLS <u>are characterized by polynomial quadratic eigenfunctions f,g,h</u> in the following way. Let

$$f = \sum_0^N f_j(x,t) E^j , \quad g = \sum_0^{N-1} g_j(x,t) E^j , \quad h = \sum_0^{N-1} h_j(x,t) E^j . \qquad (II.8a)$$

Then (recall (II.ii)),

$$f^2 - gh = \prod_1^{2N} (E - E_k), \qquad (II.8b)$$

so the recursion formulas of (II.v) become (after normalizing the linear system by the constant factor $\sqrt{f^2 - gh}$ to be consistent with (II.6a,b)),

$$\begin{cases} f_N = 1 ; \quad f_{N-1} = -\frac{1}{2} \sum_1^{2N} E_k ; \quad g_{N-1} = -q; \quad h_{N-1} = ir; \\[4pt] f_{N-2} = \frac{1}{2} [\sum_{i>j} E_i E_j - qr - \frac{1}{4} (\sum_1^{2N} E_k)^2]; \\[4pt] g_{N-2} = \frac{1}{2} q_x + \frac{i}{2} (\sum_1^{2N} E_k) q; \\[4pt] h_{N-2} = -\frac{1}{2} r_x + \frac{i}{2} (\sum_1^{2N} E_k) r . \end{cases} \qquad (II.8c)$$

(II.vii) The <u>constraints on the invariant data</u>, $\{E_k\}^{2N}$ are:

(a) E_k are distinct (otherwise the N - phase potential has degeneracies).

(b) If E_k is a zero of $f^2 - gh$, then E_k^* is also a zero (follows from (II.2a,b,c Together, these constraints imply that E_k in (II.8b) <u>are distinct and occur in conjugate pairs</u>.

(II.viii) The $\vec{\mu}$ <u>(dynamical) representation of the NLS solution</u> q is:

$$\frac{q_x}{q} = i(2 \sum_{j=1}^{N-1} \mu_j(x,t) - \sum_{k=1}^{2N} E_k),$$

$$\frac{q_t}{q} = 2i \left[\sum_{j>k} E_j E_k - \frac{3}{4} \left(\sum_1^{2N} E_k \right)^2 \right] - 4i \left[-\frac{1}{2} \left(\sum_1^{2N} E_k \right) \left(\sum_1^{N-1} \mu_j \right) + \sum_{j>k} \mu_j \mu_k \right],$$

where the entire x, t dependence of $q(x,t)$ is governed by the ODE's:

$$(\mu_k)_x = -2i\, R(\mu_k) / \prod_{\substack{j \neq k \\ j=1}}^{N-1} (\mu_k - \mu_j) \; , \qquad 1 \leq k \leq N-1$$

(II.9b)

$$(\mu_k)_t = (\mu_k)_x \cdot \left(\sum_{j=1}^{2N} E_j - 2 \sum_{\substack{j \neq k \\ j=1}}^{N-1} \mu_j \right) ,$$

and $R(\mu)$ is the hyperellipic irrationality determined by (II.8b):

$$R^2(\mu) = \prod_1^{2N} (\mu - E_k) = R_N^2(\mu) \; . \qquad (II.9c)$$

These variables $\mu_j(x,t)$ arise simply as the zeros of the polynomial g. (II.8a) Since $g_{N-1} = -iq$ by (II.8c), we factor g:

$$g(x,t) = -iq \prod_1^{N-1} (E - \mu_j(x,t)). \qquad (II.10a)$$

II(ix) It follows quite easily from the linear system (II.2) and the polynomial ansatz (II.8) that <u>the focusing NLS constraint, $r = q^*$,</u> translates to the following constraints on polynomial quadratic eigenfunctions:

$$f_j \in R, \; j = 0,\ldots, N,$$

$$f^2(\mu_k) = \prod_{j=1}^{2N} (\mu_k - E_j) \; . \qquad (II.10b)$$

$$h_j = -g_j^* \, , \; j = 0,\ldots, N-1,$$

$$h(x,t,E) = -iq^* \prod_1^{N-1} (E - \mu_j^*(x,t)). \qquad (II.10c)$$

II(x): Conservation laws in the $\vec{\mu}$ - representation

The key to averaging and modulation theory is the remarkable structure that the generating functions for conservation laws have in the $\vec{\mu}$-coordinates. Recall (II.5),

$$[f]_t - [4Ef + i(qh + rg)]_x = 0.$$

After some effort, we are able to represent the density f and flux terms f, qh, and rg as symmetric functions of $\{\mu_j\}$:

$$f = \frac{1}{2}\left(2\sum_1^{N-1}\mu_j - \sum_1^{2N} E_j\right)\prod_1^{N-1}(E - \mu_j) + E\prod_1^{N-1}(E - \mu_j),$$

$$rg = -i|g|^2 \prod_1^{2N-1}(E - \mu_j), \qquad (II.11)$$

$$|q|^2 = -i/2\, q_t/q - \frac{1}{2}q_{xx}/q, \quad q_{xx}/q = (q_x/q)_x + (q_x/q)^2,$$

qh is similar by (II.10), and (II.9a) provide $(\ln q)_t$ x. We appeal to this structure in Section III.

II.(xi) Linearization of the quasiperiodic NLS flows

As shown by the μ-representation (II.9), the entire x, t dependence (aside from a plane wave factor) of an N phase solution q is captured by the μ ode's (II.9b,c). In a minor miracle shared by the other solution equations (see [2], e.g.), these ode's <u>linearize</u> via the classical Abel-Jacobi map associated to the Riemann surface $\mathcal{R}$: $(E, R(E))$, $R^2(E) = \prod_1^{2N}(E - E_k)$, of genus N-1. In fact, the explicit angle variables $\vec{\theta} = \vec{k}x + \vec{\omega}t$ in the action-angle description of these solutions [21,22,3,4] are the images of $\vec{\mu}(x,t)$ under the Abel-Jacobi map. To be more explicit, we require some ingredients of $\mathcal{R}$.

We choose a basis of closed cycles on $\mathcal{R}$ in Figure 4 for the case N=4 (genus 3). All closed paths on $\mathcal{R}$ are linear combinations of this basis a_j, b_j, j =1, ... , N-1. Next we choose a normalized basis

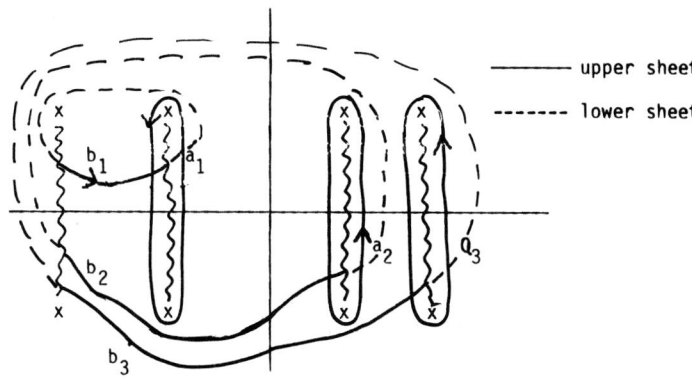

Homology basis on $\mathcal{R}$, N=4.

Figure 4

of **holomorphic** differentials on :

$$\psi_j = \sum_{k=1}^{N-1} C_{jk} \, E^{N-1-k} \, dE/R(E), \quad j = 1, \ldots, N-1, \tag{II.12a}$$

where the normalization constants C_{jk} are uniquely specified by the conditions

$$\oint_{a_i} \psi_j = \delta_{ij} \quad \text{(the Kronecker symbol).} \tag{II.12b}$$

The **period matrix** B of $\mathcal{R}$ is then defined by

$$B_{ij} = \oint_{b_i} \psi_j = B_{ji}, \quad i,j = 1, \ldots, N-1. \tag{II.12c}$$

Note that C_{ij} and B_{ij} depend only on $\{E_j\}$, and thus are invariants of $q_N(x,t)$.

FACT The NLS constraints on $\{E_j\}^{2N}$, (II.vii), yield the period constraints:

$$\begin{cases} \operatorname{Re}(C_{ij}) = 0, \\ \operatorname{Re}(B_{jj}) = 0, \\ \operatorname{Re}(B_{ij}) = -\frac{1}{2} \text{ for } i \neq j. \end{cases} \tag{II.13}$$

With these ingredients, the <u>Abel-Jacobi map</u>: $(\mu_1, \ldots, \mu_{N-1}) \to (\ell_1, \ldots, \ell_{N-1})$, <u>is defined by</u>:

$$\ell_j(\vec{\mu}) = \sum_{k=1}^{N-1} \int_{\mu_k^o}^{\mu_k} \psi_j, \quad j = 1, \ldots, N-1. \tag{II.14a}$$

By straightforward computation one can show, using the μ ode's (II.9b) and Lagrange interpolation formulae,

$$\frac{\partial}{\partial x} \ell_j(\vec{\mu}) = -2i\, C_{j1}$$

$$\frac{\partial}{\partial t} \ell_j(\vec{\mu}) = -2i\, [(\sum_1^{2N} E_k) C_{j1} + 2C_{j2}].$$

Thus, the $\ell_j(\vec{\mu})$ are indeed <u>phases</u> (linear sums in x and t):

$$\ell_j(x,t) = -2i\, [C_{j1}x + [(\sum_1^{2N} E_k) C_{j1} + 2C_{j2}]t] + \ell_j(0,0). \tag{II.14b}$$

The constraints (II.3) show that <u>the focusing NLS flows are constrained by</u>

$$\text{Im}[\ell_j(x,t) - \ell_j(0,0)] = 0. \tag{II.14c}$$

These facts clearly show the $\ell_j(x,t)$ are phases; next we want to give the precise 2π-periodic phases, $\theta_j = \kappa_j x + \omega_j t$, with explicit formulas for the wavenumbers κ_j and frequencies ω_j, and allowable values of the integration constants $\ell_j(0,0)$. These facts require delicate analysis of the Abel-Jacobi map and θ function representation of q; the details are given in [2,3,4]. For the purposes of this paper, we simply state that to write an explicit formula for the NLS solution q, (II.9a) says one needs an explicit formula for $\sum_1^{N-1} \mu_j(x,t)$. Then (II.14a) shows $\mu_1, \ldots, \mu_{N-1}$ satisfy the Jacobi-Inversion problem (invert the Abel-Jacobi map) along the restricted flows (II.14b,c). The classical solution to this problem is given by ratios of Riemann theta functions (see [4]). We will skip these details, and give the relevant phase information:

$$\ell_j(x,t) = \frac{1}{2\pi} \theta_j(x,t) + \ell_j(0,0),$$

$$\theta_j(x,t) = \kappa_j x + \omega_j t,$$

$$\kappa_j = -4\pi i \, C_{j1},$$

$$\omega_j = -4\pi i \, [(\sum_1^{2N} E_j) C_{j1} + 2C_{j2}].$$
(II.15)

Finally, $\ell_j(0,0)$ must satisfy technical constraints [4,8], which we omit, but which are fundamental to show the isospectral class M_q (of all q with $\sum^s(q) = \{E_1,\ldots, E_{2N}\}$) contains one real torus, coordinated by $\vec{\theta} \in [0,2\pi)^N$ and to produce the explicit, C^∞, quasiperiodic theta function solutions of NLS.

(II.xii) The μ-loci for focusing NLS flows

The pointwise analytic behavior of $\mu_j(x,t)$ is quite complicated. However, Ercolani and Forest [4] have shown that the Abel-Jacobi map, (II.14a), restricted to the NLS flows (II.14b,c), (II.15), imposes topological constraints on the closed cycles of $\mu_1,\ldots, \mu_{N-1}$:

$$\mu_j \sim a_j,$$
(II.16)

where ~ denotes "is homologous to", or equivalent as a path of integration on R. Refer to Figures 3 and 4. This fact, that the cycles of μ_j must stay in the same homology class for NLS flows, is responsible on one hand for providing the minimal control necessary to extend the KdV multiphase averaging results to (sG) and NLS (Sec. III), but on the other hand plants the seeds of linearized exponential instabilities in the periodic (sG) and NLS phase spaces (Sec. IV).

(II.xiii) Meromorphic Differentials on $\mathcal{R}$

We now define two unique, meromorphic differentials on $\mathcal{R}$ which are fundamentally connected to the wavenumbers $\vec{\kappa}$ and frequencies $\vec{\omega}$ of q, the Floquet discriminant $\Delta(E,q)$, the averaged conservation laws and slow modulations of q, and the linearized instabilities of q. These objects warrant attention.

Definition
$$\Omega^{(k)} = \sum_{j=0}^{N+1} D_j^{(k)} E^j \, dE/R(E), \quad k = 1, 2, \tag{II.17a}$$

$D^{(1)} = 0$, are the unique Abelian differentials of the second kind on $\mathcal{R}$, $R^2(E) = \prod_1^{2N+1\,2N}(E - E_k)$, that satisfy the following properties:

(a) $\Omega^{(k)}$ has a pole of order $k+1$ at ∞^+, ∞^- on $\mathcal{R}$, and no other poles;

(b) the expansions of $\Omega^{(1)}, \Omega^{(2)}$ in the local coordinate $E = \zeta^{-1}$ near $\infty^{\pm}$ are:

$$\Omega^{(1)} \sim \pm [-\frac{1}{\zeta^2} d\zeta + \text{holomorphic part}] \text{ near } \infty^{\pm},$$
$$\Omega^{(2)} \sim \pm [-\frac{4}{\zeta^3} d\zeta + \text{holomorphic part}] \text{ near } \infty^{\pm};$$
(II.17b)

(c) $\oint_{a_j} \Omega^{(k)} = \oint_{\mu_j} \Omega^{(k)} = 0$, $k = 1, 2$, $j = 1, \ldots, N-1$.
cycles

Uniqueness follows by the Riemann - Roch Theorem. By use of the formulas (II.15) for $\vec{\kappa}, \vec{\omega}$, and the Riemann bilinear identity for differentials of the first (ψ_j) and second ($\Omega^{(k)}$) kind, we find:

$$\kappa_j = \oint_{b_j} \Omega^{(1)}, \quad \omega_j = \oint_{b_j} \Omega^{(2)}. \tag{II.18}$$

For <u>potentials periodic in x of period L</u>, so that $\{\kappa_j\}$ are constrained by

$$\frac{L}{2\pi} = \frac{k_1}{\kappa_1} = \frac{k_2}{\kappa_2} = \ldots = \frac{k_N}{\kappa_N}, \quad k_j \in \mathbb{Z}, \tag{II.19}$$

one can derive the Hochstadt formula [23,22,6]:

$$\Omega^{(1)} = -\frac{1}{L} \frac{\Delta'(E) dE}{\sqrt{4 - \Delta^2(E)}}, \tag{II.20}$$

where $\Delta'(E) = \frac{d}{dE} \Delta(E; q)$.

The symmetries on $\mathcal{R}$, and properties of $\Omega^{(k)}$, lead to the important constraints on the coefficients $D_j^{(k)}$ of $\Omega^{(k)}$:

$$D_j^{(k)} \in \mathbb{R} \quad \forall j, k. \tag{II.21}$$

This concludes our discussion of exact NLS wavetrains. We now turn to apply this structure to stability considerations: with respect to slow modulations of parameters $(\vec{k}, \vec{\omega})$ in Sec. III; and with respect to perturbations of initial conditions (small perturbations of closed degrees of freedom) in Sec. IV.

III. Modulations of NLS Wavetrains

We consider an NLS wavetrain which appears locally as an N-phase solution, $q \simeq q_N(\theta_1, \ldots, \theta_N; E_1 \ldots, E_{2N})$, but which has physical characteristics (such as wavenumbers and frequencies), parametrized by $E_1, \ldots, E_{2N}$, that change slowly over large scales (several periods) in x and t. We call such q(x,t) <u>a modulating N-phase wavetrain</u>; the parameters $\{E_j(X,T)\}_1^{2N}$, $X = \epsilon x$, $T = \epsilon t$, model the slow modulations of the wave. (We remark here that, <u>by ansatz, this is a long wavelength theory</u>). <u>Our goal</u> is to determine the dependence of $E_j(X,T)$, perhaps in the presence of external perturbations. Following [24,1,3], we use the Whitham formalism of averaged conservation laws to derive a quasilinear system of 2N, first order, pde's for $\{E_j(X,T)\}_1^{2N}$. First-order stability is predicted if the system is hyperbolic (modulations can propagate), and instability if the system is elliptic.

We will develop the N-phase NLS modulation equations to an equal level of mathematical structure as has been attained for KdV [1], sinh-Gordon [3] and sine-Gordon [5] wavetrains: (a) An invariant representation of the modulation equations is derived in terms of Abelian differentials on the underlying Riemann surface $\mathcal{R}$ of q_N; (b) a Riemann invariant form, with explicit characteristic speeds and Riemann invariants, is achieved; and (c) the characteristics for focusing NLS are shown non-real for all q_N, so that all focusing NLS quasiperiodic wavetrains (that fit this ansatz) are unstable at first-order to such slow modulations. We close this section with a discussion of the physical implications of (c), with the upshot that with boundary conditions and external perturbations, the unstable modulated NLS wavetrains can stabilize. This theory should be compared with the more familiar linearized stability of Section IV.

Remark: We assume the existence of modulating NLS wavetrains. D. McLaughlin's paper in these Proceedings [25] is aimed at this question for KdV wavetrains. Also, the results of Lax-Levermore [26] and Venakides [20] bear on this question.

Consider a conservation law, $\tilde{\mathcal{J}}_t + \mathcal{X}_x = 0$, valid for solutions of NLS. We evaluate $\mathcal{J}_t + \mathcal{X}_x$ on a modulating N-phase wave, $q_N(\theta_1, \ldots, \theta_N; E_1, \ldots, E_{2N})$, and demand that the phase average vanish. This yields ([3]) the averaged conservation law:

$$\partial_T \langle \mathcal{J}(q_N) \rangle + \partial_X \langle \mathcal{X}(q_N) \rangle = 0, \qquad (III.1a)$$

$$T = \epsilon t, \quad X = \epsilon x,$$

where the phase average $\langle \cdot \rangle$ of any function F of q_N is defined by

$$\langle F(q_N) \rangle = \frac{1}{(2\pi)^N} \int_0^{2\pi} \cdots \int_0^{2\pi} d\theta_1 \cdots d\theta_N \, F(q_N(\vec{\theta};\vec{E})), \qquad (III.1b)$$

and is computed for frozen values of $\{E_k(X,T)\}$. (We note that the NLS densities $\mathcal{J}_j$ and fluxes $\mathcal{X}_j$ are independent of the plane wave phase, which we will call θ_N, as can be seen from inspection of (II.7a,b). Thus (III.1b) collapses to an (N-1) - fold integral in the calculations to follow.)

The averaged density, $\langle \mathcal{J} \rangle$, and averaged flux, $\langle \mathcal{X} \rangle$, therefore depend only on $\{E_k(X,T)\}_1^{2N}$. To obtain a closed system of modulation equations we must average 2N independent conservation laws.

III.A. Averages of the Generating Functions for all NLS Conservation Laws.

Recall the fundamental conservation law, (II.5),

$$[f]_t - [4Ef + i(qh + rg)]_x = 0, \qquad (III.2)$$

which generates all conservation laws, (II.7a), by expanding $E \sim \infty$. We want to average these terms, thereby averaging the entire family of conservation laws. After normalization by $\sqrt{f^2 - gh} = R(E)$, the density and flux terms have simple expressions in the μ coordinates, (II.11). Also, since $f_x = -rg + qh$ and $\langle f_x \rangle = 0$, $\langle rg \rangle = \langle qh \rangle$. Thus we only have to compute the averages of f and rg, which are given by

$$\langle f \rangle = \{(E - \frac{1}{2} \sum_{1}^{2N} E_k) \langle \prod_{1}^{N-1} (E - \mu_j) \rangle + \langle \sum_{1}^{N-1} \mu_j \cdot \prod_{1}^{N-1} (E - \mu_j) \rangle \} \frac{1}{R(E)}, \qquad (III.3)$$

$$i \langle rg \rangle = \{\frac{1}{2} (2 \sum_{k \neq j} E_k E_j - \frac{1}{2} (\sum_{1}^{2N} E_k)^2) \langle \prod_{1}^{N-1} (E - \mu_j) \rangle$$

$$- (\sum_{1}^{2N} E_k) \langle \sum_{1}^{N-1} \mu_j \cdot \prod_{1}^{N-1} (E - \mu_j) \rangle$$

$$+ 2 \langle (\sum_{k \neq j} \mu_j \cdot \mu_k + \sum_{1}^{N-1} (\mu_j)^2) \prod_{1}^{N-1} (E - \mu_j) \rangle \} \frac{1}{R(E)} .$$

These averages are computed [1,3,5] by changing variables of integration from $\vec{\theta}$ to $\vec{\mu}$; the Jacobian is computed from the Abel-Jacobi map:

$$\frac{\partial \vec{\theta}}{\partial \vec{\mu}} = \frac{1}{(2\pi)^{N-1}} \det \left(\frac{\mu_k^{N-1-\ell}}{R(\mu_k)} \right). \qquad (III.4)$$

The product structure of (III.3) and (III.4), together with the $\vec{\mu}$-homology result (II.16), allows the (N-1)-fold integrals to be factored into products of one-dimensional integrals [5]. The resulting expressions are:

$$\langle f \rangle = (\sum_{j=0}^{N} D_j^{(1)} E^j) \frac{1}{R(E)}, \qquad (III.5a)$$

where

$$D_N^{(1)} = 1, \quad D_{N-1}^{(1)} = -\frac{1}{2} \sum_{1}^{2N} E_k ,$$

$$D_j^{(1)} = \frac{1}{\det M} [\det M^{(N-j-1,N)} - \frac{1}{2} (\sum_{1}^{2N} E_k) \det M^{(N-j-1,N-1)}], \quad 0 \leq j \leq N-2,$$

$$M_{k\ell} = \oint_{a_k \sim \mu_k} E^{N-1-\ell} \frac{dE}{R(E)},$$

and $M^{(\alpha,\beta)}$ denotes the matrix M with column α replaced by the vector $\oint_a \frac{E^\beta dE}{R(E)}$;

$$\langle 4Ef + i(qh + rg) \rangle = (\sum_{j=0}^{N+1} D_j^{(2)} E^j) \frac{1}{R(E)} , \qquad (III.5b)$$

where

$$D^{(2)}_{N+1} = 4, \quad D^{(2)}_N = -2\sum_1^{2N} E_k, \quad D^{(2)}_{N-1} = 2\sum_{i \neq j} E_i E_j - \frac{1}{2}(\sum_1^{2N} E_k)^2,$$

$$D^{(2)}_j = \frac{1}{\det M}[-4 \det M^{(N-1-j,N+1)} + 2(\sum_1^{2N} E_k) \det M^{(N-1-j,N)}$$

$$+ (\frac{1}{2}(\sum E_k)^2 - 2\sum_{i \neq j} E_i E_j) \det M^{(N-j-1,N-1)}], \quad 0 < j < N-2.$$

A crucial fact for modulational stability considerations (Theorem III.1, part iii) is that the focusing NLS constraints: $E_j \Rightarrow E_j^*$, $\mu_j \sim a_j$ cycle, yield the constraints

$$D^{(1)}_j, D^{(2)}_j \in \mathbb{R} \quad \forall \ j = 0, \ldots, N+1. \quad (III.5c)$$

(In the defocusing case, $E_j \in \mathbb{R}$ also leads to the result (III.5c).)

<u>Remark</u>: The coefficients $D^{(k)}_j$ appear as ratios of determinants by solving the linear algebraic system (II.17c) of normalization conditions.

If we insert these results into (III.1), we have <u>the first-order N-phase NLS modulation equations by the method of averaging</u>:

$$\frac{\partial}{\partial T}\{\sum_0^N D^{(1)}_j \frac{E^j}{R(E)}\} + \frac{\partial}{\partial X}\{\sum_0^{N+1} D^{(2)}_j \frac{E^j}{R(E)}\} = 0. \quad (III.6)$$

The averages of the usual conservation laws (II.7) for NLS are obtained by expanding this expression in the local coordinate $\xi = E^{-1}$ near $E = \infty$. There are <u>2N</u> variables $E_1(X,T), \ldots, E_{2N}(X,T)$, so we presumably may choose any 2N of these, arriving at a system of 2N, quasilinear, first order p.d.e's.

<u>Remarks</u>: (1) There is an obvious consistency problem. Does one choice of 2N conservation laws imply all the rest are satisfied?

(2) What is the structure of these 2N modulation equations?

(3) Do these perturbation equations predict stability or instability?

The above questions are answered by the following remarkable connection: the averaged generating functions define unique, invariantly defined Abelian differentials on the Riemann surface $\mathcal{R}$ of $R^2(E) = \prod_1^{2N} (E - E_j)$.

Lemma (Invariant Representation of Averaged Generating Functions)

$$<f>\, dE \equiv \Omega^{(1)}, \qquad (III.7)$$

$$<4Ef + i(qh + rg)>\, dE \equiv \Omega^{(2)},$$

where $\Omega^{(1)}$, $\Omega^{(2)}$ are the unique Abelian differentials of the second kind defined earlier by (II.17,a,b,c).

Now form the unique differential

$$\Omega \equiv \Omega_T^{(1)} \, \Omega_X^{(2)}, \qquad (III.8a)$$

which has the expansion in the local parameter near $E = \infty^{\pm}$:

$$\Omega(E) \sim \mp \sum_{j=2}^{\infty} [<\mathcal{J}_j>_T + <\mathcal{X}_j>_X] \zeta^{j-2} \, d\zeta \text{ near } E = \zeta^{-1} = \infty^{\pm}. \qquad (III.8b)$$

The coefficients of $\Omega(E)$ near ∞ are averages of the conservation forms (II.7). Also, since $\oint_{a_j} \Omega^{(1)} = \oint_{a_j} \Omega^{(2)} = 0$, it follows that

$$\oint_{a_j} \Omega = 0, \; j = 1,\ldots, N-1. \qquad (III.8c)$$

Theorem III.1 Assume the first 2N averaged conservation laws are satisfied by $\{E_k(X,T)\}^{2N}$,

$$<\mathcal{J}_j>_T + <\mathcal{X}_j>_X = 0, \; j = 2,\ldots, 2N+2. \qquad (III.9a)$$

Then all higher averaged conservation laws are satisfied, and the modulation equations (III.9) take the equivalent form,

$$\Omega \equiv \Omega_T^{(1)} - \Omega_X^{(2)} \equiv 0. \qquad (III.9b)$$

Remark on proof: It follows from the Riemann-Roch Theorem that the differential Ω, defined by (III.8a,b,c), must vanish identically if it has a zero of degree 2N or harder at $\infty^{\pm}$. See [1,3]. This partly resolves the consistency question of Sec. III.A.

Theorem III.2 (Consequences of the Invariant Representation (III.9))

(A) Expansion of $\Omega = 0$ near $E = \infty^{\pm}$ shows all averaged conservation laws are satisfied ($\Omega \equiv 0$ implies each coefficient must vanish):

$$\Omega \sim -\sum_{2}^{\infty} \zeta^{j-2} \, d\zeta \, [\langle \mathcal{J}_j \rangle_T + \langle \mathcal{X}_j \rangle_X] \equiv 0.$$

(b) The other marked points on $\mathcal{R}$ are the branch points E_k. We expand $\Omega = 0$ in the local parameter ζ near E_k, $E - E_k = \zeta^2$, and find

$$\Omega \sim [\tilde{\Omega}_k^{(1)}(E_k)_T - \tilde{\Omega}_k^{(2)}(E_k)_X] \, \zeta^{-2} \, d\zeta + \text{holomorphic part},$$

where

$$\tilde{\Omega}_k^{(\ell)} = \sum_{j=0}^{N+1} D_j^{(\ell)} (E_k)^j \Big/ \sqrt{\prod_{\ell \neq k}^{2N} (E_k - E_\ell)}.$$

Since $\Omega \equiv 0$, the leading coefficient vanishes, and we deduce the explicit Riemann invariant form of the modulation equations, where the modulating simple spectra $\{E_k\}$ are the Riemann invariants:

$$(E_k)_T + S^{(k)}(E_k)_X = 0, \quad k = 1,\ldots,2N,$$

$$S^{(k)} \equiv \left(\sum_{j=1}^{N+1} D_j^{(2)} (E_k)^j\right) \Big/ \left(\sum_{j=0}^{N} D_j^{(1)} (E_k)^j\right).$$

(III.10)

(C) From (III.5C), the coefficients $D_j^{(1)}$, $D_j^{(2)}$ are <u>real</u>, while $E_k \not\in R$ for focusing NLS. Thus, <u>the characteristic sppeds</u> $S^{(k)}$ <u>are non-real, so the modulation equations are elliptic, and modulational instability is predicted for all quasiperiodic NLS wavetrains</u>. In the <u>defocusing</u> case, all E_k are real, so stability is predicted.

(D) From $\oint_{b_j} \Omega = 0 = (\oint_{b_j} \Omega^{(1)})_T - (\oint_{b_j} \Omega^{(2)})_X$, and formulas (II.18), we deduce the consistency condition referred to as <u>conservation of waves</u>:

$$(\kappa_j)_T = (\omega_j)_X \quad , \quad j = 1, \ldots, N-1.$$

Remarks on PHYSICAL IMPLICATIONS OF MODULATIONAL INSTABILITY

This negative result gives information about potential physical applications of focusing NLS wavetrains, but it is by no means fatal. The same result obtains for all (sG) wavetrains which contain at least one purely oscillatory mode [5] and for which this theory applies. One contention is that specific boundary conditions and/or external perturbations may be required to support and sustain the oscillations. Another approach could be to retain higher order terms (second derivatives) to see if they stabilize the modulations.

In many physical applications (e.g., Josephson junctions) and numerical simulations [27] of NLS and (sG) wavetrains, periodic boundary conditions in x of fixed period L are rigidly enforced. This imposes N commensurability constraints, equations (II.19), on the parameters $E_1, \ldots E_{2N}$:

$$\frac{L}{2\pi} = \frac{k_j}{\kappa_j} \quad , \quad j = 1, \ldots, N, \; k_j \in Z \; ,$$

reducing the system to N degrees of freedom. Clearly, κ_j must all be identically constant or these commensurability conditions are broken. In particular, $(\kappa_j)_T = 0$, $j = 1, \ldots, N$. Now, by conservation of waves, (III.11): $(\omega_j)_X = 0$, $j = 1, \ldots, N$. These impose N spatial constraints on an N parameter system, so we conclude: There is no spatial modulation in the presence of periodic boundary conditions of fixed period. Moreover, in the absence of external modulations, there are no terms to balance temporal modulations of $\{E_k\}$, $(E_k)_T = -S^{(k)}(E_k)_X \equiv 0$, so that: no modulations, temporal nor spatial, are possible for x-periodic wavetrains of fixed period in the absence of external perturbations.

However, in [28] we show how to derive generalized modulation equations in the presence of external perturbations (such as damping or driving). In this situation, (III.9b) becomes

$$\Omega = \Omega_T^{(1)} - \Omega_X^{(2)} = dF, \tag{III.11}$$

where $\Omega^{(1)}$, $\Omega^{(2)}$ are as before, but the exact differential dF is explicitly

III.B. An Invariant Representation of the Modulation Equations

The above questions are answered by the following remarkable connection: <u>the averaged generating functions define unique, invariantly defined Abelian differentials on the Riemann surface</u> $\mathcal{R}$ <u>of</u> $R^2(E) = \prod_1^{2N}(E - E_j)$.

<u>Lemma</u> (Invariant Representation of Averaged Generating Functions)

$$<f> \, dE \equiv \Omega^{(1)}, \qquad (\text{III}.7)$$

$$<4Ef + i(qh + rg)> dE \equiv \Omega^{(2)},$$

where $\Omega^{(1)}$, $\Omega^{(2)}$ are the unique Abelian differentials of the second kind defined earlier by (II.17,a,b,c).

Now form the unique differential

$$\Omega \equiv \Omega_T^{(1)} - \Omega_X^{(2)}, \qquad (\text{III}.8a)$$

which has the expansion in the local parameter near $E = \infty^{\pm}$:

$$\Omega(E) \sim \mp \sum_{j=2}^{\infty} [<\mathcal{J}_j>_T + <\mathcal{X}_j>_X]\zeta^{j-2}\, d\zeta \text{ near } E = \zeta^{-1} = \infty^{\pm}. \qquad (\text{III}.8b)$$

The coefficients of $\Omega(E)$ near ∞ are averages of the conservation forms (II.7). Also, since $\oint_{a_j} \Omega^{(1)} = \oint_{a_j} \Omega^{(2)} = 0$, it follows that

$$\oint_{a_j} \Omega = 0, \quad j = 1,\ldots, N-1. \qquad (\text{III}.8c)$$

<u>Theorem III.1</u> Assume the first 2N averaged conservation laws are satisfied by $\{E_k(X,T)\}_1^{2N}$,

$$<\mathcal{J}_j>_T + <\mathcal{X}_j>_X = 0, \quad j = 2,\ldots, 2N+2. \qquad (\text{III}.9a)$$

Then <u>all higher averaged conservation laws are satisfied</u>, and <u>the modulation equation</u> (III.9) <u>take the equivalent form</u>,

$$\Omega \equiv \Omega_T^{(1)} - \Omega_X^{(2)} \equiv 0. \qquad (\text{III}.9b)$$

computed from the average of the external perturbation of NLS.

The correction, dF, does not involve derivatives of E_j, and therefore the NLS characteristic speeds remain complex. The external perturbation does not stabilize the quasiperiodic modulation equations. But, for fixed period L in x, which prevents spatial modulations, the temporal modulations of $E_j(T)$ are now balanced by the effects, dF, of external perturbations:

$$\Omega_T^{(1)} = dF. \qquad (III.12)$$

This represents a system of N ode's for $E_1,\ldots, E_{2N}$, subject to the N constraints (II.19).

The predictions of this perturbation analysis for the driven, damped (sG) and NLS equations under periodic boundary conditions is currently under study. (See Section IV.F).

IV. Linearized Instabilities of Periodic NLS Wavetrains

The exact, x-independent plane solution of focusing NLS, $q = a \exp(2ia^2 t)$, is well-known to be linearly unstable. With decaying boundary conditions on the infinite line, this classical instability (called the Benjamin-Feir instability in water waves) is known to saturate by the formation of envelope solitons. The analogous instability obtains for the x-independent solutions of sine-Gordon.

Ercolani, Forest and McLaughlin [6] have shown that these linear, exponential instabilities are quite general and may exist in the neighborhood of arbitrary, degenerate, spatially periodic (sG) solutions. Independently, Tracy [16] used the periodic inverse theory to study these modulational instabilities for periodic NLS in the neighborhood of the plane wave. In this section, we use the geometric structure described in Sec. I,II to generalize the classical results on the NLS plane wave. We will: (A) outline the classical argument on the plane wave instabilities; (B) show how the Floquet discriminant for the plane wave, (I.7a), labels all unstable modes; (C) show that the geometry of the focusing NLS $\vec{\mu}$ cycles explains the linearized instabilities; (D) briefly discuss the rigorous analysis in [6] necessary for a complete mathematical solution; and (E) present a different proof from [6] of the linearized results, based on the method of averaging.

IV. A. The Classical Instability Argument for the NLS Plane Wave Solution.

(i) Consider the exact, x-independent solution $q_0 = ae^{2ia^2t}$, and seek

$$q(x,t) = q_0 + q_1, \quad |q_1| \ll 1,$$

$$= [a + \alpha(x,t)]e^{2ia^2t}, \quad |\alpha| \ll 1, \quad a \in \mathbb{R} .$$
(IV.1a)

The linearized NLS about an arbitrary solution q_0 is:

$$iq_{1_t} + q_{1_{xx}} + 4(q_0)^2 q_1 + 2q_0^2 q_1^* = 0.$$
(IV.1b)

For these particular choices, $q_0 = ae^{2ia^2t}$, $q_1 = \alpha(x,t)e^{2ia^2t}$, we find

$$\begin{cases} (i\partial_t + \partial_{xx} + 2a^2)\alpha = -2a^2\alpha^* \\ (-i\partial_t + \partial_{xx} + 2a^2)\alpha^* = -2a^2\alpha , \end{cases}$$

which decouples to a simple equation for α:

$$(\partial_{xx} + \partial_{xxxx} + 4a^2\partial_{xx})\alpha = 0 .$$

Now let $\alpha(x,t) = e^{\sigma t} e^{iKx} \hat{\alpha}$, and we find

the linear growth rate σ is:

$$\sigma^2 = -K^2(K^2 - 4a^2).$$
(IV.2).

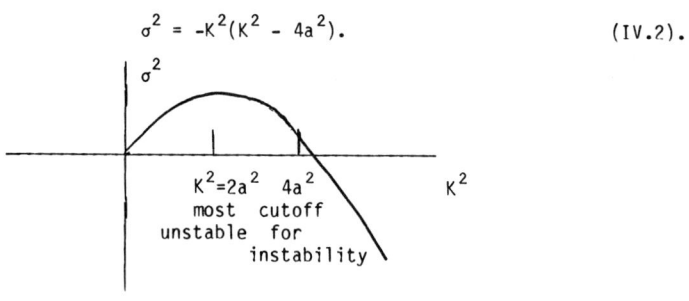

Figure 5

Conclusions: 1°) Long waves ($0 < K^2 < 4a^2$) are unstable. The $K = 0$ solution, $ae^{2ia^2 t}$, feeds the $K_*^2 = 2a^2$ longwavelength modulational instability.

2°) Boundary conditions are important. For fixed spatial period $L = 2\pi/K$, only a discrete set of unstable modes $K_n^2 = \frac{n^2 \pi^2}{L^2}$ is allowed. For L small enough ($4a^2 < K^2 = \frac{4\pi^2}{L^2}$), $q_0 = ae^{2ia^2 t}$ is linearly stable. As L increases, more unstable modes are generated; in fact,

$2n$ unstable modes exist for $ae^{2ia^2 t}$ if $\frac{n^2 \pi^2}{a^2} < L^2 < \frac{(n+1)^2 \pi^2}{a^2}$. (IV.3)

(IV.B) Connection between Linearized Stability of the Plane Wave and the Associated Floquet Discriminant.

In section I.A, we discussed the Floquet discriminant $\Delta(E;q)$ and described how the double points (E_d^j) of $\Delta(E;q)$ correspond one-to-one with all closed degrees of freedom associated to q. Of course, in studying the linearization of NLS about $q_0 = ae^{2ia^2 t}$, we are computing the number of unstable Fourier modes as a function of the spatial period L. On the other hand, we have explictly analyzed all double points (closed modes) of $\Delta(E; q_0 = ae^{2ia^2 t})$ as a function of varying L.

There is the obvious striking parallel between the number of linearly unstable Fourier modes and the number of non-real double points of Δ, as indicated in relations (I.7c) and (IV.3): they are equal.

Theorem V. (Classical Analysis of the NLS Plane Wave)

For $\frac{n^2 \pi^2}{a^2} < L^2 < \frac{(n+1)^2 \pi^2}{a^2}$, and $q_0 = ae^{2ia^2 t}$, there are exactly 2N exponentially unstable modes for NLS linearized about q_0; moreover, there are exactly 2N double points of $\Delta(E;q_0)$ off the real E axis.

Remarks. This is essentially the result of Tracy [16]. We want to state the analogous result for arbitrary periodic NLS wavetrains.

IV.C. $\vec{\mu}$-Coordinates: A Geometric Realization of Linearized Instability

As we have stated, all degeneracies of a given periodic solution $q(x,t)$ are accounted for by double points of the Floquet discriminant $\Delta(E;q)$. A generic periodic potential of (I.2a) has no degeneracies: In any neighborhood of $q(x,t)$ in the NLS periodic phase space, all degrees of freedom are excited. Correspondingly, all double points are split into a pair of simple periodic spectra.

The μ variables coordinatize the degrees of freedom for $q(x,t)$, both closed and open; recall relations (I.9) and (II.16), and Figure 3. Therefore, we can compare the geometry of the μ-cycles for a degenerate $q(x,t)$ with the $\vec{\mu}$-geometry when the double points are opened order ϵ.

We focus on the opening of a single double point, E_d^0. There are two cases to consider: $E_d^0 \in \mathbb{R}$ and $E_d^0 \notin \mathbb{R}$.

<u>Case 1</u>: $E_d^0 \in \mathbb{R}$, with the corresponding $\mu^0 \equiv E_d^0$ (Figure 6a).

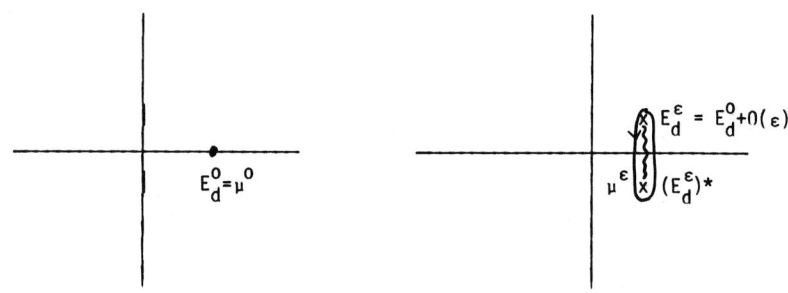

Figure 6a   Figure 6b

Now open the double point order ϵ (Figure 6b) into two conjugate simple spectra: $E_d^\epsilon = E_d^0 + O(\epsilon)$ and $(E_d^\epsilon)^*$. The corresponding μ^ϵ - cycle is homologous to an a-cycle, as indicated in Figure 6b. This picture <u>displays</u> the estimate $|\mu^\epsilon - \mu^0| = O(\epsilon)$, completely within the NLS constraints, for both the x and t flows. From the $\vec{\mu}$-representation (II.9) of solutions, the corresponding $q^0(x,t)$ and $q^\epsilon(x,t)$ do not depart more than order ϵ, which suggests linearized stability of the x and t flows (and even nonlinear stability).

<u>Case 2</u> $E_d^0 \notin \mathbb{R}$, so $(E_d^0)^*$ is also a double point, with corresponding $\mu_1^0 \equiv E_d^0$,

$\mu_2^0 \equiv (E_d^0)^*$ (Figure 7a).

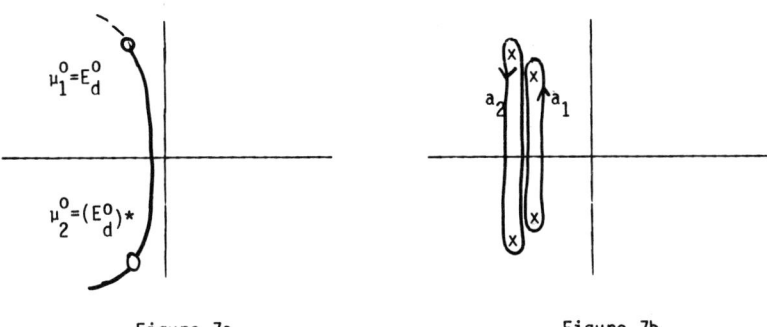

Figure 7a   Figure 7b

Now the double points $E_d^0, (E_d^0)^*$ split into four simple spectra, $E_k^\epsilon = E_d^0 + 0(\epsilon)$, $k = 1,2$, and $(E_k^\epsilon)^*$. The a-cycles created by this perturbation, a_1 and a_2 in Figure 7b, are a basis for the closed flows of μ_1^ϵ and μ_2^ϵ. Clearly, these orbits are 0(1) perturbations of $\mu_1^0 = E_d^0$, $\mu_2^0 = (E_d^0)^*$. Thus, an 0(ϵ) change in the periodic spectrum leads to an 0(1) change in the μ-spectrum, and thereby an 0(1) change in the potential $q(x,t)$. This is the modulational instability. To be precise, the instability is in the t-flow, not the x-flow. This is proved analytically in [6] and Sec. IV. to follow. The geometric evidence is achieved by taking the linear combinations of a_1, a_2 cycles corresponding to the x-flow and t-flow, as indicated in Figure 8.

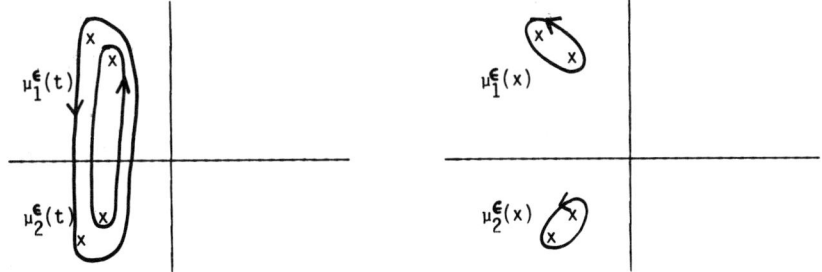

Figure 8

IV.D. Outline of the Linearized Analysis in [6]

A complete analysis of the linearization about an N-phase wavetrain is developed in [6] for the periodic (sG) equation. The same methods apply to NLS, but have not been carried out yet. At this point we will simply outline that approach in the NLS context; in the next section, we present a more limited analysis aimed at calculating the linearized growth rates.

As indicated in (II.4), quadratic eigenfunctions satisfy the linearized NLS about an arbitrary solution. Moreover, the eigenfunctions of (I.2) for periodic q (called Baker functions) have explicit representations. By evaluating the quadratic Baker functions at a countable number of periodic spectra E_j (all double points and half the simple spectra), a basis of the periodic phase space is constructed out of solutions of the linearized NLS. From the behavior of Baker functions, each basis element is analyzed, which yields explicit formulas that determine boundedness or exponential growth in the x and t flows of the basis elements. It is shown in [6] that all x-flows are stable, and for t-flows the only possible exponential growth elements correspond to double points; moreover, in the language of NLS, the non-real double points correspond to all linearized instabilities, and they are finite in number.

IV.E. Linearization with the $\vec{\mu}$-Representation and the Method of Averaging

We wish to supply the analysis that supports the heuristic arguments in IV.C. We consider a degenerate, N-phase solution of NLS, $q_N(x,t)$, and compute the linearized growth rates corresponding to opening a closed degree of freedom (double point) order ε. Again, there are two cases (Figures 6,7): E_d real or non-real, corresponding to opening $\delta = 1$ or 2 d.f.

We have the exact $\vec{\mu}$ - representation for q_N, (II.9), and the exact $\vec{\mu}$-representation for the perturbed solution $q_{N+\delta}$, $\delta=1,2$, which has the simple spectra $E_1,\ldots, E_{2N}$ of q_N and 2δ more. Therefore, we can linearize directly in the $\vec{\mu}$-representaton as follows:

(1) Begin with the $\vec{\mu}$-ode's (II.9) for $\mu_1, \ldots, \mu_{N+\delta}$, the dynamical coordinates of $q_{N+\delta}$. We have $R_{N+\delta}^2(\mu) = \prod_1^{2N+2\delta}(\mu - E_k)$, where $E_{2N+j} = E_d + \varepsilon E_{2N+j}^{(1)}$, $j = 1,2$, and the other two are their conjugates. Also, for $k=1, \ldots, 2N$, $E_k = E_k^0 + O(\varepsilon)$, where $R_N^2(\mu) = \prod_1^{2N}(\mu - E_k^0)$ is the curve for q_N. Also, expand $\mu_j = \mu_j^0 + \varepsilon \mu_j^{(1)}$, where μ_j^0, $j = 1, \ldots, N-1$, are the $\vec{\mu}$-variables for q_N, and $\mu_N = E_d + \mu_N^1$, $\mu_{N+1} = E_d^* + \varepsilon \mu_{N+1}^{(1)}$.

(2) Next, we insert these expansions into the ode's (II.9), expand in ε, and retain terms to order ε. The ε^0 terms, of course, simply reproduce the fact that $\mu_j^{(0)}(x,t)$, $j = 1, \ldots N-1$, are the $\vec{\mu}$-variables for q_N. The ε^1 terms are the linearized μ representation of $q_{N+\delta}$ about q_N, which takes the form of a linear system for $\vec{\mu}^{(1)} = (\mu_1^{(1)}, \ldots, \mu_{N+1}^{(1)})$:

$$[\vec{\mu}^{(1)}]_t = F(\vec{\mu}^{(0)}(x,t))\vec{\mu}^{(1)}. \qquad (IV.1)$$

Our concern is the behavior of the variables μ_N, μ_{N+1} that have been untied from the double points. These particular $O(\varepsilon)$ terms decouple from the others, governed by the ode's:

$$(\mu_N^{(1)})_{x \atop t} = \left[\frac{-2i \, R_N(E_d)[\sum_1^{2N} E_j^{(0)} - 2 \sum_1^{N-1} \mu_j^{(0)} + 2E_d]^1_0}{\prod_1^{N-1}(E_d - \mu_j^{(0)})} \right] \cdot \mu_N^{(1)} \qquad (IV.2)$$

and a similar equation for $\mu_{N+1}^{(1)}$.

We apply classical Floquet theory to (IV.2); the x,t means of the coefficient in (IV.2) determine the x,t Floquet exponents of $\mu_N^{(1)}$. <u>Linearized stability in x or t obtains if the x or t mean is purely imaginary; the linearized growth rate is given by the real part of the Floquet mean.</u>

We compute these Floquet means by the method of averaging described in Section III. The results are:

Theorem IV.1 (Floquet means)

The Floquet $\begin{pmatrix} x \\ t \end{pmatrix}$ - mean for $\mu_N^{(1)} = i \int_{(E_d,-R_N(E_d))}^{(E_d,R_N(E_d))} \Omega^{(1)}_{(2)}$, (IV.3)

where $\Omega_N^{(1)}$, $\Omega_N^{(2)}$ are the Abelian differentials (II.17a) associated to $q_N(x,t)$, and E_d is the unperturbed double point, either real or non-real.

The stability results follow by analysis of these integrals. Of course, the differentials $\Omega_N^{(k)}$ are familiar objects in our previous studies (equations (II.18,20,21), (III. 7,8,9,13,14)). We compute:

$$\int_{(E_d,-)}^{(E_d,+)} \Omega^{(1)} = \lim_{E_{2N+1}, E_{2N+2} \to E_d} \int_{b_N} \Omega^{(1)} = \lim_{E_{2N+j} \to E_d} \kappa_{N+1} = \kappa_{N+1} \in \mathbb{R}, \; \forall \; E_d,$$

where we have used (II.18) and the fact that periodicity in x fixes all κ_j. This proves

Theorem IV.2 (Linearized stability of x-flows)

The Floquet X-mean for $\mu_N^{(1)} = i\,\kappa_{N+1}$, which is purely imaginary for all double points E_d, real and non-real. All x-flows are linearly stable.

The analysis of the t-mean is slightly more delicate, comparable to the proof that the coefficients of $\Omega^{(k)}$ are real, (III.5c). The results of this analysis are:

Theorem IV.3 (a) The linearized t-flows of $\mu_N^{(1)}$ are <u>stable if the double point</u> E_d <u>is real</u>, and <u>unstable if the double point</u> E_d <u>is non-real</u>.

(b) There are at most a finite numer of non-real double points (**Propert** (I.V.)).

IV.F. <u>Experimental Evidence of the Modulational Instabilities; Concluding Remarks</u>

This completes our discussion of linearized stability. It should be noted that these modulational instabilities have been witnessed quite conclusively in the transient response of numerical experiments [27] on periodic (sG). When the system is integrated with initial data $q_N(x,0)$ that has non-real double points, these instabilities are immediately excited. The waveform then saturates in the formation of a structure approximated by adding the degrees of freedom created by opening up the non-real double points into simple spectra. The resulting structure has no non-real double points in its discriminant, and is at least locally stable until external perturbations carry the data far away. Locally, once the instabilities have saturated, the perturbation equations (III.12) apply. As alluded to earlier, and clarified in [6], homoclinic orbits are associated to non-real double points. These orbits in the periodic NLS phase space are currently under study for their role in temporally chaotic dynamics, with finite d.f. spatial structure, of the damped, driven NLS equation.

As alluded to at the end of Sec. III, the effectiveness of these combined tools (exact solutions, linearized stability and modulations under external perturbations) as measuring and predicting devices is the focus of our current work on near integrable equations. Surely one necessary ingredient will be the ability to track the evolution of these systems through the generation of new spatial degrees of freedom. We have indicated one mechanism that we can control, the generation through modulational instabilities. The Lax-Levermore-Venakides results on KdV indicate other mechanisms, intimately related to degenerate, characteristics of the modulation equations. The analogous theory for (sG) and NLS has yet to be resolved, but poses as a challenging learning experience.

* Supported in part by NSF Grant DMS 8411002.

References

1. Flaschka, H., Forest, M.G. and McLaughlin, D.W., Multiphase averaging and the inverse spectral solution of the KdV equation, Comm. Pure Appl. Math. 33, 739-784 (1980).

2. Forest, M.G., and McLaughlin, D.W., Spectral theory for the periodic sine-Gordon equation: a concrete viewpoint, J. Math. Phys. 23 (7), 1248-1277 (1982).

3. Forest, M.G., and McLaughlin, D.W., Modulations of sine-Gordon and sinh-Gordon wavetrains, Stud. Appl. Math. 68, 11-59 (1983).

4. Ercolani, N.M., and Forest, M.G., The geometry of real sine-Gordon wavetrains, Comm. Math. Phys. 99, 1-49 (1985).

5. Ercolani, N.M., Forest, M.G., and McLaughlin, D.W., Modulational stability of two-phase sine-Gordon wavetrains, Stud. Appl. Math. 71, 91-101 (1984).

6. Ercolani, N.M., Forest, M.G., and McLaughlin, D.W., Modulational instabilities of periodic sine-Gordon waves: a geometric analysis, to appear in Nonlinear Systems of PDE in Appl. Math., AMS-SIAM Summer Seminar, 1984.

7. Ercolani, N.M., Forest, M.G., and McLaughlin, D.W., Oscillations and instabilities in near integrable pde's, to appear in Nonlinear systems of PDE in Applied Math., AMS-SIAM Summer Seminar, 1984.

8. Lee, Jong-Eao, Ph.D. Thesis, Ohio State University, 1986.

9. Zakharov, V., and Shabat, A., Soviet Phys. JETP 34, 62 (1972).

10. Ablowitz, M., Kaup, D., Newell, A., and Segur, H., Stud. Appl. Math. 53, 249 (1974).

11. McKean, H.P., and van Moerbeke, P., The Spectrum of Hill's equation, Invent. Mat. 30, 217-274 (1975).

12. Dubrovin, B.A., Matveev, V.B., and Novikov, S.P., Usp. Math. Nauk 31, 55-136 (1976).

13. McKean, H.P., and Trubowitz, E., C.P.A.M. 29, 143 (1976).

14. Kotlyarov, V. and Its, A.R., Dop. Akad Nauk., UkRSR, A (11) 1976.

15. Ablowitz, M.J. and Ma, Y.C., Stud. Appl. Math. 65, 113 (1981).

16. Tracy, E., Ph.D. Thesis, Univ. Maryland, 1984.

17. Daté, E., Multisoliton solutions and quasi-periodic solutions of nonlinear equations of sine-Gordon type, preprint, 1981.

18. Previato, E., Ph.D. Thesis, Harvard Univ., 1984.

19. Deift, P. Lund, F., and Trubowitz, E., Nonlinear wave equations and constrained harmonic motion, Comm. Math. Phys. 74, 141-188 (1980).

20. Venakides, S., The zero dispersion limit of KdV with periodic initial data, these Proceedings.

21. Flaschka, H., and McLaughlin, D.W., Prog. Theor. Phys. 55, 438 (1976).

22. Forest, M.G., and McLaughlin, D.W., Canonical variables for the periodic sine-Gordon equation and a method of averaging, LA-UR 78-3318, Report of the Los Alamos Scientific Laboratory, 1978.

23. McKean, H.P., The sine-Gordon and sinh-Gordon equations on the circle, C.P.A.M. 34, 197-257 (1981).

24. Whitham, G.B., *Linear and Nonlinear Waves*, Wiley-Interscience, 1974.

25. McLaughlin, D.W., On the construction of a modulating N-phase KdV solution, these proceedings, preprint 1983.

26. Lax, P.D., and Levermore, C.D., The zero dispersion limit for the KdV equation, I,II,III, C.P.A.M. 36, 253 (1983); 36, 571 (1983); 36, 809 (1983).

27. Overman, E.A., McLaughlin, D.W., and Bishop, A.R., Coherence and chaos in the driven, damped sine-Gordon equation: measurement of the soliton spectrum, to appear, Physica D, 1985.

28. Forest, M.G., and McLaughlin, D.W., Modulations of perturbed KdV wavetrains, SIAM J. Appl. Math. $\underline{44}$, (2), 287 (1984).

ON HIGH-ORDER ACCURATE INTERPOLATION FOR NON-OSCILLATORY SHOCK CAPTURING SCHEMES

Ami Harten

School of Mathematical Sciences
Tel-Aviv University

and

Department of Mathematics
University of California, Los Angeles

ABSTRACT: In this paper we describe high-order accurate Godunov-type schemes for the computation of weak solutions of hyperbolic conservation laws that are essentially non-oscillatory. We show that the problem of designing such schemes reduces to a problem in approximation of functions, namely that of reconstructing a piecewise smooth function from its given cell averages to high order accuracy and without introducing large spurious oscillatons. To solve this reconstruction problem we introduce a new interpolation technique that when applied to piecewise smooth data gives high-order accuracy wherever the function is smooth but avoids having a Gibbs-phenomenon at discontinuities.

1. Introduction

In this paper we present an interpolation technique that gives rise to new high-order accurate non-oscillatory schemes for the numerical solution of scalar conservation laws:

(1.1a) $\quad u_t + f(u)_x \equiv u_t + a(u)u_x = 0, \quad -\infty < x < \infty, \ t > 0$

(1.1b) $\quad u(x,0) = u_0(x), \quad -\infty < x < \infty.$

We assume the initial data $u_0(x)$ to be piecewise-smooth functions that are either periodic or of compact support, and denote the evolution operator of the unique entropy solution by $E(t)$.

Research sponsored by the United States Army under Contract NO. DAAG29-30-C-0041 while in residence at the MRC, University of Wisconsin-Madison, and by NASA Consortium Agreement #NCA2-IR390-403 and ARO Grant #DAAG29-82-0090 while at UCLA.

Let $v_h(x,t)$ denote a numerical approximation to a weak solution $u(x,t)$ of (1.1), such that $v_j^n = v_h(x_j, t_n)$, $x_j = jh$, $t_n = n\tau$, satisfies the conservation form

(1.2a) $$v_j^{n+1} = v_j^n - \lambda(f_{j+1/2} - f_{j-1/2}) \equiv [E_h(\tau) \cdot v^n]_j.$$

Here $E_h(\tau)$ is the numerical solution operator, $\lambda = \tau/h$ and $f_{j+1/2}$, the numerical flux, is a function of $2k$ variables

(1.2b) $$f_{j+1/2} = f(v_{j-k+1}^n, \ldots, v_{j+k}^n)$$

which is consistent with (1.1a) in the sense that

(1.2c) $$f(u, u, \ldots, u) = f(u).$$

It is well known that if the total variation of the numerical solution (1.2) is uniformly bounded in h for $0 < t < T$

(1.3) $$TV(v_h(\cdot, t)) \leq C \cdot TV(u_0),$$

then any refinement sequence $h \to 0$, $\tau = O(h)$, has a subsequence $h_j \to 0$ so that

(1.4) $$v_{h_j} \xrightarrow{L_1} u$$

where u is a weak solution of (1.1). Furthermore, if all limit solutions (1.4) satisfy an entropy condition that implies uniqueness of the initial value problem (1.1), then the numerical scheme is convergent (see [4]).

Our goal in designing numerical schemes is to obtain a computer code that is both reliable and efficient. To achieve reliability we would like to use schemes that are total-variation-stable in the sense of (1.3); to get efficiency we need high-order accuracy. Unfortunately it became clear in the development of shock-capturing schemes that it is not easy to satisfy both requirements at the same time: Endowing a scheme with a property that implies automatic control over the growth of the total variation of the numerical solution may very well lead to a restriction on the order of accuracy of the scheme.

The first successful attempt to achieve nonlinear stability was to require positivity of the numerical solution operator, which led to the development of monotone schemes. However the requirement of positivity automatically implies first order accuracy of the scheme (see [2]).

The next step in the development was to consider the larger class of total-variation-diminishing (TVD) schemes, where the numerical solution operator is required to diminish the total variation of any BV function v

(1.5) $$TV(E_h(\tau) \cdot v) \leq TV(v) ;$$

these schemes trivially satisfy (1.3) with $C = 1$ (see [3]). We were able to construct TVD schemes that in the sense of local truncation error are high-order accurate everywhere except at local extrema where they necessarily degenerate into first order accuracy, (see [9], [3], [1], [7], [8]). The perpetual damping of local extrema determines the cumulative global error in the L_p-norm to be of $(1 + \frac{1}{p})$-th order, i.e. only first-order in the maximum norm but second-order in L_1.

Recently ([5]) we went one step further and introduced a larger class of <u>non-oscillatory</u> <u>schemes</u>, in which the application of the numerical solution operator to any mesh function v is required not to increase the <u>number</u> of local extrema (note that this statement does not depend on h nor on the smoothness of v). Unlike TVD schemes, which are a subset of this class, nonoscillatory schemes are not required to damp the values of each local extremum at every single time-step, but are allowed to occasionally accentuate a local extremum. Because of this last property we were able to construct non-oscillatory schemes that are second-order accurate also in the maximum norm.

In the present paper we relax the control over the possible growth of the total variation of the numerical solution even further and consequently are able to design schemes that are accurate to any finite order r. These schemes satisfy

(1.6) $$TV(E_h(\tau) \cdot u) \leq TV(u) + o(h^{r+1}), \quad u \in U$$

where U is the set of functions that are C_∞ except at a finite number of points (in any finite interval) where they may have a discontinuity in the function or some derivative (U describes the generic form of the "computable" solutions of (1.1)). We shall refer to this class of schemes as <u>essentially non-oscillatory</u> schemes. The inequality (1.6) ensures that the scheme does not have a Gibbs-like phenomenon of oscillations that are proportional to the size of the jump at a discontinuity; the permissible increase in total variation is solely due to the smooth part of the function. Unlike the second-order accurate non-oscillatory schemes of [5] which are a subset of this class, essentially non-oscillatory schemes may occasionally increase the number of local extrema. However these extra oscillations are on the level of truncation error in the smooth part and therefore can be regarded as "nonessential oscillations".

In the following we shall present high-order Godunov-type schemes that satisfy (1.6), and show that the design problem reduces to solving a problem in interpolation; the latter is the main topic of this paper.

2. High-Order Accurate Godunov-Type Schemes

Let

$$(2.1) \quad \bar{u}(x) = \frac{1}{h} \int_{-h/2}^{h/2} u(x+y) dy \equiv (A_h \cdot u)(x)$$

denote the sliding average of $u(x)$. The sliding average $\bar{u}(x,t)$ of a weak solution of (1.1) satisfies

$$(2.2a) \quad \frac{\partial}{\partial t} \bar{u}(x,t) + \frac{1}{h} [f(u(x+\frac{h}{2},t)) - f(u(x-\frac{h}{2},t))] = 0.$$

Integrating (2.2a) from t to $t + \tau$ we get

$$(2.2b) \quad \bar{u}(x,t+\tau) = \bar{u}(x,t) - \lambda[\bar{f}(x+\frac{h}{2},t;u) - \bar{f}(x-\frac{h}{2},t;u)]$$

where $\lambda = \tau/h$ and

$$(2.2c) \quad \bar{f}(x,t;u) = \frac{1}{\tau} \int_0^\tau f(u(x,t+\eta)) d\eta.$$

Thus the exact weak solution of (1.1) satisfies the following relation on the computational mesh

(2.3) $$\bar{u}_j^{n+1} = \bar{u}_j^n - \lambda[f(x_{j+1/2},t_n;u) - f(x_{j-1/2},t_n;u)];$$

where $\bar{u}_j^n = \bar{u}(x_j,t_n)$ is the cell average of the solution at time t_n.

Next we compare (2.3) to the numerical scheme in conservation form (1.2) where we set $v_j^n \equiv \bar{u}_j^n$. We see that if the resulting numerical flux $f_{j+1/2}$ satisfies

(2.4a) $$\hat{f}_{j+1/2} = \frac{1}{\tau}\int_0^\tau f(u(x_{j+1/2},t_n+\eta))d\eta + O(h^r)$$

then

(2.4b) $$v_j^{n+1} = \bar{u}_j^{n+1} + O(h^{r+1}),$$

provided that the coefficient in the $O(h^r)$ term in (2.4a) is sufficiently smooth. Relation (2.4b) shows that in the sense of cell-averages the truncation error of the scheme is $O(h^{r+1})$, i.e.

(2.5) $$\bar{u}(t+\tau) - E_h(\tau) \cdot \bar{u}(t) = O(h^{r+1}).$$

We shall refer to a scheme that satisfies (2.4)-(2.5) as an r-th order Godunov-type scheme. (Note the difference from Lax-Wendroff-type schemes that are derived by approximating a Taylor expansion of the solution and where the truncation error is made small in a pointwise sense.)

We observe that although (2.3) is a relation between the cell averages of the solution at t_n and t_{n+1}, the evaluation of the numerical flux in (2.4a) involves point values of the solution. Since

(2.6) $$\bar{u}(x) - u(x) = O(h^2)$$

we have to devise a technique to recover point values from given cell averages to a desired accuracy, in order to obtain Godunov-type schemes that are more than second-order accurate. The rest of this paper is devoted to describing an algorithm for the solution of this problem in approximation: Given $\{\bar{u}_j\}$, cell averages of $u \in U$ (i.e. piecewise smooth with a finite number of discontinuities) find $R(x;\bar{u})$ that reconstructs $u(x)$ to any finite order r

(2.7a) $$R(x;\bar{u}) = u(x) + O(h^r)$$

wherever $u(x)$ is smooth, and such that

(2.7b) $$\bar{R}(x_j;\bar{u}) = \bar{u}_j,$$

(2.7c) $$TV(\bar{R}(\cdot;\bar{u})) \leq TV(u) + O(h^r);$$

here $\bar{R}$ denotes the sliding average of R.

It is easy to see that once we solve this approximation problem, the Godunov-type scheme

(2.8) $$v^{n+1} \equiv E_h(\tau) \cdot v^n = A_h \cdot E(\tau) \cdot R(\cdot;v^n),$$

where A_h is the cell-averaging operator (2.1) and $E(\tau)$ is the evolution operator of (1.1), is r-th order accurate and essentially non-oscillatory in the sense of (1.6).

To see that the scheme (2.8) is essentially non-oscillatory we observe that both A_h and $E(\tau)$ are positive operators and therefore also TVD. Consequently

(2.9) $$TV(E_h(\tau) \cdot \bar{u}) \equiv TV(A_h \cdot E(\tau) \cdot R(\cdot;\bar{u})) \leq TV(R(\cdot;\bar{u}))$$

and (1.6) follows immediately from (2.7c).

Next we show that (2.8) is r-th order accurate in the sense of (2.4). Let us denote

(2.10a) $$v_h(\cdot,t) = E(t) \cdot R(\cdot;v^n).$$

Using (2.7b) and the fact that $v_h(x,t)$ is an exact solution of the conservation law (1.1a) we can rewrite (2.8) in the form (2.3), i.e.

(2.10b) $$v_j^{n+1} = v_j^n - \lambda(f_{j+1/2} - f_{j-1/2})$$

where

(2.10c) $$f_{j+1/2} = \frac{1}{\tau}\int_0^\tau f(v_h(x_{j+1/2},t))dt.$$

(2.4a) then follows from (2.7a) and the stability of entropy solutions of (1.1).

Remarks: (1) The scheme (2.8) is the abstract form of Godunov-type schemes. In practice we use an approximation to (2.10) which is obtained by using an appropriate numerical quadrature for the integral in (2.10c), and using an approximate solution operator to evaluate $v_h(x_{j+1/2},t)$ in (2.10a). For more details see [5] and [6].

(2) We initialize the computation by taking cell-average of the given initial data, i.e.

(2.11a) $$v_j^0 = \bar{u}_j^0 = \frac{1}{h} \int_{-h/2}^{h/2} u_0(x_j + y) dy.$$

When we apply the scheme (2.8) N times, $N \cdot \tau = t$, and the initial data are such that the solution is smooth, we expect the truncation error (2.5) to accumulate linearly. Thus at the end of the time loop we get (since $\tau = O(h)$)

(2.11b) $$v_j^N = \bar{u}(x_j,t) + O(h^r).$$

Since in general we are interested in pointwise values we output $R(x_j;v^N)$ which gives us the pointwise data to the desired accuracy:

(2.11c) $$R(x;v^N) = R(x;\bar{u}) + O(h^r) = u(x,t) + O(h^r).$$

3. Interpolation

In this section we present a new interpolation technique $H_m(x;u)$ for functions $u \in U$

(3.1a) $$H_m(x_j;u) = u(x_j).$$

The interpolant $H_m(x;u)$ is a piecewise polynomial function of x, i.e.

(3.1b) $$H_m(x;u) = q_{m,j+1/2}(x;u) \quad \text{for} \quad x_j \leq x \leq x_{j+1}$$

where $q_{m,j+1/2}$ is a polynomial in x of degree m. Wherever $u(x)$ is smooth we get that

(3.2) $$\frac{d^k}{dx^k} H_m(x;u) = \frac{d^k}{dx^k} u(x) + O(h^{m+1-k}), \quad 0 \leq k \leq m.$$

(We use here the standard convention that $k = 0$ corresponds to the function itself.)

The new feature of this interpolation technique is that, although u may be discontinuous, the interpolant $H_m(x;u)$ is essentially non-oscillatory in the sense that

(3.3) $$TV(H_m(\cdot;u)) \leq TV(u) + O(h^{m+1}).$$

This interpolation technique will be used in the following sections to develop an algorithm for the solution of the reconstruction problem (2.7).

To accomplish (3.2) we take $q_{m,j+1/2}(x;u)$ to be the m-th degree polynomial that interpolates $u(x)$ at the $m + 1$ successive points $\{x_i\}$, $i_m(j) \leq i \leq i_m(j) + m$, that include x_j and x_{j+1}, i.e.

(3.4a) $$q_{m,j+1/2}(x_i;u) = u(x_i), \quad i_m(j) \leq i \leq i_m(j) + m$$

(3.4b) $$1 - m \leq i_m(j) - j \leq 0.$$

Clearly there are exactly m such polynomials corresponding to the m different choices of $i_m(j)$ subject to (3.4b).

We have assumed that $u(x)$ has a finite number of discontinuities. Hence for h sufficiently small there are at least $m + 1$ points of smoothness between any two successive discontinuities. Consequently if (x_j,x_{j+1}) is an interval in which $u(x)$ is smooth, there is at least one choice of $i_m(j)$ such that $u(x)$ is smooth in $x_{i_m(j)} \leq x \leq x_{i_m(j)+m}$.

Next we show that any algorithm to assign $i_m(j)$ (3.4) to (x_j,x_{j+1}) that has the property:

(3.5) smoothness in (x_j,x_{j+1}) ⟹ smoothness in $(x_{i_m(j)},x_{i_m(j)+m})$,

yields $H_m(x;u)$ that satisfies both (3.2) and (3.3).

Let x_0 be a point that has a neighborhood in which u is smooth. Hence for h sufficiently small there is an interval (x_j,x_{j+1}) that includes x_0 in which u is smooth. It follows then from (3.5), (3.4) and the fact that $q_{m,j+1/2}$ is an interpolating polynomial that uses data from an interval in which u is smooth that

(3.6) $\quad \dfrac{d^k}{dx^k} q_{m,j+1/2}(x;u) = \dfrac{d^k}{dx^k} u(x) + O(h^{m+1-k})\quad$ for $\quad 0 \leq k \leq m$

and $x_j < x < x_{j+1}$;

this implies (3.2).

We turn now to show (3.3). First let us consider an interval (x_j, x_{j+1}) in which u is smooth. It follows immediately from (3.6) with $k = 0$ that

(3.7) $\quad TV_{[x_j, x_{j+1}]}(q_{m,j+1/2}) \leq TV_{[x_j, x_{j+1}]}(u) + O(h^{m+1}).$

Moreover in the intervals in which u is smooth and $\dfrac{du}{dx}$ is bounded away from zero, it follows from (3.6) with $k = 1$ that for h sufficiently small $\dfrac{d}{dx} q_{m,j+1/2} \neq 0$ in such an interval and consequently

(3.8) $\quad TV_{[x_j, x_{j+1}]}(q_{m,j+1/2}) = TV_{[x_j, x_{j+1}]}(u).$

Next let us consider an interval (x_j, x_{j+1}) that contains a single discontinuity of u. It turns out that for h sufficiently small $q_{m,j+1/2}$, for any choice of $i_m(j)$ in (3.4), is monotone in (x_j, x_{j+1}); this implies (3.8). To simplify the argument let us consider the case that u is a step-function with the discontinuity located in (x_j, x_{j+1}). In this case $u(x_i) = u(x_{i+1})$ for all $i \neq j$ and therefore

(3.9a) $\quad q_{m,j+1/2}(x_i;u) = q_{m,j+1/2}(x_{i+1};u)$

for all indices i such that

(3.9b) $\quad i \neq j$ and $0 \leq i - i_m(j) \leq m - 1.$

From (3.9a) it follows that $\dfrac{d}{dx} q_{m,j+1/2}$ has a root in each of the $m - 1$ intervals (x_i, x_{i+1}) with indices i satisfying (3.9b). Since $\dfrac{d}{dx} q_{m,j+1/2}$ is a polynomial of degree $m - 1$ it cannot have an additional root in (x_j, x_{j+1}) and therefore $q_{m,j+1/2}$ is strictly monotone there.

We conclude from the above analysis that an increase in total variation is possible only in those intervals in which u has a smooth local extremum. Since there is only a finite number of such intervals and since the increase there (3.7) is on the level of interpolation error, (3.3) follows (see [6] for a more detailed analysis).

In the following we present an algorithm to assign $i_m(j)$ to (x_j, x_{j+1}). In order to satisfy (3.5) this algorithm makes use of the information about smoothness contained in a table of divided differences of u. The latter can be defined recursively by

(3.10a) $$u[x_i] = u(x_i)$$

(3.10b) $$u[x_i,\ldots,x_{i+k}] = (u[x_{i+1},\ldots,x_{i+k}] - u[x_i,\ldots,x_{i+k-1}])/(x_{i+k} - x_i).$$

It is well known that if u is C^∞ in $[x_i, x_{i+k}]$ then

(3.11a) $$u[x_i,\ldots,x_{i+k}] = \frac{1}{k!} \frac{d^k}{dx^k} u(\xi_{i,k}), \quad x_i < \xi_{i,k} < x_{i+k}.$$

However if u has a jump discontinuity in the p-th derivative in this interval, $0 < p < k$, then

(3.11b) $$u[x_i,\ldots,x_{i+k}] = O(h^{-k+p}[\frac{d^p}{dx^p} u]);$$

here $[\frac{d^p}{dx^p} u]$ denotes the jump in the p-th derivative.

Our algorithm is recursive: It arrives at $H_r(x;u)$, which amounts to assigning $i_r(j)$, by successively evaluating $i_k(j)$, $k = 1,\ldots,r$. We start by setting

(3.12a) $$i_1(j) = j,$$

i.e. $q_{1,j+1/2}$ is the first degree polynomial interpolating u at x_j and x_{j+1}. Let us assume that we have already defined $i_k(j)$, i.e. $q_{k,j+1/2}$ is the k-th degree polynomial interpolating u at

(3.13) $$x_{i_k(j)},\ldots,x_{i_k(j)+k}.$$

We consider now as candidates for $q_{k+1,j+1/2}$ the two (k+1)-th degree interpolating polynomials obtained by adding to (3.13) the neighboring point to the left or the one to the right; this corresponds to setting $i_{k+1}(j) = i_k(j) - 1$ or $i_{k+1}(j) = i_k(j)$, respectively. We choose the one that gives a $(k+1)$-th order divided difference that is smaller in absolute value:

$$(3.12b) \quad i_{k+1}(j) = \begin{cases} i_k(j) - 1 & \text{if } |u[x_{i_k(j)-1}, \ldots, x_{i_k(j)+k}]| \\ & \quad < |u[x_{i_k(j)}, \ldots, x_{i_k(j)+k+1}]| \\ i_k(j) & \text{otherwise} \end{cases}$$

Using Newton's form of interpolation it is easy to see that

$$(3.14) \quad q_{k+1,j+1/2} - q_{k,j+1/2} = u[x_{i_{k+1}(j)}, \ldots, x_{i_{k+1}(j)+k+1}] \cdot \prod_{i=i_k(j)}^{i_k(j)+k} (x - x_i).$$

Since the product in the RHS of (3.14) is the same for both choices in (3.12b), we see that our algorithm selects as $q_{k+1,j+1/2}$ this $(k+1)$-th polynomial that deviates the least from $q_{k,j+1/2}$ in (x_j, x_{j+1}). Clearly if h is sufficiently small it follows from (3.11) that the algorithm (3.12) satisfies the requirement (3.5), and consequently has the desired properties (3.2) and (3.3). However if h is not small, (3.14) shows that the algorithm attempts to find the "least oscillatory" polynomial (subject to the restricted choice in (3.12b)), and thus is meaningful also in this case.

In the next two sections we describe two different techniques to solve the reconstruction problem (2.7) in terms of interpolation. The importance of using the particular interpolation described in this section is that its nonoscillatory nature goes over to the reconstruction algorithm. The nonoscillatory nature of $R(x;\bar{u})$ (2.7c) is demonstrated in section 7 by numerical examples (analysis is presented in [6]).

4. Reconstruction Via The Primitive Function

In this section we apply a technique frequently used in area-preserving approximations[1] in order to solve the reconstruction problem in terms of interpolation.

---

[1] I thank Nira Dyn from Tel-Aviv University for pointing this out. A similar approach was taken by Woodward [10] and Zalesak [11].

Given cell-averages $\bar{u}_j$ of a piecewise-smooth function $u \in U$

$$(4.1) \qquad \bar{u}_j = \frac{1}{h_j} \int_{x_j - 1/2}^{x_j + 1/2} u(y)dy, \qquad h_j = x_{j+1/2} - x_{j-1/2}$$

we can immediately evaluate the point-values of the primitive function $U(x)$

$$(4.2) \qquad U(x) = \int_{x_0}^{x} u(y)dy$$

by

$$(4.3) \qquad U(x_{j+1/2}) = \sum_{i=i_0}^{j} h_j \bar{u}_j .$$

Since

$$(4.4) \qquad u(x) \equiv \frac{d}{dx} U(x)$$

we can apply interpolation to the point values of the primitive function (4.3) and then obtain an approximation to $u(x)$ by defining

$$(4.5) \qquad R(x;\bar{u}) = \frac{d}{dx} H_r(x;U).$$

We note that this procedure does not require uniformity of the mesh.

The primitive function $U(x)$ is smoother than $u(x)$ (by one extra derivative) and therefore $U \in U$. Hence we get from (3.2) that wherever $u(x)$ is smooth

$$(4.6) \qquad \frac{d^k}{dx^k} H_r(x;U) = \frac{d^k}{dx^k} U(x) + O(h^{r+1-k}), \qquad 0 \leq k \leq r.$$

Using (4.4) and (4.5) we can rewrite (4.6) as

$$(4.7) \qquad \frac{d^\ell}{dx^\ell} R(x;\bar{u}) = \frac{d^\ell}{dx^\ell} u(x) + O(h^{r-\ell}),$$

which implies (2.7a) for $\ell = 0$.

We turn now to study $\bar{R}(x;\bar{u})$, the sliding average of (4.5):

$$(4.8) \qquad \bar{R}(x;\bar{u}) = \frac{1}{h} \int_{-h/2}^{h/2} R(x+y)dy = \frac{1}{h} [H_r(x + \frac{h}{2};U) - H_r(x - \frac{h}{2};U)].$$

Denoting the interpolation error by

$$(4.9) \qquad e(x) = H_r(x;U) - U(x)$$

we can rewrite (4.8) as

$$\bar{R}(x;\bar{u}) = \frac{1}{h}[U(x+\frac{h}{2}) - U(x-\frac{h}{2})] + \frac{1}{h}[e(x+\frac{h}{2}) - e(x-\frac{h}{2})]$$

(4.10)
$$= \frac{1}{h}\int_{x-h/2}^{x+h/2} u(y)dy + \frac{1}{h}[e(x+\frac{h}{2}) - e(x-\frac{h}{2})]$$

$$= \bar{u}(x) + \frac{1}{h}[e(x+\frac{h}{2}) - e(x-\frac{h}{2})].$$

Since

(4.11a) $$e(x_{j+1/2}) = 0,$$

(4.11b) $$e(x) = O(h^{r+1}),$$

we get from (4.10) that (4.5) satisfies (2.7b), i.e.

(4.12a) $$\bar{R}(x_j;\bar{u}) = \bar{u}_j$$

and that wherever $e(x)$ is smooth

(4.12b) $$\bar{R}(x;\bar{u}) = \bar{u}(x) + O(h^{r+1}).$$

It follows from (4.12) that $\bar{R}(x;\bar{u})$ is a piecewise-polynomial interpolation of degree r of $\bar{u}$. Note that (4.12b) is one order more accurate than (4.7) with $\ell = 0$, which is $R(x;\bar{u}) = u(x) + O(h^r)$.

5. Reconstruction Via Deconvolution

In this section we describe another technique to reconstruct a piecewise smooth function $u \in U$ from its given cell-averages $\bar{u}_j$. We assume that the mesh is uniform and consider the cell-averages $\bar{u}_j$ to be point values of $\bar{u}(x)$,

(5.1a) $$\bar{u}(x) = \frac{1}{h}\int_{-h/2}^{h/2} u(x-y)dy$$

(5.1b) $$\bar{u}_j = \bar{u}(x_j).$$

The function $\bar{u}(x)$ (5.1a) was referred to earlier as the sliding average of

u. Here we consider it to be the convolution of $u(x)$ with the characteristic function of a cell $\psi_h(x)$

(5.2a) $$\psi_h(x) = \begin{cases} 1/h & \text{for } |x| < h/2 \\ 0 & \text{for } |x| > h/2 \end{cases},$$

i.e.

(5.2b) $$\bar{u}(x) = \int_{-\infty}^{\infty} u(x-y)\psi_h(y)dy = (u*\psi_h)(x).$$

In the following we describe a procedure to reconstruct $u(x)$ up to $O(h^r)$ for any finite r; this will be referred to as "finite-order deconvolution". Expanding $u(x-y)$ in (5.2b) around $y = 0$, we get

(5.3a) $$\bar{u}(x) = \sum_{k=0}^{\infty} \frac{u^{(k)}(x)}{k!} \frac{(-1)^k}{h} \int_{-h/2}^{h/2} y^k dy \equiv \sum_{k=0}^{\infty} \alpha_k h^k u^{(k)}(x)$$

where

$$\alpha_k = \begin{cases} 0 & k \text{ odd} \\ \frac{2^{-k}}{(k+1)!} & k \text{ even} \end{cases}.$$

Multiplying both sides of (5.3a) by $h^\ell \dfrac{d^\ell}{dx^\ell}$ and then truncating the expansion in the RHS at $O(h^r)$ we get

(5.4a) $$h^\ell \bar{u}^{(\ell)}(x) = \sum_{k=0}^{r-\ell-1} \alpha_k h^{k+\ell} u^{(k+\ell)}(x) + O(h^r).$$

Writing the relations (5.4a) for $\ell = 0,\ldots,r-1$ in matrix form, we obtain

(5.4b) $$\begin{vmatrix} \bar{u}(x) \\ h\bar{u}'(x) \\ h^2\bar{u}''(x) \\ \vdots \\ h^{r-1}\bar{u}^{(r-1)}(x) \end{vmatrix} = \begin{bmatrix} 1 & 0 & \alpha_2 & 0 & \alpha_4 & \cdots & \alpha_{r-1} \\ & \ddots & & \ddots & & & \\ & & & & & & \alpha_4 \\ & & & & & & 0 \\ & 0 & & & & & \alpha_2 \\ & & & & & & 0 \\ & & & & & & 1 \end{bmatrix} \begin{vmatrix} u(x) \\ hu'(x) \\ h^2u''(x) \\ \vdots \\ h^{r-1}u^{(r-1)}(x) \end{vmatrix} + O(h^r).$$

Let us denote the coefficient matrix in the RHS of (5.4b) by C. This matrix is upper triangular and diagonally dominant. Multiplying both sides of (5.4b) by C^{-1} from the left we get

(5.4c)
$$\begin{bmatrix} u(x) \\ hu'(x) \\ \vdots \\ h^{r-1}u^{(r-1)}(x) \end{bmatrix} = C^{-1} \begin{bmatrix} \bar{u}(x) \\ h\bar{u}'(x) \\ \vdots \\ h^{r-1}\bar{u}^{(r-1)}(x) \end{bmatrix} + O(h^r).$$

Relation (5.4c) is the essence of finite-order deconvolution.

Given $\bar{u}_j$ (5.1) we apply the interpolation technique of section 3 to $\bar{u}$ and compute $O(h^r)$ approximations to $h^\ell \bar{u}^{(\ell)}$ at x_j by taking appropriate derivatives of $H_m(x;\bar{u})$, $m \geq r - 1$; denote these approximations by $\bar{D}_{\ell,j}$, $0 \leq \ell \leq r-1$:

(5.5a) $$\bar{D}_{0,j} = \bar{u}_j$$
(5.5b) $$\bar{D}_{\ell,j} = h^\ell \bar{u}^{(\ell)}(x_j) + O(h^r), \quad \ell = 1,\ldots,r-1.$$

Using $\bar{D}_{\ell,j}$ in the RHS of (5.4c) we obtain $O(h^r)$ approximations to $h^\ell u^{(\ell)}(x_j)$ which we denote by $D_{\ell,j}$, i.e.

(5.6a) $$D_{0,j} = u(x_j) + O(h^r)$$
(5.6b) $$D_{\ell,j} = h^\ell u^{(\ell)}(x_j) + O(h^r), \quad 1 \leq \ell \leq r - 1.$$

Since C is an upper triangular matrix, $D_{\ell,j}$ are computed by back-substitution: We set

(5.7a) $$D_{r-1,j} = \bar{D}_{r-1,j}$$

and then compute backwards for $k = r - 2,\ldots,0$

(5.7b) $$D_{k,j} = \bar{D}_{k,j} - \sum_{\ell=k+1}^{r-1} \alpha_\ell D_{\ell,j}.$$

Finally we define

(5.8) $$R(x;\bar{u}) = \sum_{k=0}^{r-1} \frac{1}{k!} D_{k,j} [(x - x_j)/h]^k \quad \text{for} \quad |x - x_j| < h/2.$$

It follows immediately from (5.6) that

$$(5.9) \qquad R(x;\bar{u}) = \sum_{k=0}^{r-1} \frac{1}{k!} u^{(k)}(x_j)(x-x_j)^k + O(h^r) = u(x) + O(h^r),$$

which implies (2.7a).

To see that the reconstruction (5.8) satisfies property (2.7b)

$$(5.10a) \qquad \bar{R}(x_j;\bar{u}) = \bar{u}_j$$

we evaluate

$$\bar{R}(x_j;\bar{u}) = \frac{1}{h} \int_{-h/2}^{h/2} R(x_j - y;\bar{u}) dy$$

to get

$$(5.10b) \qquad \bar{R}(x_j;\bar{u}) = \sum_{k=0}^{r-1} \frac{1}{k!} \frac{D_{k,j}}{h^k} \frac{(-1)^k}{h} \int_{-h/2}^{h/2} y^k dy = D_{0,j} + \sum_{k=1}^{r-1} \alpha_k D_{k,j},$$

where α_k are (5.3b). Relation (5.10a) then follows immediately from comparing the RHS of (5.10b) to (5.7b) with $k = 0$

$$\bar{D}_{0,j} = D_{0,j} + \sum_{k=1}^{r-1} \alpha_k D_{k,j}$$

and then using (5.5a). We note that this result is independent of the particular values assigned to $\bar{D}_{\ell,j}$ for $1 < \ell < r - 1$.

<u>Remarks</u>: (1) $u(\bar{x})$ is smoother than $u(x)$, and therefore $u \in U \Rightarrow \bar{u} \in U$. Consequently

$$(5.11a) \qquad \frac{d^\ell}{dx^\ell} H_m(x;\bar{u}) = \frac{d^\ell}{dx^\ell} \bar{u}(x) + O(h^{m+1-k}).$$

Since H_m is only continuous at x_j, for $k > 1$

$$(5.11b) \qquad \frac{d^k}{dx^k} H_m(x_j + 0;\bar{u}) \neq \frac{d^k}{dx^k} H_m(x_j - 0;\bar{u}).$$

In the numerical experiments presented in this paper we have taken

$$(5.12a) \qquad \bar{D}_{0,j} = \bar{u}_j$$

(5.12b) $$\overline{D}_{K,j} = h^k M\left(\frac{d^k}{dx^k} H_m(x_j + 0; \overline{u}), \frac{d^k}{dx^k} H_m(x_j - 0; \overline{u})\right)$$

where $M(x,y)$ is the min mod function

(5.13) $$M(x,y) = \begin{cases} s \cdot \min(|x|,|y|) & \text{if } \operatorname{sgn}(x) = \operatorname{sgn}(y) = s \\ 0 & \text{otherwise.} \end{cases}$$

Clearly this choice satisfies (5.5) for $m \geq r-1$.

(2) The finite order deconvolution extends very easily to the case where the mollifier $\psi_h(y)$ is replaced by a smoother function or a function with a different support.

6. The Constant Coefficient Case

In this section we consider the Godunov-type scheme (2.8) in the constant coefficient case

(6.1a) $$u_t + au_x = 0, \quad a = \text{const.}$$

(6.1b) $$u(x,0) = u_0(x).$$

The exact solution in this case is just pure translation with a constant speed a

(6.2) $$u(x,t) = [E(t) \cdot u_0](x) = u_0(x - at).$$

Consequently we can express the numerical scheme explicitly in the following form:

(6.3) $$v_j^{n+1} = \overline{R}(x_j - a\tau; v^n);$$

here $\overline{R}$ is the sliding average of R.

The truncation error in the sense of cell-averages (2.5) is then

(6.4) $$\overline{u}(x - a\tau) - \overline{R}(x - a\tau; \overline{u}),$$

which measures how well $\overline{R}(\cdot; \overline{u})$ approximates $\overline{u}$. We have noted that generally if $R(\cdot; \overline{u})$ approximates u to $O(h^r)$, then $\overline{R}(\cdot; \overline{u})$ approximates $\overline{u}$ to $O(h^{r+1})$, which corresponds to an r-th order accurate scheme. To have this gain of one extra order of accuracy we need special smoothness of the reconstruction error; this is evident from (2.4a) and more directly from (4.10)-(4.11). When we do not have this smoothness we find that the scheme is r-th order accurate in the L_1-norm but not in the maximum norm (see section 7).

We note that although the problem to be solved (6.1) is linear, the numerical scheme (6.3) is highly nonlinear. The nonlinearity enters through the interpolation, where the stencil (3.12) is chosen differently at each point and each time level, depending on local smoothness of the numerical solution. In this respect (6.3) is conceptually different from standard finite-difference schemes where the stencil is arbitrarily predetermined.

We would also like to point out that this scheme breaks away from the somewhat artificial notion of "upstream differencing": The decision what stencil to use in the reconstruction is made on the basis of smoothness considerations and has nothing to do with the "direction of the wind". The latter enters only when applying $E(\tau)$ in (2.8). The resulting stencil is a combination of the two.

7. Numerical Examples

In this section we present some numerical examples in order to illustrate the nature of the approximations described in this paper.

The first set of examples deals with smooth data. In Figs. 1a to 1f we show approximations to $u(x) = \sin \pi x$, $-1 \leq x \leq 1$; these were reconstructed via deconvolution (see section 5) from the cell-averages input

(7.1a) $$\bar{u}_j = \bar{u}(x_j) = \frac{\sin(\pi h/2)}{(\pi h/2)} \cdot \sin(\pi x_j)$$

(7.1b) $$x_j = -1 + j \cdot \frac{2}{N}, \quad 0 \leq j \leq N - 1.$$

In this example we took $N = 6$ and extended the data outside $[-1,1]$ by periodicity. The cell-averages input is shown in Figs. 1a-1f by circles. Both the reconstructed approximation $R(x;\bar{u})$ (5.8) and $u(x) = \sin \pi x$ are shown by solid lines.

Let us denote the polynomial degree of $H(x;\bar{u})$ (5.12) by P_I and that of

$R(x;\bar{u})$ by (5.8) by P_R:

(7.2) $\quad\quad\quad\quad P_I = \deg[H(x;\bar{u})], \quad P_R = \deg[R(x,\bar{u})].$

Fig. 1a shows the reconstruction associated with the original first-order accurate Godunov scheme: $P_I = P_R = 0$.

Fig. 1b shows the reconstruction associated with the "second-order" TVD schemes [3]: $P_I = P_R = 1$. Comparing it to Fig. 1a we see that the local extrema are flattened in exactly the same way - this demonstrates the degeneracy to first-order accuracy there.

Fig. 1c shows the reconstruction associated with the second-order accurate non-oscillatory scheme of [5]: $P_I = 2$, $P_R = 1$. Comparing it to Fig. 1b we observe a considerable improvement in the quality of the approximation. We also note that the reconstruction increases the total variation of the input data; however, this increase is small - it is of the size of the truncation error.

Figs. 1d, 1e and 1f show the reconstruction corresponding to $P_I = k$, $P_R = k - 1$ and $k = 3, 4$ and 5, respectively. We observe that unlike the previous approximations, the reconstruction here does not go through the circles; this is a consequence of (2.6).

The reconstruction $R(x;\bar{u})$ is discontinuous at $x_{j+1/2}$; the size of the jump is proportional to the reconstruction error. As we increase the accuracy of the reconstruction going from Fig. 1a to Fig. 1f, we observe that the size of the jumps become smaller and smaller; in Fig. 1f the jumps are hardly noticeable.

In Table 1 we list the L_∞-norm of the interpolation error $|H(x;\bar{u}) - \bar{u}(x)|$ and of the reconstruction error $|R(x;\bar{u}) - u(x)|$ for a refinement sequence

(7.3) $\quad\quad\quad\quad N = 4, 8, 16, 32, 64$ in (7.1).

We turn now to examine the performance of the resulting Godunov-type scheme (2.8) in the constant coefficient case

(7.4) $\quad\quad\quad\quad u_t + u_x = 0, \quad u(x,0) = \sin \pi x, \quad -1 < x < 1.$

We input the initial data in the form of the cell-averages (7.1). Again we assume periodicity in space, which implies a period of 2 in time as well.

In Table 2 we list both the L_∞- and the L_1-error at $t = 2$ of the scheme (2.10), (6.3) where $R(x;\bar{u})$ is reconstruction via deconvolution with P_I and P_R defined by (7.2). The calculations in this table were performed for the refinement sequence (7.3) with $\lambda = \tau/h = 0.8$. In addition to the actual error we also list the quantity r which is the computational order of accuracy. Assuming the error to be exactly $e_h = \text{const} \cdot h^r$ we evaluate r for any two successive calculations in the refinement sequence by

(7.5) $$r = \log_2(e_h/e_{h/2}).$$

In Table 3 we repeat the calculations in Table 2 for the scheme (2.10), (6.3) where now $R(x;\bar{u})$ is reconstruction via the primitive function (4.5). Here P_U denotes the polynomial degree of the interpolation $H(x;U)$ of the primitive function (4.2), i.e.

(7.6) $$P_U = \deg[H(x;U)].$$

We observe from Table 2 that the schemes with $P_I = k$ and $P_R = k - 1$ are (at least) k-th order accurate in both L_∞ and L_1. The TVD scheme with $P_I = P_R = 1$ is a little bit better than first-order in L_∞ and a little bit worse than second-order in L_1.

We observe from Table 3 that the schemes with the odd order $P_U = 3$ and $P_U = 5$ seem to be third and fifth-order accurate, respectively, in both the L_∞ and L_1 norms.

However when P_U is an even number, $P_U = 2,4,6$, this order is realized only in the L_1-sense. In the L_∞-norm it seems to be one order lower in accuracy.

Comparing Table 2 and Table 3 for the fine mesh $N = 64$, we find that the scheme from Table 2 with $P_I = k$ and the scheme with $P_U = k+1$ from Table 3 seem to give comparable accuracy.

We turn now to examine the performance of the various approximations when applied to the discontinuous function

$$(7.7) \quad u(x) = \begin{cases} -x \sin(3\pi s^2/2) & \text{for } -1 < x < -\frac{1}{3} \\ |\sin(2\pi s)| & \text{for } |x| < \frac{1}{3} \\ 2x - 1 - \frac{1}{6} \sin(3\pi x) & \text{for } \frac{1}{3} < x < 1 \end{cases}$$

which we extend periodically outside $[-1,1]$.

The circles in both Figs. 2a and 2b show the given cell-average of (7.7) at $N = 40$ uniformly spaced mesh points. The two solid lines in Figure 2a are $\bar{u}(x)$ and $H(x;\bar{u})$ with $P_I = 6$. The two solid lines in Fig. 2b are $u(x)$ and $R(x;\bar{u})$ with $P_R = 5$. We see in both figures that the approximations give good accuracy in the smooth part and are essentially nonoscillatory. This is as to be expected since $N = 40$ provides sufficient resolution of the problem in the sense that there are at least 7 points of smoothness inbetween discontinuities.

In Figs. 3a and 3b we repeat the calculation in Fig. 2 for $N = 15$. It is interesting to note that in spite of the lack of resolution in this case both approximations are essentially non-oscillatory.

To compare the two reconstruction techniques we repeat in Fig. 3c the same case described in Fig. 3b, but now using reconstruction via primitive function with $P_U = 6$. The main difference is in the rounding of local extrema at discontinuities present in Fig. 3b; this is due to the min mod operation in (5.12b).

Next we apply the Godunov-type scheme (2.10), (6.3) to the solution of $u_t + u_x = 0$ with the initial data (7.7). As before we initialize the computation by taking cell-averages of the initial data and use $\lambda = \tau/h = 0.8$. In Figs. 4, 5 and 6 we present calculations performed with reconstruction via deconvolution using $P_I = P_R = 1$ in Fig. 4; $P_I = 2$, $P_R = 1$ in Fig. 5 and $P_I = 4$, $P_R = 3$ in Fig. 6. The results are presented at: (a) $t = 2$, (b) $t = 4$. The circles in these figures show the reconstructed numerical approximation at the mesh points; the solid line shows the exact solution.

We observe that in all cases we get non-oscillatory approximations, and that the schemes are dissipative in time. The quality of the results improves with increasing formal order of accuracy.

In Fig. 7 we compare the two reconstruction techniques. In Fig. 7b we repeat the same calculation as in Figs. 4 to 6 using $P_I = 5$ and $P_R = 4$. In Fig. 7a we use reconstruction via primitive function with $P_U = 6$ in (7.6). Unlike the situation in Figs. 3b-3c, here the two schemes produce very similar results.

Acknowledgement

I would like to thank Sukumar Chakravarthy, Bjorn Engquist and Stan Osher for many stimulating discussions and for various contributions to this research (which is part of a joint project).

References

[1] P. Colella and P.R. Woodward, The piecewise-parabolic method (PPM) for gas-dynamical simulations, J. Comp. Phys. v. 54, (1984), 174-201.

[2] A. Harten, J.M. Hyman and P.D. Lax (with appendix by B. Keyfitz), "On finite-differnce approximations and entropy conditions for shocks", Comm. Pure Appl. Math., v. 29, (1976), 297-322.

[3] A. Harten, High resolution schemes for hyperbolic conservation laws, J. Comp. Phys., 49 (1983), 357-393.

[4] A. Harten, On a class of high resolution total-variation-stable finite-difference schemes, SINUM, v. 21, (1984), 1-23.

[5] A. Harten and S. Osher, "Uniformly high-order accurate non-oscillatory schemes, I.", MRC Technical Summary Report #2823, May 1985.

[6] A. Harten, S. Osher, B. Engquist and S. Chakravarthy, "Uniformly high-order accurate non-oscillatory schemes, II", in preparation.

[7] S. Osher and S.R. Chakravarthy, "High-resolution schemes and the entropy condition", SINUM, v. 21, (1984), 955-984.

[8] S. Osher and S.R. Chakravarthy, "Very high order accurate TVD schemes", ICASE Report #84-44, (1984).

[9] B. Van Leer, Towards the ultimate conservative scheme, II. Monotonicity and conservation combined in a second order scheme, J. Comp. Phys. 14 (1974), 361-376.

[10] P. Woodward, in Proceedings of the NATO Advanced Workshop in Astrophysical Radiation Hydrodynamics, Munich, West Germany, August 1982; also Lawrence Livermore Lab. Report #90009.

[11] S.T. Zalesak, "Very high-order and pseudo-spectral flux-corrected transport (FCT) algorithms for conservation laws", in "Advances in computer methods for partial differential equations", Vol. 4 (R. Vichnevetsky and R.S. Stepleman, eds.) IMACS, Rutgers University, 1981.

Table 1. Interpolation and reconstruction L_∞-errors for $u(x) = \sin \pi x$.

	N	$P_I=1, P_R=1$	$P_I=2, P_R=1$	$P_I=3, P_R=2$	$P_I=4, P_R=3$	$P_I=5, P_R=4$	$P_I=6, P_R=5$
Interpolation Error $\|H(\cdot;\bar{u}) - u\|_\infty$	4	1.86×10^{-1}	1.86×10^{-1}	7.39×10^{-2}	7.39×10^{-2}	3.17×10^{-2}	3.17×10^{-2}
	8	6.85×10^{-2}	2.84×10^{-2}	7.63×10^{-3}	3.16×10^{-3}	9.37×10^{-4}	3.88×10^{-4}
	16	1.87×10^{-2}	3.72×10^{-3}	5.36×10^{-4}	1.07×10^{-4}	1.70×10^{-5}	3.39×10^{-6}
	32	4.78×10^{-3}	4.71×10^{-4}	3.45×10^{-5}	3.40×10^{-6}	2.76×10^{-7}	2.72×10^{-8}
	64	1.20×10^{-3}	5.91×10^{-5}	2.17×10^{-6}	1.07×10^{-6}	4.36×10^{-9}	2.14×10^{-10}
Reconstruction Error $\|R(\cdot;\bar{u}) - u\|_\infty$	4	2.57×10^{-1}	2.57×10^{-1}	1.07×10^{-1}	1.07×10^{-1}	4.69×10^{-2}	4.69×10^{-2}
	8	1.64×10^{-1}	6.28×10^{-2}	1.52×10^{-2}	5.34×10^{-3}	1.74×10^{-3}	5.85×10^{-4}
	16	4.87×10^{-2}	1.37×10^{-2}	1.13×10^{-3}	2.15×10^{-4}	3.28×10^{-5}	6.22×10^{-6}
	32	1.27×10^{-2}	3.27×10^{-3}	7.54×10^{-5}	7.18×10^{-6}	5.37×10^{-7}	5.23×10^{-8}
	64	3.20×10^{-3}	8.07×10^{-4}	7.88×10^{-6}	2.52×10^{-7}	8.49×10^{-9}	4.16×10^{-10}

Table 2. Solution of $u_t + u_x = 0$, $u(x,0) = \sin \pi x$ at $t = 2$ using reconstruction via deconvolution with $\lambda = \tau/h = 0.8$.

	N	$P_I = 1$, $P_R = 1$	r	$P_I = 2$, $P_R = 1$	r	$P_I = 3$, $P_R = 2$	r	$P_I = 4$, $P_R = 3$	r	$P_I = 5$, $P_R = 4$	r	$P_I = 6$, $P_R = 5$	r
L_∞-ERROR	4	5.62×10^{-1}		4.23×10^{-1}		2.88×10^{-1}		1.82×10^{-1}		1.20×10^{-1}		7.92×10^{-2}	
	8	2.59×10^{-1}	1.12	8.77×10^{-2}	2.27	3.35×10^{-2}	3.10	7.95×10^{-3}	4.52	4.38×10^{-3}	4.78	9.56×10^{-4}	6.37
	16	1.12×10^{-1}	1.21	1.60×10^{-2}	2.45	4.34×10^{-3}	2.95	2.74×10^{-4}	4.86	1.45×10^{-4}	4.92	8.19×10^{-6}	6.87
	32	4.74×10^{-2}	1.24	3.51×10^{-3}	2.19	4.92×10^{-4}	3.14	9.95×10^{-6}	4.78	4.21×10^{-6}	5.11	6.74×10^{-8}	6.92
	64	1.97×10^{-2}	1.27	7.99×10^{-4}	2.14	5.52×10^{-5}	3.16	5.42×10^{-7}	4.20	1.20×10^{-7}	5.13	8.08×10^{-10}	6.38
L_1-ERROR	4	6.57×10^{-1}		4.61×10^{-1}		3.31×10^{-1}		2.28×10^{-1}		1.51×10^{-1}		1.03×10^{-1}	
	8	2.37×10^{-1}	1.47	1.10×10^{-1}	2.07	2.97×10^{-2}	3.48	9.97×10^{-3}	4.52	3.47×10^{-3}	5.44	1.15×10^{-3}	6.48
	16	9.39×10^{-2}	1.34	2.14×10^{-2}	2.36	3.05×10^{-3}	3.28	3.95×10^{-4}	4.66	1.01×10^{-4}	5.10	1.13×10^{-5}	6.67
	32	2.95×10^{-2}	1.67	4.47×10^{-3}	2.26	2.03×10^{-4}	3.91	1.44×10^{-5}	4.78	1.83×10^{-6}	5.79	9.33×10^{-8}	6.92
	64	8.37×10^{-3}	1.82	1.02×10^{-3}	2.13	1.44×10^{-5}	3.82	5.46×10^{-7}	4.72	3.18×10^{-8}	5.85	7.85×10^{-10}	6.89

Table 3. Solution of $u_t + u_x = 0$, $u(x,0) = \sin \pi x$ at $t = 2$, using reconstruction via primitive function and $\lambda = \tau/h = 0.8$.

N	$p_U = 2$	r	$p_U = 3$	r	$p_U = 4$	r	$p_U = 5$	r	$p_U = 6$	r
4	4.877×10^{-1}		3.193×10^{-1}		2.391×10^{-1}		1.535×10^{-1}		1.103×10^{-1}	
8	2.442×10^{-1}	0.99	5.221×10^{-2}	2.61	3.072×10^{-2}	2.96	6.664×10^{-3}	4.52	4.026×10^{-3}	4.78
16	1.065×10^{-1}	1.20	7.310×10^{-3}	2.84	4.050×10^{-3}	2.92	2.223×10^{-4}	4.91	1.401×10^{-4}	4.84
32	4.636×10^{-2}	1.20	9.384×10^{-4}	2.96	4.593×10^{-4}	3.14	7.112×10^{-6}	4.97	4.083×10^{-6}	4.95
64	1.952×10^{-2}	1.25	1.171×10^{-4}	3.00	5.132×10^{-5}	3.16	2.228×10^{-7}	5.00	1.177×10^{-7}	5.12
	L_∞-ERROR									
4	8.233×10^{-1}		5.353×10^{-1}		3.665×10^{-1}		2.328×10^{-1}		1.620×10^{-1}	
8	2.373×10^{-1}	1.79	6.870×10^{-2}	2.96	2.800×10^{-2}	3.71	8.267×10^{-3}	4.82	3.864×10^{-3}	5.39
16	9.176×10^{-2}	1.37	8.731×10^{-3}	2.98	3.271×10^{-3}	3.10	2.729×10^{-4}	4.92	1.053×10^{-4}	5.20
32	3.024×10^{-2}	1.60	1.096×10^{-3}	2.99	2.225×10^{-4}	3.88	8.631×10^{-6}	4.98	1.864×10^{-6}	5.82
64	8.506×10^{-3}	1.83	1.368×10^{-4}	3.00	1.582×10^{-5}	3.82	2.701×10^{-7}	5.00	3.164×10^{-8}	5.88
	L_1-ERROR									

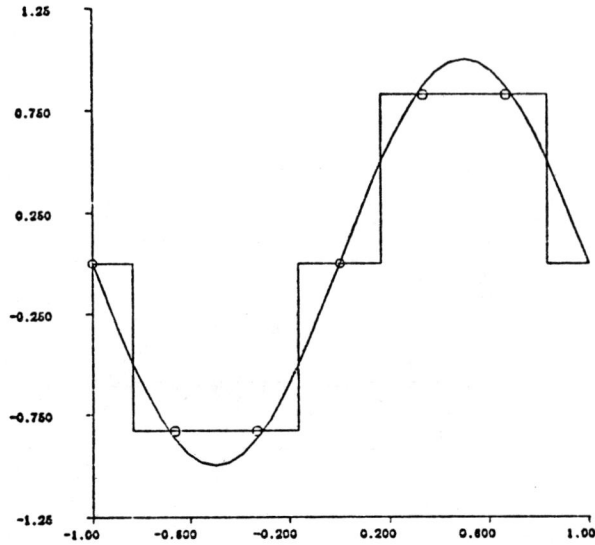

Fig. 1a. Reconstruction of $\sin \pi x$. $P_I = P_R = 0$.

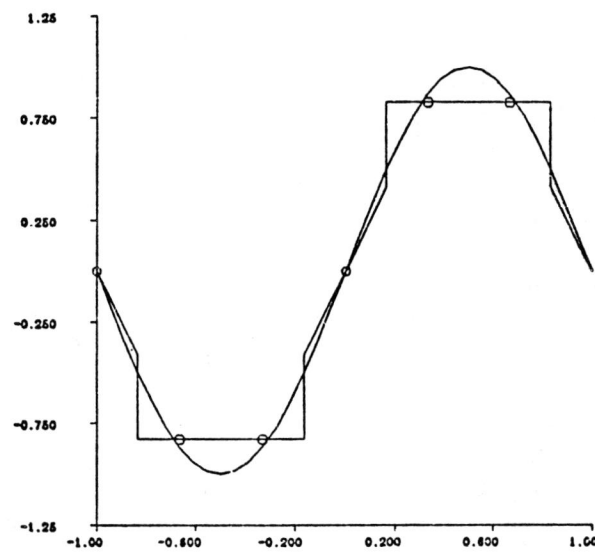

Fig. 1b. $P_I = P_R = 1$.

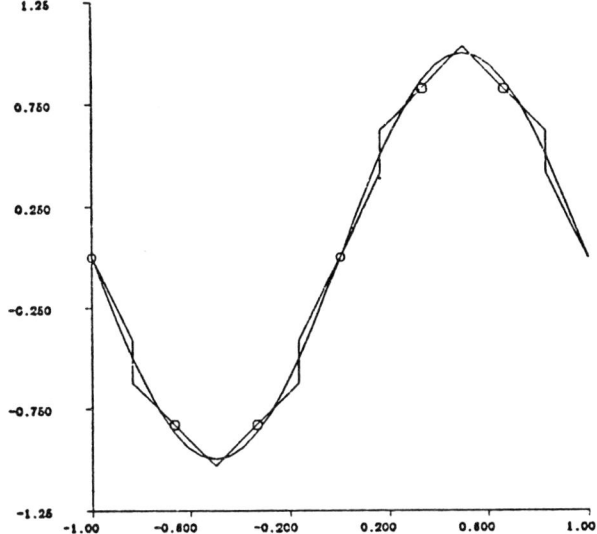

Fig. 1c. $P_I = 2$, $P_R = 1$.

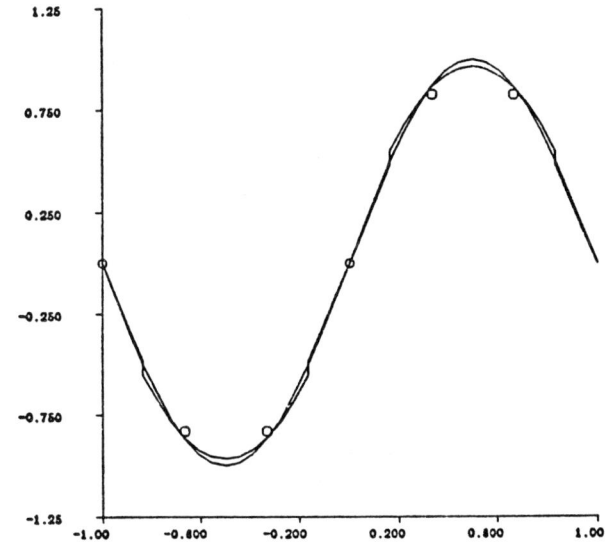

Fig. 1d. $P_I = 3$, $P_R = 2$.

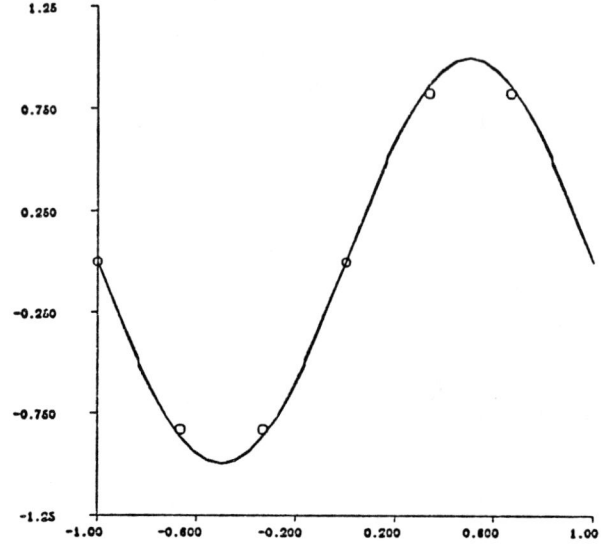

Fig. 1e. $P_I = 4$, $P_R = 3$.

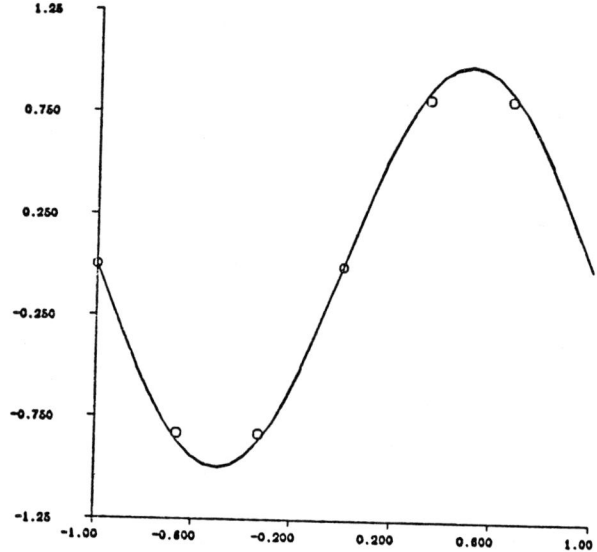

Fig. 1f. $P_I = 5$, $P_R = 4$.

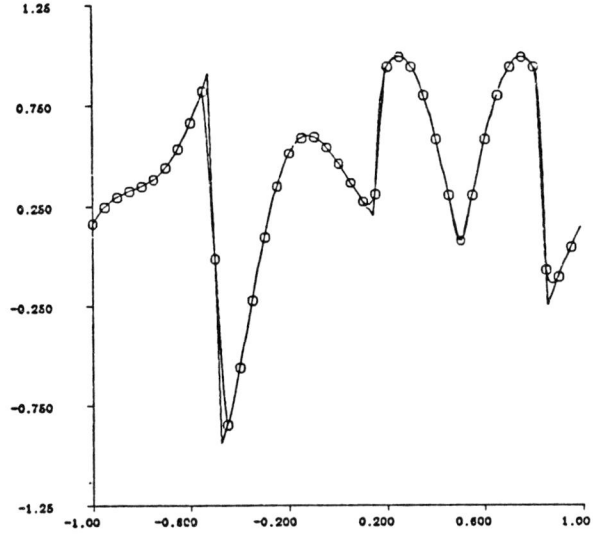

Fig. 2a. $H(x;\bar{u})$ with $P_I = 6$ vs. $\bar{u}(x)$. $N = 40$.

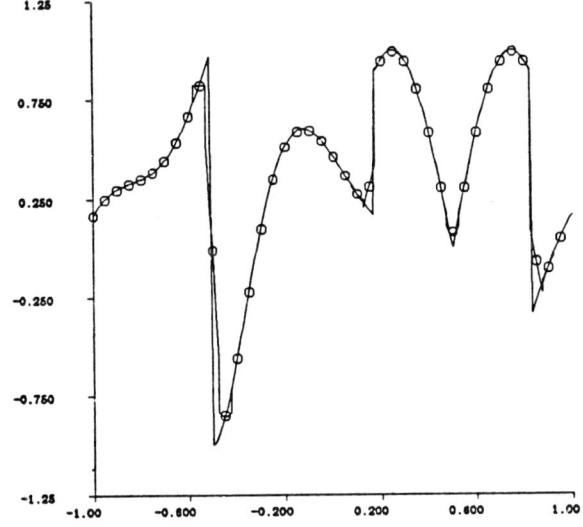

Fig. 2b. $R(x;\bar{u})$ with $P_I = 6$, $P_R = 5$ vs. $u(x)$. $N = 40$.

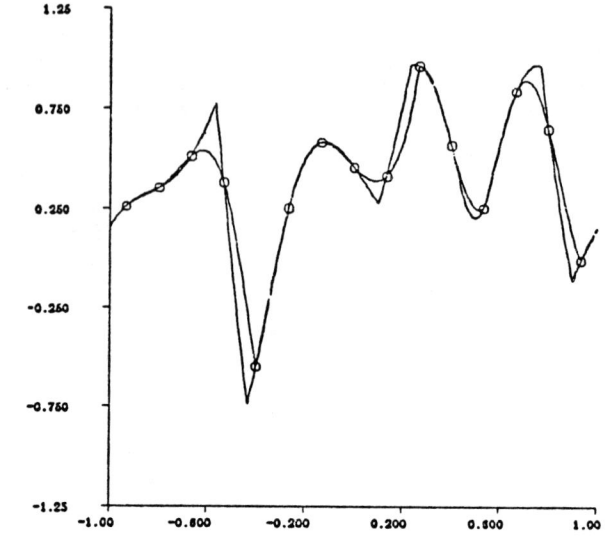

Fig. 3a. $H(x;\bar{u})$ with $P_I = 6$ vs. $\bar{u}(x)$. $N = 15$.

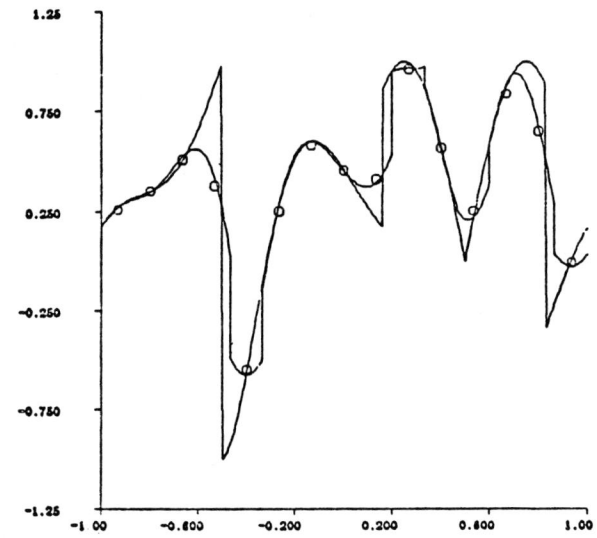

Fig. 3b. $R(x;\bar{u})$ with $P_I = 6$, $P_R = 5$ vs. $u(x)$. $N = 15$.

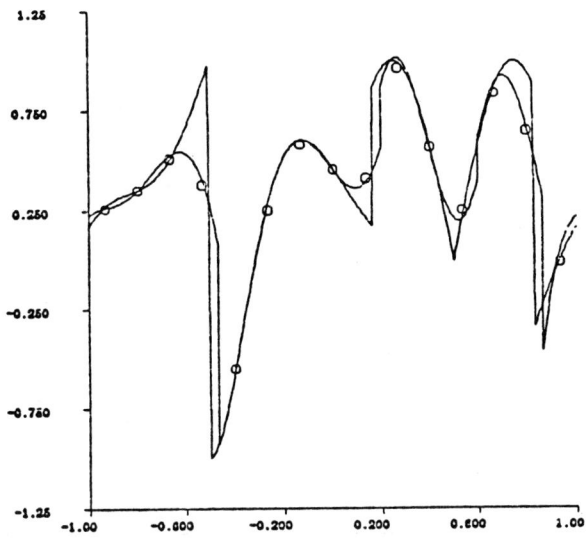

Fig. 3c. Reconstruction via Primitive function. $R(x;\bar{u})$ with $P_U = 6$ vs. $u(x)$. $N = 15$.

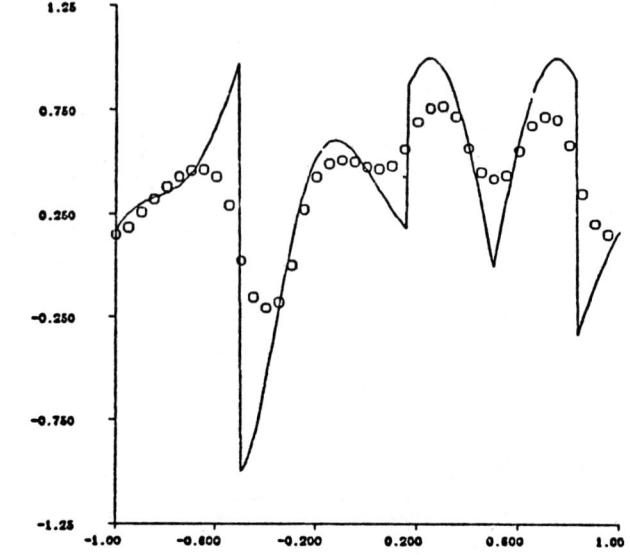

Fig. 4a. Solution with $P_I = P_R = 1$ and $N = 40$ at $t = 2$.

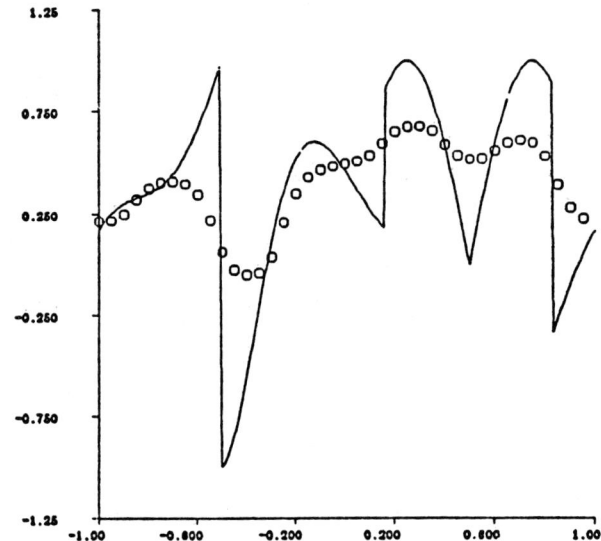

Fig. 4b. Same as Fig. 4a at $t = 4$.

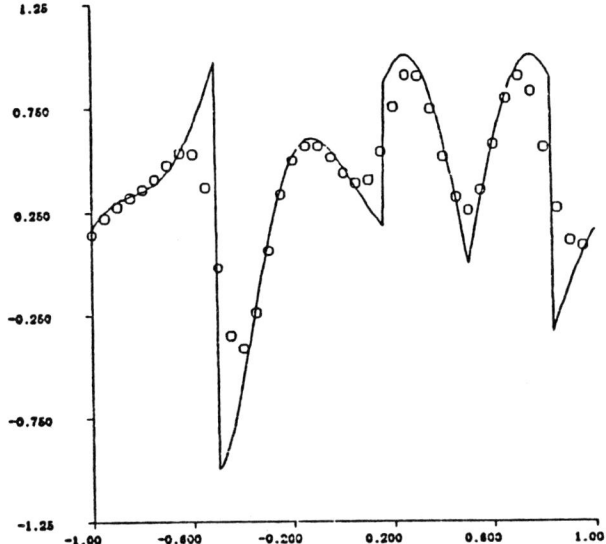

Fig. 5a. Solution with $P_I = 2$, $P_R = 1$ and $N = 40$ at $t = 2$.

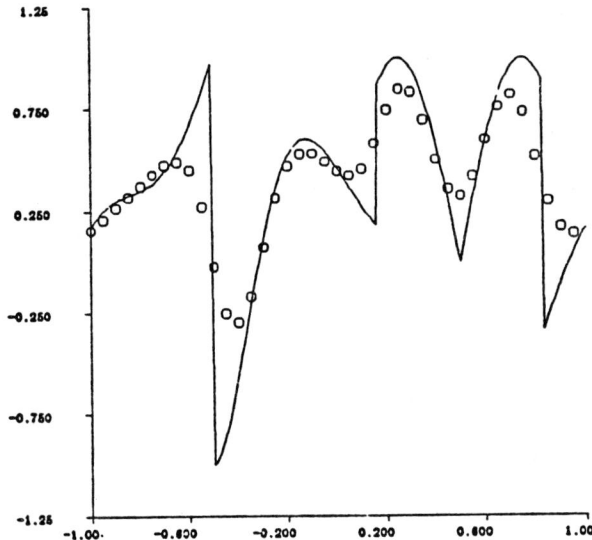

Fig. 5b. Same as Fig. 5a at $t = 4$.

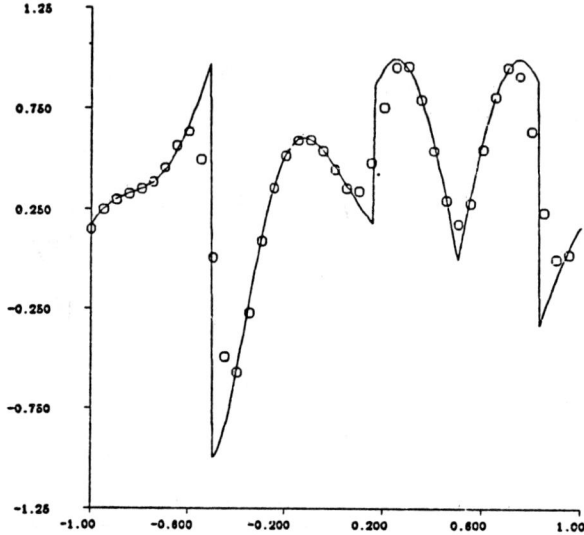

Fig. 6a. Solution with $P_I = 4$, $P_R = 3$ and $N = 40$ at $t = 2$.

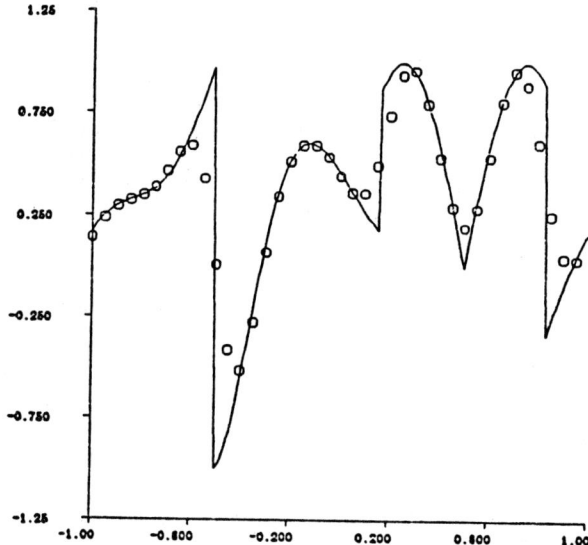

Fig. 6b. Same as Fig. 6a at $t = 4$.

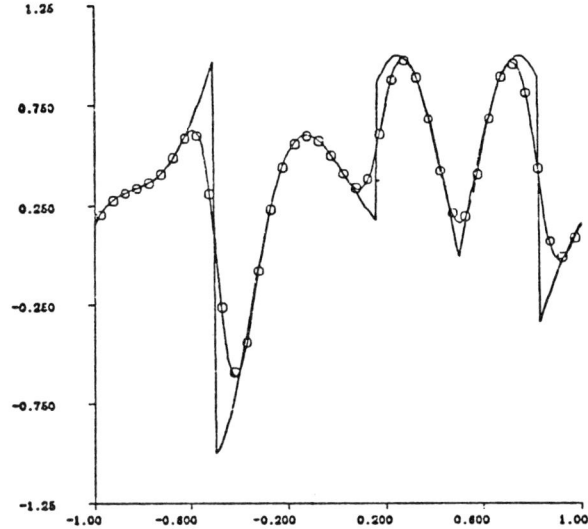

Fig. 7a. Solution using reconstruction via primitive function with $P_U = 6$ at $t = 2$.

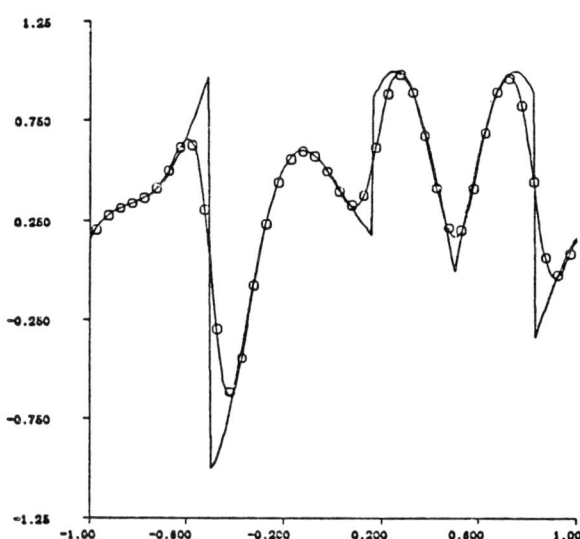

Fig. 7b. Solution using reconstruction via deconvolution with $P_I = 5$, $P_R = 4$ at $t = 2$.

ON THE WEAK CONVERGENCE OF DISPERSIVE DIFFERENCE SCHEMES

Peter D. Lax
Courant Institute
251 Mercer St.
New York, NY 10012

1. Introduction.

In a report written in 1944, [2], von Neumann described a computational method for calculating flows admitting shocks with abitrary prescribed initial data. The method was a finite difference discretization of the equations of compressible flow in which space and time derivatives were differenced systematically, and viscosity and heat conduction were set equal to zero. Calculations at the Aberdeen Proving Ground, using punched card equipment, produced approximations that contained post-shock mesh scale oscillations see Fig. 1. V. Neumann suggested that these oscillations represented the heat energy created by the irreversible action of the shock, and that as Δx, Δt tend to zero, the approximate solutions tend in the <u>weak sense</u> to the correct discontinuous solutions of the fluid equations: "These considerations suggest the surmise that (25) is always a valid approximation of the hydrodynamical motion, but with this qualification: "It is not the $x_a(t)$ which approximate the $x(q,t)$ but the averages of the x_a over an interval of sufficient length of contiguous a'-s. The x_a themselves perform oscillations around these averages and these oscillations do not tend to zero, but they make finite contributions to the total energy".

"In the mathematical terminology, the surmise means that the quasi-molecular kinetic solutions converge to the hydrodynamical one but in the weak sense" "A mathematical proof of this surmise would be most important, but it seems to be very difficult, even in the simplest special cases. The procedure to be followed will therefore be a different one: We shall test the surmise experimentally by carrying out the necessary computations....".

2. The Scheme

V. Neumann used Lagrange coordinates to describe the flow of a compressible:
e and V denote internal energy and volume per unit mass, p is pressure,

- given by an equation of state,

1) $$p = p(e,V)$$

u is velocity, y Lagrangian label, t time.

2) $$u_t + p_y = 0,$$

3) $$V_t - u_y = 0,$$

4) $$e_t + pu_y = 0$$

(2) and (3) are the laws of conservation of momentum and mass. Conservation of energy is obtained by multiplying (2) by u and adding it to (4):

5) $$(e + K)_t + (up)_y = 0,$$

where K abbreviates specific kinetic energy:

6) $$K = \frac{1}{2} u^2.$$

The thermodynamic variables are associated with cell centers, located at (integer + $1/\lambda$) multiples of Δy, and at times which are integer multiples of Δt:

$$(\varepsilon, p, V)_{k+1/2}^{n}$$

Velocity is associated with cell boundaries, located at integer multiples of Δy, and at times which are (integer + $1/2$) multiples of Δt:

$$u_k^{n+1/2}$$

The differential equations are then approximated by replacing dervatives with centered difference quotients:

7) $$u_k^{n+1/2} = u_k^{n-1/2} - \lambda(p_{k+1}^{n} - p_k^{n}),$$

8) $$V_{k+1/2}^{n+1} = V_{k+1/2}^{n} + \lambda(u_{k+1}^{n+1/2} - u_k^{n+1/2}),$$

9) $$e_{k+1/2}^{n+1} = e_{k+1/2}^{n} - \lambda p_{k+1/2}^{n+1/2} (u_{k+1}^{n+1/2} - u_{k}^{n+1/2}),$$

where

10) $$p_{k+1/2}^{n+1/2} = \frac{1}{2} (p_{k+1/2}^{n} + p_{k+1/2}^{n+1})$$

and

11) $$\lambda = \frac{\Delta t}{\Delta y}.$$

Equations (7) and (8) are in conservation form. In [4] J.G. Trulio and K.R. Trigger have shown how to derive a difference equation for energy that is in conservation form. Average the momentum equation (7) over n and $n+1$, and multiply it by u_k^{n+1}:

12) $$K_k^{n+1} = K_k^{n} - \lambda(p_{k+1/2}^{n+1/2} - p_{k-1/2}^{n+1/2}) u_k^{n+1/2}$$

where

13) $$K_k^n = \frac{1}{2} u_k^{n-1/2} u_k^{n+1/2}$$

Average (12) over k and $k+1$ and add it to (9):

14) $$E_{k+1/2}^{n+1} = E_{k+1/2}^{n} - \lambda(p_{k+1/2}^{n+1/2} u_{k+1}^{n+1/2} - p_{k}^{n+1/2} u_{k}^{n+1/2})$$

where

15) $$E_{k+1/2}^{n} = e_{k+1/2}^{n} + K_{k+1/2}^{n},$$

where

16) $$K_{k+1/2}^{n} = \frac{1}{2} (K_k^n + K_{k+1}^n),$$

and

17) $$p_k^{n+1/2} = \frac{1}{2} (p_{k+1/2}^{n+1/2} + p_{k-1/2}^{n+1/2}).$$

Extend the lattice functions defined above to all y, t by setting

18) $\quad u_\Delta(y,t) = u_k^{n+1/2} \quad \text{for} \quad |y - k\Delta y| < \frac{1}{2}\Delta y, \quad n\Delta t < t < (n+1)\Delta t,$

and similarly for the other variables. Denote by $\mathcal{J} = \mathcal{J}_\Delta$ space translation by $\Delta y/2$, and by $\mathcal{T} = \mathcal{T}_\Delta$ time translation by $\Delta t/2$. Then (12) and (15) lead to

19) $\quad K_\Delta = \frac{1}{2} \frac{\mathcal{T} + \mathcal{T}^{-1}}{2} (\mathcal{J} u_\Delta)(\mathcal{J}^{-1} u_\Delta),$

while, (10) and (17) can be combined into

20) $\quad P_\Delta = (\frac{\mathcal{T}+\mathcal{T}^{-1}}{2})(\frac{\mathcal{J}+\mathcal{J}^{-1}}{2})P_\Delta .$

The energy conservation law (15) can be abbreviated as

21) $\quad (\mathcal{T}-\mathcal{T}^{-1})((e_\Delta + K_\Delta) + \lambda(\mathcal{J} - \mathcal{J}^{-1})P_\Delta u_\Delta = 0$

Let us assume that the scheme under discussion is stable, i.e. that the variables u_Δ, e_Δ, V_Δ remain uniformly bounded in the maximum norm as $\Delta \to 0$, then there exist subsequence for which u_Δ, p_Δ, V_Δ, e_Δ, K_Δ, P_Δ, $P_\Delta u_\Delta$ tend to limits in the sense of distribution theory. We denote these limits by

22) $\quad \bar{u}, \bar{p}, \bar{v}, \bar{e}, \bar{K}, \bar{P}, \overline{Pu}.$

Equations (7), (8) and (21) are in conservation form; it follows as shown in [1], that the distribution limits (22) satisfy, in the sense of distributions, the differential equations

23) $\quad \bar{u}_t + \bar{p}_y = 0$

24) $\quad \bar{v}_t - \bar{u}_y = 0$

and

25) $\quad (\bar{e}+\bar{K})_t + (\overline{Pu})_y = 0$

Equations (23) and (24) are the laws of conservation of momentum and mass (2) and (3). Equation (25) differs from the energy conservation equation (4):

25)' $$(\bar{e} + \tfrac{1}{2}\bar{u}^2)_t + (\bar{p}\,\bar{u})_y = 0 ;$$

we now calulate the deviation.

Since P_Δ is linearly related to p_Δ, it follows that

26) $$\bar{P} = \bar{p}.$$

Next we set

27) $$u_\Delta = \bar{u} + v_\Delta,\ P_\Delta = \bar{p} + q_\Delta$$

By definition of the quantities (22), the functions v_Δ and q_Δ tend to zero in the sense of distributions. Setting (27) into (19) and (20) we obtain

28) $$\bar{K} = \tfrac{1}{2}\bar{u}^2 + \bar{H},$$

where

28)' $$\bar{H} = \tfrac{1}{4}\overline{(\mathcal{J}+\mathcal{J}^{-1})(\mathcal{J}v)(\mathcal{J}^{-1}v)},$$

and

29) $$\overline{Pu} = \bar{p}\,\bar{u} + \bar{w},$$

where

29)' $$\bar{w} = \tfrac{v}{4}\overline{(\mathcal{J}+\mathcal{J}^{-1})(\mathcal{J}+\mathcal{J}^{-1})q}.$$

Set (28) and (29) into (25):

30) $$(\bar{e} + \tfrac{1}{2}\bar{u}^2)_t + (\bar{p}\,\bar{u})_y + \bar{H}_t + \bar{w}_y = 0$$

This differs from the energy conservation equation (25)' by additional terms that have the character of <u>Reynolds stress</u> of turbulence theory. Thus the weak limit satisfies the energy equation iff the numerical Reynolds stress is zero:

31) $$\bar{H}_t + \bar{w}_y = 0.$$

When does (31) hold? Certainly when both $\bar{H}$ and $\bar{w}$ are zero. For instance, when q_Δ consists of mesh oscillations, $(J+J^{-1})q_\Delta$ would tend to zero in the L^1 sense, which according to (29)' would make $\bar{w} = 0$. Similarly, when v_Δ executes mesh oscillations, (28)' shows that $\bar{H} = 0$. So it could turn out that V. Neumann's surmise is correct after all, but for reasons quite subtle.

Jonathan Goodman and the author are investigating a somewhat analogous semi-discrete approximation to the model equation

$$u_t + (\tfrac{1}{2} u^2)_y = 0,$$

which turns out to be completely integrable. Here too one has weak but not strong convergence; we hope that the complete integrability makes it possible to analyze precisely the numerical Reynolds stress that arises in this simplified model.

We conclude by remarking on a striking difference between weak limits of solutions of differential equations and difference equations. It boils down to this:

If a uniformly bounded sequence of functions u_n converges weakly but __not__ strongly to $\bar{u}$, then

$$\overline{u^2} \neq \bar{u}^2.$$

On the other hand, if the uniformly bounded sequence u_Δ converges weakly but not strongly to u, then $(J_\Delta u_\Delta)(J_\Delta^{-1} u_\Delta)$ could converge weakly to $\bar{u}^2$. It follows that a theory of a Young measure suitable for solutions of difference equations would have novel features.

Bibliography

1. P.D. Lax and Burton Wendroff, Systems of Conservation Laws, Comm. Pure and Applied Mathematics III (1960) 217-237.

2. J.v. Neumann, Proposal and Analysis of a New Numerical Method for the Treatment of Hydro-dynamical Shock Problems. Vol VI, Collected Works.

3. J.v. Neumman and R.D. Richtyer. A Method for the Numerical Calculation of Hydrodynamical Shocks. J. Appl. Phys. 21 (1950) 380-385.

4. J.G. Trulio and K.R. Trigger. Numerical Solution of One Dimensional Hydrodynamical Shock Problem, UCRL Report 6522, 1961.

5. P.D. Lax, On Dispersion Difference Schemes 1985, Physica 18 D (1986) 250-254, North-Holland, Amsterdam

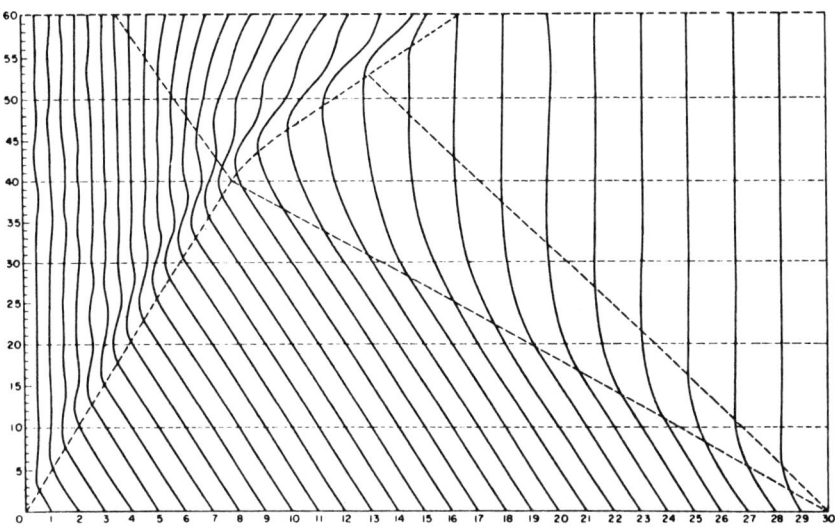

FIGURE 1

NONLINEAR GEOMETRIC OPTICS FOR HYPERBOLIC SYSTEMS OF CONSERVATION LAWS

Andrew Majda[*]

Department of Mathematics
and
Applied and Computational Mathematics Program
Princeton University
Princeton, New Jersey 08544

Table of Contents

I. Introduction

II. The Formal Theory: Weakly Nonlinear Hyperbolic Waves in 1-D

  2.1: The Single Wave Expansion in 1-D
  2.2: Multi-Wave Non-Resonant Asymptotic Solutions in 1-D
  2.3: Periodic Resonant Wave Asymptotics in 1-D
  2.4: An Application to Periodic Resonant Waves for Compressible Fluid Flow in 1-D

III. The Formal Theory: Weakly Nonlinear Hyperbolic Waves in Several Space Dimensions

  3.1: The Single Wave Expansion in Multi-D
  3.2: Resonant Oblique Small Amplitude Plane Wave Interactions in Multi-D
  3.3: The Asymptotic Equations for Resonant Oblique Waves for 2-D Compressible Flow

IV. The Rigorous Theory of Nonlinear Geometric Optics for Weak Solutions of Conservation Laws in 1-D

  4.1: The Validity of Nonlinear Geometric Optics for Weak Solutions of 1-D Conservation Laws
  4.2: The Proof of the Theorem for Scalar Convex Conservation Laws with Periodic Initial Data
  4.3: Some Accessible Open Problems in the Rigorous Theory of Weakly Nonlinear Geometric Optics

I. Introduction

Here we give a detailed discussion of recent developments, both formal and rigorous, in the theory of weakly nonlinear geometric optics for constructing asymptotic solutions of quasi-linear hyperbolic systems in one and several space variables. This method is the main perturbation technique used in analyzing nonlinear wave motion for hyperbolic systems. The ideas for this method originated in the late 1940's and early 1950's in pioneering work of Landau [8],

---

[*] Partially supported by N.S.F. Grant #DMS84-03223 and A.R.O. and O.N.R. Grants.

Lighthill [10], and Whitham [18]. However, it is only in very recent work [4], [5], [7], [1] that these methods have been developed through systematic self-consistent perturbation schemes for resonant and nonresonant wave problems in one and several space dimensions. Sections II and III of this paper give a detailed discussion and description of these formal perturbation methods applied to problems in 1-D and multi-D, respectively. The reader can consult the survey in [16] which reviews the literature on weakly nonlinear hyperbolic waves before 1981 and compare this treatment with the one described in sections II and III to see the recent progress in the field in constructing such formal perturbation expansions.

One of the goals of the use of such asymptotic methods is to reduce extremely complex problems to simpler but often non-trivial problems which are more readily understood. In fact one of the main themes of this paper is that for weakly nonlinear hyperbolic wave motions, extremely complicated wave motions for systems in 1-D and multi-D are well approximated through solutions of much simpler equations such the inviscid Burgers equation. As we will see in section 2, the tacit assumptions used in the formal derivation of the expansions of weakly nonlinear geometric optics are that solutions of the underlying hyperbolic system remain smooth; nevertheless, in a variety of applied contexts these methods are often used <u>after shock waves have formed</u> in general weak solutions of conservation laws and yield a very good approximation. In section 4 we give a leisurely discussion of the recent paper of DiPerna and the author [3] which contains rigorous results on the use of weakly nonlinear geometric optics for systems in 1-D. These results indicate that the method is even better for weak solutions then could be anticipated from the formal predictions of the perturbation theory!! For example, within <u>errors that are of order</u> ε^2 <u>uniformly for all time</u>, the <u>weak solution</u> of a general initial value problem for a <u>general</u> $M \times M$ <u>system</u> of <u>genuinely nonlinear conservation</u> laws <u>with initial data</u> with <u>amplitude</u> ε is <u>approximated by</u> the <u>weak solutions</u> of completely <u>decoupled Burgers equations</u>.

There are several recent multi-D applications of the formal methods presented here to complex physical problems including the development of simplified models in reacting gas flow [17], the regular reflection of weak shocks [6], and the formation of Mach stems in reacting shock fronts [13], [14]. We will not discuss any of these applications in detail here but briefly mention one accessible open theoretical problem which arises from the work in [13], [14]. An application of weakly nonlinear geometric optics to the (free-surface) shock fronts in multi-D reacting gases yields the simplified governing integro-differential conservation law (see [13], [14]) for a scalar quantity, $\sigma(x,t)$, given by

$$(1.1) \quad \sigma_t + a_1 (\tfrac{1}{2} \sigma^2)_x + a_2 [\int_0^\infty \sigma(x + \beta s) \sigma_x(x + s) ds]_x = 0$$

with $a_1 - a_2 \neq 0$ and the parameter β satisfying $\beta > 1$. We remark that if $\beta \equiv 1$, the equation in (1.1) reduces to the inviscid Burgers equation. The numerical experiments with (1.1) strongly predict that smooth solutions of (1.1) develop shocks (see [14]) although the nonlocal terms have some prominent and unusual effects in this breakdown process in various parameter regimes. The accessible open problem for (1.1) which we present here is the following:

Problem: Find some (or any) initial data for (1.1) for which one can prove rigorously that shock waves occur.

A brief discussion of a number of other accessible open problems in the rigorous theory of weakly nonlinear geometric optics is given at the end of section 4 of this paper. The applications of the ideas presented here to understanding complex multi-D problems is only beginning and possible applications to complex Mach bifurcations, instability in supersonic jets, and the detailed structure of multi-D reacting shock fronts are currently being developed.

Acknowledgement: The author thanks his scientific collaborators, R. Rosales, J. Hunter, and R. DiPerna for their explicit and implicit contributions to this paper.

II. The Formal Theory: Weakly Nonlinear Hyperbolic Waves in 1-D

We consider solutions of the general $M \times M$ strictly hyperbolic system in a single space dimension given by the M equations,

$$(2.1) \quad (F_0(u))_t + (F_1(u))_x = 0, \quad -\infty < x < \infty \quad t > 0.$$

The requirement of strict hyperbolicity at a constant vector, $u_0 \in R^M$, with t, a time-like direction, means that

$$(2.2a) \quad \det(A_0(u_0)) \neq 0$$

and the generalized eigenvalue problems

$$(2.2b) \quad \begin{array}{c} (A_1(u_0) - \lambda A_0(u_0)) \cdot r = 0 \\ \ell \cdot (A_1(u_0) - \lambda A_0(u_0)) = 0 \end{array}$$

have M-distinct real eigenvalues $\lambda_1 < \lambda_2 < \ldots < \lambda_M$ with corresponding right eigenvectors, r_j, $1 < j < M$ and left eigenvectors, ℓ_K, $1 < K < M$ satisfying the normalization conditions,

$$(2.2c) \quad \ell_K \cdot A_0 r_j = \delta_{Kj}$$

with δ_{Kj} the Kronecker delta symbol. Here $A_0(u)$, $A_1(u)$ are the $M \times M$ Jacobian matrices of the corresponding smooth mappings, $F_0(u)$, $F_1(u)$, i.e. $A_j(u) = \frac{\partial F_j(u)}{\partial u}$, $j = 0,1$. We also assume that $F_j(u)$ admits the Taylor expansion at u_0 given by

$$(2.3) \quad F_j(u_0 + \epsilon v) = F_j(u_0) + \epsilon A_j v + \frac{\epsilon^2}{2} B_j(v,v) + O(\epsilon^3) \quad j = 0,1$$

where the B_j are the corresponding Hessian matrices of $F_j(u)$ at u_0.

Our objective here is to construct formal asymptotic approximate solutions of the hyperbolic system in (2.1) with small amplitude rapidly oscillating initial data with the form,

$$(2.4) \quad u_0^\epsilon(x) = u_0 + \epsilon u_1^0(x, \frac{x}{\epsilon})$$

with $u_1^0(x,\tilde{x})$ a smooth structure function. To clarify and simplify the presentation we will often assume that $u_1^0(x,\tilde{x}) = u_1(\tilde{x})$ and either

<u>Hypothesis #1</u> $u_1^0(\tilde{x})$ has compact support

or

<u>Hypothesis #2</u> $u_1^0(\tilde{x})$ is a periodic function with period one and mean zero, i.e.,

$$\int_0^1 u_1^0(\tilde{x} + a)d\tilde{x} = 0 \quad \text{for any} \quad a \in R \ .$$

Under <u>Hypothesis #1</u>, the <u>basic asymptotic expansions</u> which we construct below <u>are non-resonant</u> and to the formal degree of approximation, nonlinear <u>wave fields can be superimposed</u> and do not interact as in the linear case. On the other hand, <u>Hypothesis #2 describes</u> the <u>simplest situation where different nonlinear wave fields</u> can <u>interact and resonate</u> to <u>leading order</u>.

To summarize, in the remainder of this section we will build formal asymptotic approximate solutions of the hyperbolic system in (2.1). These solutions approximate the solution, $u^\varepsilon(x,t)$, which satisfies

(2.5) $$\begin{array}{l} F_0(u^\varepsilon)_t + F_1(u^\varepsilon)_x = 0, \quad -\infty < x < \infty, \ 0 < t \\ u^\varepsilon(x,0) = u_0 + \varepsilon u_1^0(x, \frac{x}{\varepsilon}) \ . \end{array}$$

The methods which we present below in developing the formal constructions are less explicit and somewhat more complicated than those already used in [15]. The advantages of these methods is that they generalize readily to many interesting situations in several space dimensions which we describe in the next section. The fourth section of this paper contains a discussion of rigorous work by DiPerna and the author which justifies many of the expansions presented in this section, even after shock waves have formed in the given weak solution in (2.5). Nevertheless, many interesting theoretical questions remain regarding these approximations. Finally, we refer the reader to [15] for a complete discussion of the formal constructions presented below for more general initial data via explicit solution of the second order perturbation equations.

2.1: <u>The Single Wave Expansion in 1-D</u>

Here we develop the simplest asymptotic approximation for the solution in

(2.5). The result of this expansion will be a formal asymptotic solution, $\tilde{u}_W^\epsilon(x,t)$, with the special initial data,

(2.6) $$u_0^\epsilon(x) = u_0 + \epsilon \, \sigma_j^0(x, \tfrac{x}{\epsilon})r_j$$

for any fixed j with $1 \leq j \leq M$. Here r_j is the corresponding right eigenvector and $\sigma_j^0(x,\tilde{x})$ is a smooth function which is bounded with bounded first derivatives.

By analogy with linear geometric optics, we attempt to construct, formal single wave asymptotic solutions of the equations in (2.1) with the form,

(2.7) $$\tilde{u}_W^\epsilon(x,t) = u_0 + \epsilon u_1(x,t,\tfrac{\phi}{\epsilon}) + \epsilon^2 u_2(x,t,\tfrac{\phi}{\epsilon})$$

where $\phi(x,t)$ is a phase function to be determined. We require $\tilde{u}_W^\epsilon(x,t)$ defined in (2.7) to be a smooth formal solution of the strictly hyperbolic system in (2.1) to two orders in ϵ, i.e., $\tilde{u}_W^\epsilon(x,t)$ should satisfy

(2.8) $$\frac{\partial}{\partial t} F_0(\tilde{u}_W^\epsilon) + \frac{\partial}{\partial x} F_1(\tilde{u}_W^\epsilon) = o(\epsilon).$$

The requirement in (2.8) means that after inserting the ansatz from (2.7) into (2.1), we must be able to choose $u_1(x,t,\tfrac{\phi}{\epsilon})$, $u_2(x,t,\tfrac{\phi}{\epsilon})$ so that

(2.9)
A) The terms of order zero in ϵ vanish
B) The terms of order one in ϵ vanish.

Obviously, in order to have the function, $u_1(x,t,\tfrac{\phi}{\epsilon})$, consistently describe the leading order asymptotic behavior on some region G, in space time, we need to require additionally that

(2.10) $$|u_2(x,t,\tfrac{\phi}{\epsilon})| = o(\epsilon^{-1}) \quad \text{for} \quad (x,t) \in G,$$

i.e. $u_2(x,t,\theta)$ should grow <u>sublinearly</u> in θ for $x,t \in G$ and $\theta = \tfrac{\phi}{\epsilon}$. The requirement that the initial data have the special form in (2.6) for these given single wave approximate solutions will arise after we impose the requirements in (2.9) and (2.10) on the solution $\tilde{u}_W^\epsilon(x,t)$ with the single wave ansatz in (2.7).

By substituting the ansatz from (2.7) into (2.1) and expanding the nonlinear terms via (2.3), we compute that the requirements in (2.9a) are satisfied provided that we can choose $u_1(x,t,\theta)$, $\phi(x,t)$ so that,

Order zero

(2.11) $\qquad (A_0 \phi_t + A_1 \phi_x) \dfrac{\partial}{\partial \theta} u_1(x,t,\theta)\big|_{\theta=\frac{\phi}{\varepsilon}} = 0$.

Similarly, the requirement in (2.9b) is satisfied provided that we can choose $u_2(x,t,\theta)$ so that

Order one

(2.12) $\qquad -(A_0 \phi_t + A_1 \phi_x) \dfrac{\partial}{\partial \theta} u_2(x,t,\theta)\big|_{\theta=\frac{\phi}{\varepsilon}} =$

$$[A_0(u_1)_t + A_1(u_1)_x + \phi_t B_0((u_1),(u_1)_\theta) + \phi_x B_1((u_1),(u_1)_\theta)]\big|_{\theta=\frac{\phi}{\varepsilon}}.$$

We use the familiar method of multiple scales and regard the variable θ in (2.11) and (2.12) as an additional independent variable linked with (x,t) only in the final asymptotic form through the evaluation, $\theta = \dfrac{\phi}{\varepsilon}$.

The equation in (2.11) is satisfied provided that

(2.13a) $\qquad \phi^j$ satisfies the eikonal equation,

$$\phi^j_t + \lambda_j \phi^j_x = 0 \quad \text{for some } j, \ 1 \leq j \leq M$$

with λ_j an eigenvalue from (2.2b) and

(2.13b) $\qquad\qquad u_1(x,t,\theta)$ satisfies
$$u_1(x,t,\theta) = \sigma_j(x,t,\theta) r_j$$

with r_j the corresponding right eigenvector from (2.2c). We recognize the equation in (2.13a) as the familiar eikonal equation of linear geometric optics. For a single space variable, we choose the simplest phase function solution of this equation, namely

(2.13c) $\qquad \phi^j = x - \lambda_j(u_0) t, \quad \text{any } j, \ 1 \leq j \leq M.$

At this state in the argument, $\sigma_j(x,t,\theta)$ is an arbitrary amplitude function.

This completes the solution of the order zero equations.

Now, we turn to the equations in (2.12) which represent the order one contribution in powers of ε. First, we substitute the form of $u_1(x,t,\theta)$ as determined in (2.13b) into the right hand side of (2.12) and regard the equation in (2.12) to be satisfied for all θ following the ideas of the method of multiple scales. With ϕ^j satisfying (2.13), we see from (2.2c) that $\ell_j \cdot (A_0 \phi_t^j + A_1 \phi_x^j) = 0$ and the ordinary differential equation in θ defined by (2.12) with the form

$$(2.14a) \qquad (A_0 \phi_t^j + A_1 \phi_x^j) \frac{\partial}{\partial \theta} u_2 = F(\theta)$$

has a solution, $u_2(\theta)$, if and only if

$$(2.14b) \qquad \ell_j \cdot F(\theta) = 0.$$

By imposing this condition on the explicit inhomogeneous terms defined by the right hand side of (2.12), we obtain the much simpler differential equation for $\sigma_j(x,t,\theta)$,

$$(2.15) \qquad (\sigma_j)_t + \lambda_j (\sigma_j)_x + b_j \left(\frac{1}{2} \sigma_j^2 \right)_\theta = 0$$

with b_j given by the formula,

$$(2.16) \qquad b_j = \ell_j \cdot (-\lambda_j B_0(r_j, r_j) + B_1(r_j, r_j))$$

for any j with $j = 1,\ldots,M$.

The final step in the formal derivation of the leading order asymptotics is to find a solution, $u_2(x,t,\theta)$, satisfying the ordinary differential equation in (2.14) with $F(\theta)$ defined in (2.12) and with <u>sub-linear</u> growth in θ for (x,t) in a space-time region G; i.e. we need to satisfy the requirement in (2.10). For these special simple wave forms this is easily achieved. We seek the solution, $u_2(\theta)$, for (2.14) in the form, $u_2(\theta) = \sum_{p \neq j} \tilde{\sigma}_p r_p$ and compute that (2.14) is satisfied if and only if

$$(2.17) \qquad \frac{\partial \tilde{\sigma}_p}{\partial \theta} = (\lambda_j - \lambda_p)^{-1} c_{pj} (\sigma_j^2)_\theta$$

for $p \neq j$, $1 \leq p \leq M$ with c_{pj} explicit constants. We remark that one consequence of the conservation form in (2.1) is that the inhomogeneous right hand side of the equations in (2.17) is an exact θ derivative. Thus, u_2 has the explicit form,

$$u_2 = \sum_{p \neq j} (\lambda_j - \lambda_p)^{-1} c_{pj} \sigma_j^2 r_j$$

and u_2 is <u>bounded</u> <u>automatically</u> on any space-time region G where σ_j is smooth and bounded - thus, the sublinear growth condition in (2.10) is trivially satisfied in this case.

What has been achieved by this expansion? To summarize, we have demonstrated that solutions u^ε of the general quasi-linear hyperbolic initial value problem

(2.18)
$$F_0(u^\varepsilon)_t + F_1(u^\varepsilon)_x = 0, \quad x \in R^1, \, t > 0$$
$$u^\varepsilon(x,0) = u_0 + \varepsilon \sigma_j^0(x, \frac{x}{\varepsilon}) r_j$$

are uniformly approximated in regions of smoothness within terms of order ε^2 by the solution u_W^ε with the form,

(2.19)
$$u_W^\varepsilon = u_0 + \varepsilon \sigma_j(x, t, \frac{\phi_j}{\varepsilon}) r_j$$

where σ_j satisfies the much simpler scalar conservation law,

(2.20)
$$(\sigma_j)_t + \lambda_j (\sigma_j)_x + b_j (\tfrac{1}{2} \sigma_j^2)_\theta = 0$$
$$\sigma_j(x, t, \theta)|_{t=0} = \sigma_j^0(x, \theta).$$

First, we remark that if the initial data has the form in Hypothesis #1 or #2, i.e. σ_j^0 is a function of θ alone, then the solution of (2.20) is independent of x and reduces to the famous inviscid Burgers' equation,

(2.21)
$$(\sigma_j)_t + b_j (\tfrac{1}{2} \sigma_j^2)_\theta = 0$$
$$\sigma_j(t, \theta)|_{t=0} = \sigma_j^0(\theta)$$

provided that $b_j \neq 0$. We remark that in the special case that $F_0(u) = u$, the condition

(2.22) $\quad b_j \neq 0 \quad$ is equivalent to Lax's genuine nonlinearity condition at u_0

(see [12]). The more general problem in (2.20) is also easily solved exactly by reduction to the inviscid Burgers equation through the use of characteristic coordinates for the operator, $\frac{\partial}{\partial t} + \lambda_j \frac{\partial}{\partial x}$.

2.2: Multi-Wave Non-Resonant Asymptotic Solutions in 1-D

Here we consider the asymptotic approximation of the solution $u^\epsilon(x,t)$ to the initial value problem with general initial data,

(2.23)
$$F_0(u^\epsilon)_t + F_1(u^\epsilon)_x = 0 \;,\; x \in R^1,\; t > 0$$
$$u^\epsilon(x,0) = u_0 + \epsilon u_1^0(\frac{x}{\epsilon}).$$

If the problem in (2.23) were linear, we would simply decompose the data, $u_1^0(\frac{x}{\epsilon})$, according to the right eigenvectors r_j and superimpose the single wave asymptotic solutions defined in (2.19) and (2.12). Thus, to solve the asymptotic problem in (2.23), our first naive guess is to define $u_w^\epsilon(x,t)$ by superposition as

(2.24)
$$u_w^\epsilon(x,t) = u_0 + \epsilon \sum_{j=1}^{M} \sigma_j(t,\frac{\phi^j}{\epsilon}) r_j$$

where $\phi^j = x - \lambda_j t$ and each σ_j solves the much simpler initial value problem for the inviscid Burgers' or linear advection equation given by

(2.25)
$$(\sigma_j)_t + b_j(\frac{1}{2}\sigma_j^2)_\theta = 0,\; \theta \in R^1,\; t > 0$$
$$\sigma_j(t,\theta)|_{t=0} = \ell_j \cdot u_1^0(\theta)$$

for $1 \leq j \leq M$. When such a simple multi-wave expansion from (2.24) and (2.25) generates a self-consistent formal asymptotic solution, we say that the multi-wave approximation is nonresonant. Of course the requirements of generating a formal asymptotic solution are analagous to those already discussed in (2.8) - (2.10). We must find a function, $u_2(x,t,\vec{\theta})$ with $\vec{\theta} \in R^M$ so that with $\tilde{u}_w^\epsilon$ given by

(2.26)
$$\tilde{u}_w^\epsilon = u_0 + \epsilon \sum_{j=1}^{M} \sigma_j(t,\frac{\phi^j}{\epsilon}) r_j + \epsilon^2 u_2(x,t,\frac{\vec{\phi}}{\epsilon})$$

both of the conditions in (2.9) are satisfied and in addition,

(2.27) $$|u_2(x,t,\tfrac{\vec{\phi}}{\varepsilon})| = o(\varepsilon^{-1})$$

for $(x,t) \in G$, some space-time domain; then the multi-wave approximation from (2.24) and (2.25) is uniformly valid on G and we have achieved a remarkable simplification by approximating the general small amplitude initial value problem in (2.23) by the much simpler equations in (2.24) and (2.25) with exact solutions. Obviously, with the form in (2.24) and (2.13), we have already guaranteed that the terms of order zero in powers of ε vanish; the crux of the matter is to choose $u_2(x,t,\vec{\theta})$ so that simultaneously the terms of order one in ε vanish, i.e. (2.9b) is satisfied, and u_2 satisfies the sublinear growth conditions. This will not be possible always and resonances can occur, this is the topic of the next subsection. Here we will derive some simple sufficient conditions for the form validity of the nonresonant wave expansions including a careful formal analysis of the behavior of u_2; these sufficient conditions always apply under hypothesis #1, i.e. when $u_0^1(x)$ is smooth with compact support.

By following the calculations in (2.12) it is tedious but straightforward to compute with the ansatz in (2.26) that the order one terms in ε vanish provided that

| Order one |

(2.28) $$-\sum_{\ell=1}^{M} \mathcal{a}_\ell \frac{\partial}{\partial \theta_\ell} u_2(x,t,\vec{\theta})\Big|_{\vec{\theta}=\frac{\vec{\phi}}{\varepsilon}} = G(t,\vec{\theta})\Big|_{\vec{\theta}=\frac{\vec{\phi}}{\varepsilon}}.$$

Here the matrix coefficients, $\mathcal{a}_\ell$ are given by

(2.29) $$\mathcal{a}_\ell = \phi_t^\ell A_0 + \phi_x^\ell A_1 = -\lambda_\ell A_0 + A_1$$

for $1 \leq \ell \leq M$. The term $G(t,\vec{\theta})$ is a sum of simple and binary wave interaction terms,

(2.30a) $$G(t,\theta) = \sum_{j=0,1}^{M} g_j(t,\theta_j) + \sum_{\substack{1 \leq p,q \leq M \\ p \neq q}} g_{pq}(t,\theta_p,\theta_q)$$

with the single wave terms given by

(2.30b) $\quad g_j(t,\theta_j) = A_0((\sigma_j)_t \, r_j + [-\lambda_j B_0(r_j,r_j) + B_1(r_j,r_j)]\sigma_j(\sigma_j)_{\theta_j}$

and the binary wave interaction terms given by

(2.30c) $\quad g_{pq}(t,\theta_p,\theta_q) = \frac{1}{2}[-\lambda_p B_0(r_p,r_q) + B_1(r_p,r_q)]\sigma_q(t,\theta_q)(\sigma_p(t,\theta_p))_{\theta_p}$.

Below, we use the notation $\sigma'(t,\theta)$ for $\sigma(t,\theta)_\theta$. Clearly, the single wave contributions in (2.30b) are handled exactly as we discussed earlier in (2.14)-(2.16) provided that the Burgers equations from (2.25) are satisfied. To handle the contributions from the binary wave interaction terms in (2.28), we use the method of multiple scales and regard the $\vec{\theta}$ variables as independent variables. By superposition, we only need to construct a solution $u_2(x,t,\theta_p,\theta_q)$ satisfying the sublinear growth condition from (2.27) and the linear P.D.E.,

(2.31) $\quad (-\lambda_p A_0 + A_1)\frac{\partial}{\partial \theta_p} u_2 + (-\lambda_p A_0 + A_1)\frac{\partial}{\partial \theta_q} u_2 = g_{pq}(\theta_p,\theta_q)$

for $p \neq q$, $1 \leq p,q \leq M$. Once again, in a single space variable, we solve for u_2 explicitly through the basis expansion,

$$u_2 = \sum_{j=1}^{M} \tilde{\sigma}_j r_j.$$

The problem in (2.31) is equivalent to the M scalar advection equations,

(2.32) $\quad (\lambda_j - \lambda_p)\frac{\partial}{\partial \theta_p} \tilde{\sigma}_j + (\lambda_j - \lambda_q)\frac{\partial}{\partial \theta_q} \tilde{\sigma}_j = \ell_j \cdot g_{pq}(\theta_p,\theta_q)$

for $p \neq q$, $1 \leq j \leq M$.

Thus, we only need to decide the conditions on the inhomogeneous terms $\ell_j \cdot g_{pq}(\theta_p,\theta_q)$ with g_{pq} given explicitly in (2.30c) which guarantee that the scalar advection equations in (2.32) have solutions $\tilde{\sigma}_j$ with sublinear growth in $|\theta_p| + |\theta_q|$. The cases when $j = p,q$ are special and we handle these first. From (2.30c), we compute that when $j = p$, (2.32) becomes with some constant, c_p^{pq},

(2.33) $\quad (\lambda_p - \lambda_q)\frac{\partial}{\partial \theta_q} \tilde{\sigma}_p = c_p^{pq} \sigma_q(t,\theta_q) \frac{\partial}{\partial \theta_p}(\sigma_p(t,\theta_p))$

and the solution $\tilde{\sigma}_p$ within a function of θ_p is given by integration,

(2.33b) $\quad \tilde{\sigma}_p(\theta_p,\theta_q) = (\lambda_p - \lambda_q)^{-1} c_p^{pq} \sigma'(t,\theta_p) \int_{\theta_q^0}^{\theta_q} \sigma_q(t,\theta) d\theta.$

We see that under Hypothesis #1 or Hypothesis #2 on the initial data, $\sigma_p(\theta_p, \theta_q)$ is always bounded on any region where $\sigma_p'(t, \theta_p)$ is bounded and

(2.33c) $\quad |\tilde{\sigma}_p(t, \theta_p, \theta_q)| \leq |c_p^{pq}||\lambda_p - \lambda_q|^{-1}|\sigma_p'(t,\theta)|_\infty (\max_{-\infty < \theta_q < \infty} |\int_{\theta_q}^{\theta_q} \sigma_q(t,\theta)d\theta|)$

with $|f(t,\theta)|_\infty = \max_{\theta \in R^1}|f(t,\theta)|$. Similarly, when $j = q$, $\tilde{\sigma}_q(t, \theta_p, \theta_q)$ can be chosen explicitly as a bounded function of (θ_p, θ_p) under Hypothesis #1 or #2 in any region where $\sigma_p(t,\theta)$, $\sigma_q(t,\theta)$ remain smooth and $\tilde{\sigma}_q$ satisfies the estimate,

(2.34) $\quad |\tilde{\sigma}_q(t, \theta_p, \theta_q)| \leq |c_p^{pq}||\lambda_p - \lambda_q|^{-1}|\sigma_p(t,\theta)|_\infty|\sigma_p(t,\theta)|_\infty.$

The treatment of the scalar advection equation in (2.32) is quite different when $j \neq p$ and $j \neq q$, i.e. for $M \geq 3$. Here the nonresonant expansion will always be valid under Hypotheis #1 on the initial data; on the other hand, this will not be true for periodic data satisfying Hypothesis #2 due to the appearance of resonant wave interactions (see the next subsection). Within a bounded function, the solution of the inhomogeneous scalar advection equation in (2.32) is given by

(2.35) $\quad \tilde{\sigma}_j = c_j^{pq}(\lambda_j - \lambda_p)^{-1} \int_0^{\theta_p} \sigma_p'(t,s)\sigma_q(t, \theta_q - h_j^{pq}(\theta_p - s))ds$

with $h_j^{pq} = \frac{\lambda_j - \lambda_q}{\lambda_j - \lambda_p}$. Next, we give some simple explicit conditions which guarantee that $\tilde{\sigma}_j$ is uniformly bounded. For a smooth function $\sigma_p(t,\theta)$, the θ-variation of σ_p is the L^1-norm of the first derivative defined by

(2.36) $\quad \mathrm{Var}(\sigma_p(t,\theta)) = \int_{-\infty}^{\infty} |\sigma_p'(t,s)|ds.$

With this definition we see that $\tilde{\sigma}_j(t, \theta_p, \theta_q)$ satisfies the estimate,

(2.37) $\quad |\tilde{\sigma}_j(t, \theta_p, \theta_q)| \leq c_j^{pq}|\lambda_j - \lambda_p|^{-1}|\sigma_p(t,\theta)|_\infty \mathrm{Var}\,\sigma$

for $1 \leq j \leq M$, $j \neq p, q$.

Let's summarize the conditions which guarantee that we can find $u_2(x,t,\vec{\theta})$ which is bounded on some space time region G and so that $u_2(x,t,\vec{\theta})$ satisfies (2.28). From (2.17), (2.34), and (2.37), we need

(2.38a) $$\sum_{j=1}^{M} (|\sigma_j(t,\theta)|_\infty + \text{Var}\,\sigma_j(t,\theta)) < C_0$$

while the conditions in (2.33) require

(2.38b) $$\sum_{\substack{q,j=1 \\ q \neq j}}^{M} (|\sigma_j'(t,\theta)|_\infty \times \max_{-\infty < \theta_q < \infty} |\int_0^{\theta_q} \sigma_q(t,\theta)d\theta|) < C_0 \ .$$

Finally, we check the self-consistency of the asymptotic expansion. The functions $\sigma_j(t,\theta)$ satisfy the simple scalar conservation laws in (2.25) with initial data, $\ell_j \cdot u_1^0(\theta)$ for each $j = 1,\ldots,M$. Since scalar conservation laws do not increase the maximum norm, the variation of a solution, or the functional, $\max_{\theta_1,\theta_2 \in R^1} |\int_{\theta_1}^{\theta_2} f(s)ds|$, we see that all of the quantities in (2.38) except $|\sigma'(t,\theta)|_\infty$ are <u>uniformly bounded in time</u> provided that initially, we have

(2.39) $$|u_1^0(x)|_\infty + \text{Var}\,u_1^0 + \max_{-\infty < x_1 < x_2 < \infty} |\int_{x_1}^{x_2} u^0(s)ds| < \infty.$$

Thus, under the assumption in (2.39) for the smooth initial data, the formal nonresonant expansion is valid with a formal error of order ε^2 for all (x,t) with $0 < t < T$ where T is any interval of existence of smooth solutions for the initial value problem for the Burgers equations in (2.25) where

(2.40) $$\max_{1 \leq j \leq M} |\sigma_j'(t,\theta)|_\infty < \infty$$

for $0 < t < T$, i.e. until shock waves have formed in these simple solutions. For many (rarefaction) initial data, $T \equiv +\infty$!!

We have "proved" the following:

<u>Proposition 1</u>: (<u>Formal Validity of the Nonresonant Multi-Wave Expansion</u>)

Assume that the initial data for the initial value problem in (2.23) satisfies the condition in (2.39), then the nonresonant multi-wave expansion defined in (2.24) and (2.25) is a formal uniformly valid approximate solution of (2.23) with errors uniformly of order ε^2 on any space-time region G defined by $-\infty < x < \infty$ and $0 < t < T$ with T restricted by (2.40). In particular,

A) For smooth data satisfying hypothesis #2 so that $u_1^0(x)$ has compact support, the general solution of the general hyperbolic system in (2.23) is always formally approximated uniformly within terms of order ε^2 in regions of smoothness by the much simpler decoupled inviscid Burgers' solutions satisfying (2.24) and (2.25).

B) The same remark as in A) applies for solutions with periodic initial data provided that $M \leq 2$.

In the above, we have carried out the formal assessment of the validity of the nonresonant asymptotic expansions in great detail. In section 4 we will develop a rigorous analysis which both confirms these formal calculations in regions of smoothness and somewhat surprisingly extends the validity of the asymptotics to regions where shock waves have formed!! This is not predicted by the formal asymptotic analysis but is a consequence of the rigorous analysis of nonlinear wave interactions.

2.3 Periodic Resonant Wave Asymptotics in 1-D

Here we consider the problem of constructing uniformly valid formal asymptotic approximate solutions for the initial value problem,

$$(2.41) \quad \begin{aligned} F_0(u^\varepsilon)_t + (F_1(u^\varepsilon))_x &= 0, \quad x \in \mathbb{R}^1, \ t > 0 \\ u^\varepsilon(x,0) &= u_0 + \varepsilon u_1^0\left(\frac{x}{\varepsilon}\right) \end{aligned}$$

where $u_1^0(x)$ is a smooth periodic function of period one with mean zero, i.e. Hypothesis #2 is satisfied. We will discover that typically for $M \geq 3$, the simple decoupled nonresonant multi-wave expansions are not a uniformly valid asymptotic approximation; in fact, the inviscid Burgers equations need to be coupled through terms incorporating nonlocal resonant wave interactions in order to develop a uniform approximation. We will develop the asymptotic expansions by a more general but less explicit approach than the one used earlier in [15]. The advantage of the more general approach that we develop here is that the same method can be used in studying the multi-dimensional planar oblique wave interactions which we describe in section three (see [7]).

In a familiar fashion, we begin with the ansatz

(2.42) $$\tilde{u}_w^\varepsilon = u_0 + \sum_{j=1}^{M} \varepsilon\sigma_j(t,\frac{\phi_j}{\varepsilon})r_j + \varepsilon^2 u_2(t,\frac{\vec{\phi}}{\varepsilon})$$

with $\phi_j = x - \lambda_j t$ and $\sigma_j(0,\theta) = \ell_j \cdot u_1^0(\theta)$ for $1 \leq j \leq M$. As in (2.13), the form of the leading terms automatically guarantees that the terms of order zero in ε from (2.9a) vanish but here we do not impose the equations in (2.25) at the outset. The equation for the terms of order one in ε is the same one as given in (2.28) with the same definitions for the coefficients as given in (2.29) and (2.30). Now, in order to guarantee that a solution $u_2(t,\frac{\vec{\phi}}{\varepsilon})$ of (2.28) can be found with sublinear growth for the periodic case, we need to investigate (as in (2.31)) the auxilliary constant coefficient P.D.E.,

(2.43) $$(-\lambda_p A_0 + A_1)\frac{\partial u}{\partial \theta_p} + (-\lambda_q A_0 + A_1)\frac{\partial u}{\partial \theta_q} = \tilde{g}(\theta_p,\theta_q)$$

with $\tilde{g}$ double periodic. We remark that the use of $\tilde{g}(\theta_p,\theta_q)$ on the right hand side of (2.43) is not a typographical error since we will ultimately use solutions of (2.43) when $\tilde{g}(\theta_p,\theta_q)$ is <u>not</u> the function defined on the right hand side of (2.31) from our earlier argument. We have the following preliminary fact:

<u>Lemma 1</u>: Assume that $\tilde{g}(\theta_p,\theta_q)$ is a doubly periodic function with $\int_0^1 \tilde{g}(s,\theta_q)ds = \int_0^1 \tilde{g}(\theta_p,s)ds = 0$, then the P.D.E. in (2.43) has a solution $u(\theta_p,\theta_q)$ with sublinear growth in $|\theta_p| + |\theta_q|$ if and only if

(2.44) $$\lim_{T\to\infty} \frac{1}{2T}\int_{-T}^{T} \ell_j \cdot \tilde{g}((\theta_p,\theta_q) + \vec{a}_{pq}^j s)ds = 0$$

for all $j \neq p,q$ and $1 \leq j \leq M$ with

$$\vec{a}_{pq}^j = (\lambda_j - \lambda_p, \lambda_j - \lambda_q)$$

<u>Remark</u>: Since $\tilde{g}(\theta_p,\theta_q)$ is a periodic function of two variables, $\tilde{g}((\theta_p,\theta_q) + \vec{a}_{pq}^j s)$ is an almost periodic function of the scalar variable s. Standard properties of almost periodic functions (see [22]) guarantee that the mean of the terms on the right hand side exists for a general doubly periodic function and is uniform for

arbitrary translations of the argument s. We postpone the proof of Lemma 1 until the end of this section and continue the argument. With the explicit binary wave interaction terms from (2.30c) defining $g_{pq}(t,\theta_p,\theta_q)$ in the periodic case, we can apply the above lemma to obtain sufficient (and <u>essentially necessary</u>) conditions for the nonresonant multi-wave expansions from the last subsection to be uniformly valid for periodic waves. These conditions require that

$$(2.45) \qquad 0 = r_{pq}^j \lim_{T \to \infty} \frac{1}{2T} \int_{-T}^{T} \sigma_p'(t, \theta_p + u_j - \lambda_p)s) \sigma_q(t, \theta_q + (\lambda_j - \lambda_q)s) ds$$

$$j \neq p \neq q$$

with r_{pq}^j, the asymmetric binary interaction coefficients defined by

$$(2.46) \qquad r_{pq}^j = \ell_j \cdot (-\lambda_p B_0(r_q, r_p) + B_1(r_q, r_p)).$$

In general, as the reader can see by inspection for periodic wave patterns, the conditions in (2.45) are not satisfied and the extremely simple nonresonant multi-wave pattern from (2.24) cannot describe the leading order asymptotics through decoupled inviscid Burgers' equations.

The lemma and the conditions in (2.45) suggest an obvious strategy for satisfying the additional requirements in (2.45) by incorporating additional non-local terms in the periodic case. With g_{pq} given by the formula in (2.30c), we consider the function $g_{pq}(t,\theta_p,\theta_q)$ uniquely defined by the conditions,

$$(2.47) \qquad \ell_j \cdot g_{pq}(t,\theta_p,\theta_q) = 0, \quad j = p,q$$

$$\ell_j \cdot g_{pq} = r_{pq}^j \lim_{T \to \infty} \frac{1}{2T} \int_{-T}^{T} \sigma_p'(t, \theta_p + (\lambda_q - \lambda_p)s) \sigma_q(t, \theta_q + (\lambda_j - \lambda_q)s)$$

$$j \neq p, q, \quad 1 \leq j \leq M.$$

The function $\tilde{g}(t,\theta_p,\theta_q)$ defined by

$$(2.48) \qquad \tilde{g}(t,\theta_p,\theta_q) = g_{pq}(t,\theta_p,\theta_q) - \bar{g}_{pq}(t,\theta_p,\theta_q)$$

satisfies all of the conditions from Lemma #1 and has a solution with sublinear growth. The terms, $g_{pq}(t,\theta_p,\theta_q)$ yield new contributions to the leading order asymptotics besides those from the single wave contributions in (2.30b) already discussed earlier in (2.14)-(2.16). Straightforward addition of these terms results in the following leading order asymptotic equations

(2.49)
$$0 = (\sigma_j)_t + b_j(\frac{1}{2}\sigma_j^2)_{\theta_j} + \sum_{\substack{p\neq j \\ q\neq j}} I_{pq}^j \lim_{T\to\infty} \frac{1}{2T} \int_{-T}^{T} \sigma_p'(t, \theta_p + (\lambda_j - \lambda_p)s)\sigma_q(t, \theta_p + (\lambda_j - \lambda_q)s)ds .$$

Unfortunately, such a straightforward approach fails for the following reason: The tacit assumption in the ansatz from (2.42) is that σ_j is a function of only two variables, (t, θ_j) while, when $\Gamma_{pq}^j \neq 0$, the integro-differential terms on the right hand side of (2.49) are functions of the additional variables (θ_p, θ_q) so these equations are <u>not</u> a self-consistent closed system of equations. How can we treat this difficulty and obtain a closed self-consistent system of equations? Here we present an idea which we call the principle of "exchange of phase functions" which uses some additional flexibility in the method of multiple scales which is usually ignored in other applications of this method ----- this idea will enable us to obtain a closed system of equations with a similar form as given in (2.47).

The idea behind the principle of "exchange of phase functions" is an extremely simple one. The order one perturbation equation from (2.28) only needs to be satisfied when $\vec{\theta}$ is restricted to the values, $\vec{\theta} = \frac{\vec{\phi}}{\varepsilon}$; thus, to satisfy the equation in (2.28), we can replace $G(t, \vec{\theta})$ by any other function $\tilde{G}(t, \vec{\theta})$ satisfying

(2.50)
$$\tilde{G}(t,\vec{\theta})|_{\vec{\theta}=\frac{\vec{\phi}}{\varepsilon}} = G(t,\vec{\theta})|_{\vec{\theta}=\frac{\vec{\phi}}{\varepsilon}}$$

in order to continue the asymptotics. We consider $\hat{h}_{pq}(t, \theta_j)$ the function of θ_j alone defined by

(2.51)
$$\lim_{T\to\infty} \frac{1}{2T} \int_{-T}^{T} \sigma_p'(t, \theta_j + (\lambda_j - \lambda_p)s)\sigma_q(t, \sigma_j + (\lambda_j - \lambda_q)s)ds$$

$$= \hat{h}_{pq}(t, \theta_j).$$

We denote the nonlocal averages appearing on the right hand side of (2.49) by

$h_{pq}(t,\theta_p,\theta_q)$, i.e.

(2.52) $\quad h_{pq}(t,\theta_p,\theta_q) = \lim_{T\to\infty} \frac{1}{2T} \int_{-T}^{T} \sigma'_p(t,\theta_p + (\lambda_j - \lambda_p)s)\sigma_q(t,\theta_q + (\lambda_j - \lambda_q)s)ds.$

The reader can readily verify that with $\phi^\ell = x - \lambda_\ell t$, $\ell = 1,\ldots,M$,

(2.53) $\quad\quad\quad\quad\quad\quad \hat{h}_{pq}(t,\frac{\phi_j}{\epsilon}) = h_{pq}(t,\frac{\phi_p}{\epsilon},\frac{\phi_q}{\epsilon}).$

Therefore, if we define $\overline{g}_{pq}$ by

(2.54) $\quad\quad\quad \begin{array}{l} \ell_j \cdot \overline{g}_{pq} = 0, \quad j = p,q \\ \ell_j \cdot \overline{g}_{pq} = \Gamma_{pq}\hat{h}_{pq}, \quad j \neq p, j \neq q, \end{array}$

we use the terms involving $\overline{g}_{pq}(t,\theta_j)$ in the leading order asymptotics but invoke the principle of "exchange of phases" guaranteed by (2.53) to replace $\overline{g}_{pq}(t,\theta_j)$ by $g_{pq}(t,\theta_p,\theta_q)$ in (2.48) to guarantee that $u_2(t,\theta_p,\theta_q)$ can be found with sublinear growth. Thus, we replace the non-local terms on the right hand side of (2.49) by those in (2.51) and obtain for the M-functions, $\{\sigma_j(t,\theta)\}_{j=1}^{M}$, the closed system of <u>Periodic Resonant Wave Asymptotic Equations</u>

(2.55a) $\quad 0 = (\sigma_j)_t + b_j(\frac{1}{2}\sigma_j^2)_\theta$

$\quad\quad\quad\quad\quad + \sum_{\substack{p\neq q \\ p\neq j \\ q\neq j}} \Gamma_{pq}^{j} \lim_{T\to\infty} \frac{1}{2T} \int_{-T}^{T} \sigma'_p(t,\theta+(\lambda_j-\lambda_p)s)\sigma_q(t,\theta+(\lambda_j-\lambda_q)s)ds$

for $j = 1,\ldots,M$ with the initial data

(2.55b) $\quad\quad\quad\quad\quad\quad \sigma_j(t,\theta)|_{y=0} = \ell_j \cdot u_1^0(\theta).$

In the above argument we have developed the following:

<u>Proposition 2 (Formal Validity of the Resonant Asymptotic Equations for Periodic Waves)</u>

The solution u^ϵ of the general small amplitude periodic initial value problem in (2.41) has a formal uniformly valid asymptotic approximation within terms of order $o(\epsilon)$ given by

(2.56)
$$u_w^\varepsilon = u_0 + \varepsilon \sum_{j=1}^{M} \sigma_j(t, \frac{\phi^j}{\varepsilon}) r_j$$

where $\{\sigma_j(t,\theta)\}_{j=1}^{M}$ solves the initial value problem for the coupled system of resonant asymptotic equations in (2.55) with r_{pq}^j defined in (2.46). This expansion has a region of formal validity within terms that are $o(\varepsilon)$ at least on any region of the form $R^1 \times [0,T]$ where the solution in (2.55) remains smooth.

In the next subsection, we will display these equations explicitly for the equations of compressible fluid flow and describe some ongoing research and further developments using these equations. We still owe the reader a sketch of the proof of Lemma 1. We expand the solution u of (2.43) as $u = \sum_{j=1}^{M} \tilde{u}_j r_j$; for $j \neq p$ and $j \neq q$, we obtain the scalar convection equation;

$$(-\lambda_p + \lambda_j) \frac{\partial \tilde{u}_j}{\partial \theta_p} + (-\lambda_q + \lambda_j) \frac{\partial \tilde{u}_j}{\partial \theta_q} = \ell_j \cdot \tilde{g}.$$

With the characteristic co-ordinate, $(\tilde{\theta}_1, \tilde{\theta}_2)$, defined by

$$\theta_p = (\lambda_j - \lambda_p) \tilde{\theta}_1$$
$$\theta_q = \tilde{\theta}_2 + (\lambda_j - \lambda_q) \tilde{\theta}_1$$

the above equation reduces to

$$\frac{\partial}{\partial \tilde{\theta}_1} \tilde{u}_j = \ell_j \cdot g(\vec{a}_{pq}^j \tilde{\theta}_1 + (0, \tilde{\theta}_2)).$$

The Lemma follows immediately from direct integration of this equation and standard properties of almost periodic functions. In particular, one property of almost periodic functions needed to complete the proof of the lemma and already used implicitly in the proof of (2.53) is the fact that for any almost periodic function, $\sigma(s)$ the limit,

$$\lim_{T \to \infty} \frac{1}{2T} \int_{-T}^{T} \sigma(s + h) ds = M$$

is a number M independent of $h \in R^I$ and the convergence is uniform in h (see Chapter 6 of [22]).

2.4: An Application to Periodic Resonant Waves for Compressible Fluid Flows in 1-D

In this section; we apply the theory of periodic resonant waves developed in the previous subsection to the inviscid 1-D compressible fluid equations. These equations are the system of three conservation laws, expressing conservation of mass, momentum, and total energy, given by

(2.57)
$$\rho_t + (\rho v)_x = 0$$
$$(\rho v)_t + (\rho v^2 + p)_x = 0$$
$$(\rho e + \tfrac{1}{2}\rho v^2)_t + (\rho e v + \tfrac{1}{2}\rho v^3 + pv)_x = 0$$

where ρ is the mass density, v is the flow velocity, S is the entropy, and $p(\rho,S)$ is the pressure given as a specified function of (ρ,S) through thermodynamic considerations - for an ideal gas, $p = R\rho^\gamma \exp(s/c_v)$ for some γ with $\gamma > 1$. We assume that at the constant background value of interest, (ρ_0,S_0), $p_\rho(\rho_0,S_0) > 0$ so that the system in (2.57) is strictly hyperbolic. To conform with the general notation in the last sections, we set $u = {}^t(u_1,u_2,u_3) = {}^t(\rho,v,S)$ and we concentrate on the asymptotic approximation of solutions of (2.57) with small amplitude periodic initial data satisfying Hypothesis #2 with the form,

(2.58)
$$u_0^\varepsilon(x) = \begin{pmatrix} \rho_0 \\ 0 \\ S_0 \end{pmatrix} + \varepsilon u_0^1(o\,\tfrac{x}{\varepsilon})$$

with ρ_0, S_0 given constants.

Much is known rigorously about the behavior of small amplitude weak solutions of the 1-D compressible fluid equations for data with compact support and finite total variation as a consequence of ideas based on Glimm's method including shock formation, structure of singularities, and large-time behavior (see the bibliography of [12] for some recent references). On the other hand, very little is known rigorously about the behavior of weak solutions of the compressible fluid equations with fixed small amplitude periodic initial data (the transformation $x' = \tfrac{x}{\varepsilon}$, $t' = \tfrac{t}{\varepsilon}$ converts the problem in (2.57) and (2.58) into this problem). In fact, even at a formal level, very little is known regarding the important conceivable large time asymptotic behavior. The compressible fluid equations in 1-D are the simplest system with $M \geq 3$ where resonance occurs and the formal asymptotic theory from the previous subsection applies. With the rescaling $x' = \frac{x}{\varepsilon}$, $t' = \frac{t}{\varepsilon}$, Proposition 2 guarantees that for fixed small amplitude periodic initial data, the resonant asymptotic equations in (2.55) yield a formally valid approximation for large times of order $O(\varepsilon^{-1})$. Thus, from a theoretical viewpoint, one might investigate the structure and large-time asymptotic behavior for the approximating system of equations in (2.55) and use this information to formulate conjectures regarding the rigorous behavior of solutions of (2.57) and (2.58). Incidentally, the validity of such an approach has been confirmed rigorously for initial data of compact support in 1-D through the recent theoretical work of DiPerna and the author [3] described in section 4. From a more applied point of view, the problem in (2.57) and (2.58) is the simplest model problem in 1-D where natural resonances couple and drive the behavior of solutions of conservation laws; the asymptotic equations in (2.55) provide a uniformly valid leading order approximation for these effects. For an example of a multi-dimensional problem of extreme complexity where similar effects as we describe through asymptotics below seem to play a prominent role, we refer the reader to the recent beautiful numerical simulations by Woodward [19], [30] of instabilities in supersonic jets.

In the remainder of this subsection, we record the resonant asymptotic equations from (2.55) for the specific system in (2.57) and then briefly describe some of the work in progress by Rosales, Schonbek, and the author ([21]) in analyzing this sytem. For the full details of the explicit calculations and more information, we refer the reader to [21]. At the background state, ${}^t(\rho_0, 0, S_0)$, the system of equations in (2.57) has the three eigenvalues,

(2.59) $$\lambda_1 = -c_0, \quad \lambda_2 = 0, \quad \lambda_3 = c_0$$

with corresponding right eigenvectors,

(2.60) $$r_1 = \begin{pmatrix} \rho_0 \\ -c_0 \\ 0 \end{pmatrix}, \quad r_2 = \begin{pmatrix} (p_s)_0 \\ 0 \\ -c_0^2 \end{pmatrix}, \quad r_3 = \begin{pmatrix} \rho_0 \\ c_0 \\ 0 \end{pmatrix}$$

where $c_0 = (p_\rho(\rho_0, S_0))^{1/2}$ is the speed of sound and the subscript zero, here and subsequently, denotes evalution at the background state, (ρ_0, S_0). The uniformly valid asymptotic approximation from the last subsection has the form,

$$(2.61) \qquad u_w^\varepsilon = \begin{pmatrix} \rho_0 \\ 0 \\ S_0 \end{pmatrix} + \varepsilon \sum_{j=1}^{3} \sigma_j(t, \frac{\phi^j}{\varepsilon}) r_j$$

where $\phi^j = x - \lambda_j t$ and $\{\sigma_j(t,\theta)\}_{j=1}^{3}$ satisfy the resonant wave asymptotic equations from (2.55) for the special system in (2.57). For this physical system, these equations are remarkably simple and are given by

The Resonant Wave Equations for Periodic 1-D Compressible Flow

$$(\sigma_1)_t + \alpha(\tfrac{1}{2}\sigma_1^2)_\theta + \beta \lim_{T \to \infty} \frac{1}{2T} \int_{-T}^{T} \sigma_2'(\tfrac{1}{2}\theta + \tfrac{1}{2}s)\sigma_3(t,s)ds = 0$$

$$(2.62) \qquad (\sigma_2)_t = 0$$

$$(\sigma_3)_t - \alpha(\tfrac{1}{2}\sigma_3^2)_\theta - \beta \lim_{T \to \infty} \frac{1}{2T} \int_{-T}^{T} \sigma_2'(\tfrac{1}{2}\theta + \tfrac{1}{2}s)\sigma_1(t,s)ds.$$

With $T(\rho, S)$ the temperature, the coefficients α and β are given by

$$\alpha = -\frac{1}{2\rho_0 c_0}[(\rho^2 c^2)_\rho]_0$$

(2.63)

$$\beta = \rho_0^2 c_0^3[(c^{-2} T_\rho)_\rho]_0.$$

Since the second wave field of gas dynamics associated with the eigenvalue λ_2 is linearly degenerate, it is not surprising that the second equation in (2.62) is trivial and $\sigma_2(t,\theta) = \sigma_2^0(\theta)$, the initial amplitude for the entropy wave. On the other hand, for gas dynamics, the two sound waves are typically genuinely nonlinear and the requirement $\alpha \neq 0$ is a disguised version of this condition at the state (ρ_0, S_0). Since $\sigma_2(t,\theta) = \sigma_2^0(\theta)$, the resonant asymptotic equations in (2.62) reduce to two Burgers' equations coupled through a skew-symmetric linear integral operator with a known kernel given by the derivative of the entropy field. Thus, for the 1-D compressible fluid equations, all leading order resonances can be described by the coupling of two Burgers' equations by a known linear integral operator - a remarkable simplification through asymptotics. Next, we simplify the integral operators on the right-hand side of (2.62) even further assuming that Hypothesis #2 is satisfied.

It will be convenient to expand the known derivative of the smooth initial entropy field $\sigma_0'(\theta)$ in the real Fourier series,

$$(2.64) \qquad \sigma_0'(\theta) = \sum_{n=1}^{\infty} (a_n e^{2\pi i n \theta} + \bar{a}_n e^{-2\pi i n \theta})$$

the projection of $\sigma_0'(\theta)$ onto the even harmonics, $P\sigma_0'(\theta)$, is given by

$$(2.65) \qquad P\sigma_0'(\theta) = \sum_{n=1}^{\infty} (a_{2n} e^{4\pi i n \theta} + \bar{a}_{2n} e^{-4\pi i n \theta}).$$

We substitute (2.64) into the nonlocal averages on the right hand side of (2.62) and compute explicitly (by expanding $\sigma(t,s)$ in a Fourier series) that

$$(2.66) \qquad \lim_{T \to \infty} \frac{1}{2T} \int_{-T}^{T} \sigma_0'(\tfrac{1}{2}\theta + \tfrac{1}{2}s)\sigma(t,s)ds$$

$$= \int_0^1 P\sigma_0'(\tfrac{1}{2}\theta + \tfrac{1}{2}s)\sigma(t,s)ds.$$

Thus, if we denote by $K(\theta)$, the 1-periodic function defined by $K(\theta) = P\sigma_0'(\tfrac{1}{2}\theta)$, we see that the equations in (2.62) can be reduced to the

Simplified Resonant Wave Equations

$$(2.67) \qquad \begin{aligned} (\sigma_1)_t + \alpha(\tfrac{1}{2}\sigma_1^2)_\theta + \beta \int_0^1 K(\theta + s)\sigma_3(t,s)ds &= 0 \\ (\sigma_3)_t - \alpha(\tfrac{1}{2}\sigma_3^2)_\theta - \beta \int_0^1 K(\theta + s)\sigma_1(t,s)ds \end{aligned}$$

with $K(\theta) = P\sigma_0'(\tfrac{1}{2}\theta)$. Thus, we see that only even harmonics in the initial entropy field are expected to produce resonances while an initial entropy distribution, $\sigma_0(\theta)$, while only odd harmonics will never resonate to leading order. In section four, we will show that resonant planar oblique wave interactions for isentropic compressible fluid flow in 2-D will yield a similar set of equations as in (2.67) although the physical interpretation of the variables $\{\sigma_j\}_{j=1}^3$ will be quite different. In the calculations by Woodward already mentioned, there is a preference of odd harmonics over the even ones in producing resonances in that extremely complex physical problem; perhaps, the preference for half the harmonics in producing resonances in the simpler problem discussed here is not merely a coincidence.

We conclude this section with a few remarks about the simplified asymptotic coupled system in (2.67) which will appear in detail in a forthcoming paper of Rosales, Schonbek, and the author [21]. First, global existence of weak solutions is easily established as a consequence of two formal estimates for solutions of the system in (2.67)

#1) $\text{Var}(\sigma_1(\cdot,t),\sigma_3(\cdot,t))) \leq \exp(Ct\|K(\theta)\|_1)\text{Var}(\sigma_1^0(\theta),\sigma_3^0(\theta))$

#2) $\frac{\partial}{\partial t}\int_0^1 (\sigma_1^2(\theta,t) + \sigma_3^2(\theta,t))d\theta \leq 0.$

(Actually, only #1) is needed for existence but #2) is an important fact.) Secondly, in [21], numerical computations for the system in (2.67) will be presented which elucidate the large-time asymptotic behavior of solutions of the system in (2.67). Due to the effects of artificial viscosity, such meaningful numerical computations for small amplitude periodic waves via a direct simulation of the compressible 1-D fluid equations in (2.57) would be a rather difficult undertaking. However, it is not difficult to design an efficient numerical scheme without artificial viscosity for the numerical solution of the asymptotic equations in (2.67). The scheme that has been developed for the calculations to be reported in [21] combines, through fractional steps, the deterministic Glimm scheme for the periodic inviscid Burgers solutions to avoid any artificial viscosity with a Fourier-spectral step which efficiently solves the nonlocal integral operators exactly.

III. The Formal Theory: Weakly Nonlinear Hyperbolic Waves in Several Space Dimensions

In this section, we present some of the ideas and results which extend the formal theory described in section II to several space variables. We both briefly review some of the ideas of Choquet-Bruhat [1] and Hunter and Keller [5] and describe several new results on resonantly interacting waves in multi-D which will appear in a forthcoming paper of Hunter, Rosales, and the author [7]. Unlike the detailed treatment presented in section II, here we will only discuss the ideas and make several remarks - most details will be omitted. However, many of the details only involve a mild extension of the ideas already developed in detail in section II and we will attempt to point out the essential differences and new difficulties. We end this section with a discussion of oblique resonant wave interactions for isentropic 2-D compressible fluid flow. The asymptotics will yield essentially the same simplified system as we discussed earlier in (2.67).

Here we will discuss the asymptotic approximation of solutions to the initial value problem for an $M \times M$ system of hyperbolic conservation laws in $N+1$ variables (N-space variables) given by

$$(3.1a) \qquad \frac{\partial}{\partial x_0} F_0(u^\varepsilon) + \sum_{j=1}^{N} \frac{\partial}{\partial x_j} F_j(u^\varepsilon) = 0, \quad x_0 > 0$$
$$x' \in R^N$$

with $x = (x_0, x')$, $x' = (x_1, \ldots, x_N)$, and small amplitude rapidly oscillating initial data,

$$(3.1b) \qquad u^\varepsilon(x_0, x')|_{x_0=0} = u_0 + \varepsilon u_1^0(x', \frac{\phi^0(x')}{\varepsilon}).$$

Here x_0 denotes the time variable, $u = {}^t(u_1, \ldots, u_M) \in R^M$, and $\{F_j(u)\}_{j=0}^{N}$ are smooth functions of u defined on an open set containing the constant vector, u_0, with $A_j = \frac{\partial F_j}{\partial u}\big|_{u_0}$, $0 \le j \le N$, the corresponding $M \times M$ Jacobian matrices at u_0. We assume that $F_j(u)$ has the Taylor expansion,

$$(3.2) \qquad F_j(u_0 + \varepsilon v) = F_j(u_0) + \varepsilon A_j v + \varepsilon^2 \frac{B_j(v,v)}{2} + O(\varepsilon^3)$$

for $j = 0, 1, \ldots, N$ where $B_j(v, w)$ is the symmetric M-vector valued bilinear form defined by the Hessian matrix of each component evaluated at u_0. The phase functions in the initial data, $\vec{\phi}^°(x')$, belong to R^ℓ so that there are ℓ distinct phase functions initially. We always assume that x_0 is a time-like direction for (3.1) and for simplicity in exposition, we require that the system in (3.1a) is strictly hyperbolic at u_0, thus,

(3.3)
A) $\det A_0(u_0) \neq 0$

B) $\det(-\lambda A_0(u_0) + \sum_{j=1}^{N} A_j(u_0)\xi_j) = 0$ has M distinct roots,

$\{\lambda_\rho(\xi)\}_{\rho=1}^{M}$, for any $\xi \in R^N$ with $\xi \neq 0$.

We denote by $R_\rho(\xi)$, $L_K(\xi)$, the corresponding non-zero left and right eigenvectors satisfying

(3.4a)
$$(-\lambda_\rho(\xi)A_0 + \sum_{j=1}^{N} A_j\xi_j)R_\rho(\xi) = 0$$

$$L_K(\xi) \cdot (-\lambda_K(\xi)A_0 + \sum_{j=1}^{N} A_j\xi_j) = 0$$

and the normalization,

(3.4b)
$$L_K(\xi) \cdot A_0 R_\rho(\xi) = \delta_{K\rho}$$

with $\delta_{K\rho}$, the Kronecker delta. For the initial data, $u_1(x', \frac{\vec{\phi}^°(x')}{\varepsilon})$, we require that $u_1^0(x', \vec{\theta})$ is a smooth function of $\vec{\theta} \in R^\ell$ and that each separate generalized mean of u_1^0 with respect to θ_j vanishes, i.e.

(3.5)
$$\lim_{T \to \infty} \frac{1}{2T} \int_{-T}^{T} u_1^0(x', \theta_1, \ldots, \theta_{j-1}, s, \theta_{j+1}, \ldots, \theta_\ell) = 0$$

for any j with $1 \leq j \leq \ell$.

Before beginning a discussion of the formal asymptotic solutions, we remark that equations of the form in (3.1) with nonlinear source terms, $S(u)$, are also easily handled by the methods described below. For the simple wave expansions which we discuss below in subsection 3.1, it is inessential that (3.5) is satisfied and in the general case, there is a mean field correction of order ε (see [51]). Furthermore, the background state can be any smooth function $u_0(x)$. On the other hand, for the general multi-dimensional theory of planar oblique binary wave interactions as presented in subsection 3.2, at the present time, it is quite essential that u_0 be a constant state and that each $\phi^j = \vec{x}\cdot\vec{\omega}^j$ be a plane wave phase function for the formal asymptotics to be justified in the fashion presented here. The theory as presented requires only minor changes to incorporate hyperbolic systems with roots of constant multiplicity (see [5]) - this is an important remark for applications to the complete 3-D Euler equations of compressible fluid flow.

3.1: The Single Wave Expansion in Multi-D

Here we describe the construction of asymptotic solutions of the equation in (3.1a) with the single-wave ansatz,

$$(3.6) \qquad \tilde{u}_w^\varepsilon = u_0 + \varepsilon u_1(x, \tfrac{\phi}{\varepsilon}) + \varepsilon^2 u_2(x, \tfrac{\phi}{\varepsilon})$$

where ϕ is a single phase function. Later, we will identify the initial data compatible with this ansatz. In an analogous fashion as in a single space variable, we must be able to choose $u_1(x,\theta), u_2(x,\theta)$ so that simultaneously both (2.9) and (2.10) are satisfied. The terms of order zero in ε vanish provided that

Order zero

$$(3.7) \qquad \left(\sum_{j=0}^{N} A_j \phi_{x_j}\right) \frac{\partial}{\partial \theta} u_1(x,\theta)\Big|_{\theta=\frac{\phi}{\varepsilon}} = 0.$$

The requirement that the power of order one in ε vanish in the formal solution of (3.1) requires that we choose u_2 so that

Order one

$$(3.8) \qquad -\left(\sum_{j=0}^{N} A_j \phi_{x_j}\right) \frac{\partial u_2}{\partial \theta}(x,\theta)\Big|_{\theta=\frac{\phi}{\varepsilon}} = \left(\sum_{j=0}^{N} A_j (u_1)_{x_j} + \sum_{j=0}^{N} \phi_{x_j} B_j(u_1, (u_1)_\theta)\right)\Big|_{\theta=\frac{\phi}{\varepsilon}}.$$

The solution of the equation in (3.7) is identical to the solution of the leading terms of the equations of geometric optics at the background state, u_0. With the notation, $\nabla \phi = \left(\frac{\partial \phi}{\partial x_1}, \ldots, \frac{\partial \phi}{\partial x_N}\right)$, the equation in (3.7) is satisfied only if there exists a p so that for some wave speed λ_p,

(3.9a)
$$\phi_{x_0} + \lambda_p(\nabla\phi) = 0, \quad 0 < x_0 < X_0$$
$$\phi(x_0, x')|_{x_0=0} = \phi^°(x'),$$

and $u_1(x,\theta)$ has the form,

(3.9b)
$$u_1(x,\theta) = \sigma(x,\theta) R_p(\nabla\phi)$$

where $R_p(\xi)$ is the right eigenvector from (3.4a) associated with the wave speed, $\lambda_p(\xi)$ here $\sigma(x,\theta)$ is an arbitrary scalar multiplier at this stage in the argument. In satisfying the equation of order one in (3.8), as in (2.14) above, we see that

(3.10)
$$L_p(\nabla\phi) \cdot \sum_{j=0}^{N} A_j \phi_{x_j} = 0$$

with $L_p(\xi)$ the left eigenvector from (3.4) and the equation in (3.8) has a solution, $u_2(x,\theta)$, by the Fredholm alternative, if and only if the inner product of the right hand side of (3.8), regarded as a function of θ, vanishes with $u_1(x,\theta)$ defined in (3.9b). This yields the scalar nonlinear differential equation for $\sigma(x,\theta)$ given by

(3.11a)
$$D_p \sigma(x,\theta) + b_p (\tfrac{1}{2} \sigma^2(x,\theta))_\theta = 0, \quad 0 < x_0$$

with initial data,

(3.11b)
$$\sigma(x,\theta)|_{x_0=0} = \sigma_0(x',\theta)$$

and $\sigma_0(x',\theta)$ a given initial wave-form at $x_0 = 0$. Here D_p is the linear transport operator of geometric optics associated with the phase function, ϕ, defined by

(3.12a)
$$D_p = \frac{\partial}{\partial x_0} + \vec{a}_p \cdot \nabla + c_p$$

with $\vec{a}_p = (a_p^j(x))$, $c_p(x)$, given by

(3.12b)
$$a_p^j = L_p(\nabla\phi) \cdot A_j R_p(\nabla\phi), \quad 1 \leq j \leq N$$

$$c_p(x) = L_p(\nabla\phi) \cdot \sum_{j=0}^{N} A_j (R_p)_{x_j}.$$

The coefficient, b_p, is a multi-dimensional generalization of the coefficient in (2.16) and is defined by

(3.13)
$$b_p = L_p(\nabla\phi) \cdot \left(\sum_{j=0}^{N} \phi_{x_j} B_j(R_p, R_p) \right).$$

We briefly recall the reasons for calling the operator in (3.12), the transport operator of linear geometric optics associated with the phase function, ϕ. If one takes the identity

$$(-\lambda_p(\xi) A_0 + \sum_{j=1}^{N} A_j \xi_j) R_p(\xi) = 0,$$

differentiates with respect to ξ and computes the dot product with $L_p(\xi)$, after evaluation at $\xi = \nabla\phi$, the result is the formula,

$$\nabla_\xi \lambda_p(\xi) \big|_{\nabla\phi} = \vec{a}_p(x).$$

If one considers the integral curves $X(x_0, x') \in R^N$ defined by the O.D.E.

$$\frac{dX}{dx_0} = \vec{a}_p(x_0, X), \quad X(x_0, x') \big|_{x_0=0} = x',$$

we use the above formula and (3.9a) to compute that

$$\frac{d}{dx_0} \phi(x_0, X(x_0, x')) = \phi_{x_0} + \lambda_p(\nabla\phi) = 0,$$

i.e. $x_0, X(x_0, x')) = \phi_0(x')$. Thus, the direction of differentiation defined by the transport operator in (3.12) is always tangent to the surfaces of constant phase for the phase function, ϕ. With the condition in (3.5), it is completely straightforward to follow the reasoning in section (2.1) to verify that $u_2(x, \theta)$ be chosen with sublinear growth in θ once the Fredholm alternative condition from (3.10) is satisfied.

To summarize, we have verified the following:

Proposition 3: For any initial wave form, $\sigma_0(x',\theta)$ with $\lim_{T\to\infty} \frac{1}{2T} \int_{-T}^{T} \sigma_0(x',s)ds = 0$ and any initial plane function, $\phi_0(x')$ with $\nabla\phi_0 \neq 0$ on the x-support of $\sigma_0(x',\theta)$, there is a uniformly valid formal asymptotic solution of (3.1) with the initial data,

$$u_0 + \varepsilon\sigma_0(x',\frac{\phi_0}{\varepsilon})R_p(\nabla\phi_0)$$

for any p with $1 < p < M$. The leading order approximation is given by

(3.14)
$$u_W^\varepsilon = u_0 + \varepsilon\sigma(x,\frac{\phi}{\varepsilon})R_p(\nabla\phi)$$

where ϕ solves the eikonal equations of geometric optics from (3.9) while the amplitude, σ, solves the much simpler nonlinear transport equation defined by the scalar equation,

(3.15)
$$D_p\sigma + b_p(\frac{1}{2}\sigma^2)_\theta = 0 \quad x_0 > 0$$
$$\sigma(x_0,x',\theta)|_{x_0=0} = \sigma_0(x',\theta)$$

with the coefficients for D_p and b_p defined in (3.12) and (3.13). The region of formal validity of this approximate solution is restricted in another way beyond the conditions imposed in the 1-D case described in section 2.1 - the expansion obviously ceases to be valid in regions where the eikonal solutions in (3.9a) form multi-dimensional caustics.

Remark: The equation in (3.15) is a scalar equation in several space variables. However, since the coefficients of D_p and b_p do not depend on θ, once characteristic coordinates are used to straighten out this linear operator, the equation in (3.15) essentially simplifies to a parametrized family of inviscid Burgers equation with lower order terms with the form,

(3.16)
$$\sigma_\tau + c_p(\tau)\sigma + b_p(\tau)(\sigma^2)_\theta = 0.$$

One significant difference in this equation when compared with the asymptotic equations in one space dimension is the appearance of the linear lower order term, $c_p(\tau)$. When the linear geometric optic light rays are expanding in multi-D, we have $c_p(\tau) > 0$ while $c_p(\tau) < 0$ corresponds to local focusing of these light rays. Thus, the equations account for both linear geometric distortion and nonlinear distortion of the wave form and amplitude.

For brevity, we will not discuss non-resonant wave patterns in multi-D although the conditions described in section 2.2 can be generalized easily to yield sufficient conditions for non-resonance and superposition of the simple wave patterns from (3.14). Instead, we describe some new work of Hunter, Rosales, and the author to appear in [7] where we study some very interesting multi-D problems with resonant wave interactions.

3.2: Resonant Oblique Small Amplitude Plane Wave Interactions in Multi-D

We consider two different highly oscillatory small amplitude wave packets which do not interact for times $x_0 < X_0$, i.e. so that the simple wave expansions discussed in section 3.1 are uniformly valid approximations for $0 < x_0 < X_0$. This is easily arranged by guaranteeing that for example, the bicharacteristic rays emanating from the supports of the initial wave forms do not intersect until times, $x_0 \geq X_0$. Thus,

$$(3.17) \quad u_w^\varepsilon(x) = u_0 + \varepsilon\sigma_1(x,\frac{\phi_1}{\varepsilon})R_1(\nabla\phi_1) + \varepsilon\sigma_2(x,\frac{\phi_2}{\varepsilon})|R_2(\nabla\phi_2) + o(\varepsilon)$$

is a uniformly valid approximation for times x_0 with $0 < x_0 < X_0$) where the functions ϕ_j, $\sigma_j(x,\theta)$. satisfy the respective decoupled asymptotic equations in (3.9), (3.11), (3.12). We assume that for times $x_0 \geq X_0$, the bicharacteristic rays associated with ϕ_1 and ϕ_2 intersect; then generally one might expect, for periodic or almost periodic wave forms, that the simple wave expansion from (3.17) ceases to be uniformly valid for times $x_0 \geq X_0$ because new oblique waves are generated from the collision and interaction of these two different wave fronts. Thus, the following problem arises naturally:

(3.18) Describe the quantitative locations and strengths of the different oblique wave patterns generated after the collision time, X_0.

The problem posed in (3.18) explains the meaning of the title of this section.

Below, we describe the uniformly valid formal asymptotic solution for the problem of oblique wave interactions posed in (3.18) under two hypotheses:

A) The two wave fronts in (3.18) are plane wave fronts with linear phase functions, $\phi_j = \sum_{r=0}^{N} x_r w_r^j = x \cdot \vec{w}^j$, $j = 1, 2$ with $\det(\sum_{K=0}^{N} A_k \omega_k) = 0$

(3.19)

B) The generic geometric condition to be defined in (3.24) is satisfied for $\vec{w}^1$ and $\vec{w}^2$ with $\vec{w}^j = ((\phi_j)_{x_0}, \nabla \phi_j)$.

With the hypotheses in (3.19), the problem given in (3.18) is solved by finding new directions, $\{\vec{w}^j\}_{j=1}^{\ell}$, with ℓ a well-determined number satisfying $\ell \leq M$ ($\ell < M$ quite often for $M \geq 4$) so that for times beyond the interaction time, X_0, the asymptotic solution in (3.17) is continued by the more complex uniformly valid asymptotic solution given by

(3.20) $\quad u_w^\varepsilon(x) = u_0 + \varepsilon \sum_{j=1}^{\ell} \sigma_j(x, \frac{\phi_j}{\varepsilon}) R_j + \varepsilon^2 u_2(x, \frac{\vec{\phi}}{\varepsilon})$

with the linear phase functions $\phi_j = \vec{w}^j \cdot x$, $1 \leq j \leq \ell$ and $|u_2(x, \frac{\vec{\phi}}{\varepsilon})| = o(\varepsilon^{-1})$. For times $x_0 > X_0$, the nonlinear amplitude functions from (3.20), $\{\sigma_j(x, \theta)\}_{j=1}^{\ell}$, no longer decouple but instead solve the $\ell \times \ell$ system,

The Resonant Plane Wave Asymptotic Interaction Equations

(3.21) $\quad D^j \sigma_j + b_j \frac{\partial}{\partial \theta}(\frac{1}{2}\sigma_j) + \sum_{\substack{1 \leq p,q \leq \ell \\ p \neq q \\ q \neq p}} \Gamma_{pq}^j \lim_{T \to \infty} \frac{1}{2T} \int_{-T}^{T} \sigma_q'(x, s) \sigma_p(x, s + h_{pq}^j \theta) ds = 0,$

$1 \leq j \leq \ell$

with $\sigma_q'(x, s) = \frac{\partial \sigma_q}{\partial s}(x, s)$. The coefficients Γ_{pq}^j and h_{pq}^j have an explicit form generalizing the formulae in (2.46) and (2.55) but these expressions are too cumbersome and not important enough for the subsequent presentation to display here. The transport operators, D^j, and the coefficients b_j are given by the formulae in (3.12) and (3.13) specialized to the planar phase functions, $\phi_j = x \cdot \vec{w}^j$. Of course, to solve the wave interaction problem from (3.18), the equations in (3.21) should be supplemented by the initial conditions,

(3.22) $$\sigma_j(x,\theta)|_{x_0=X_0} = 0, \quad j = 3,\ldots,\ell$$

while the initial data for $\sigma_1(x,\theta)$, $\sigma_2(x,\theta)$ on $x_0 = X_0$ are computed by evaluating the leading order terms from (3.17) on $x_0 = X_0$.

How are these new oblique wave directions $\vec{w}^j$ computed? In general, given the two vectors $\vec{w}^1$ and $\vec{w}^2$, there are ℓ distinct directions with $\ell \leq M$ and parametrized by linearly independent vectors, $h^j = (h_1^j, h_2^j)$, $1 \leq j \leq \ell$, so that $h^1 = (1,0)$, $h^2 = (0,1)$, and the $\vec{w}^j$ from (3.20) and (3.21) are defined by the requirements,

(3.23a) $$\vec{w}^j = h_1^j \vec{w}^1 + h_2^j \vec{w}^2, \quad 1 \leq j \leq \ell$$

and

(3.23b) $$\det(A \cdot (h_1(\vec{w}^1) + h_2(\vec{w}^2))) = 0$$

if and only if (h_1, h_2) is a multiple of some (h_1^j, h_2^j) with $1 \leq j \leq \ell$.

Here we use the shorthand notation, $A \cdot \vec{w} = \sum_{\ell}^{N} A_\ell \vec{w}^\ell$. We define $Q(n_1, n_2)$ by $Q(n_1, n_2) = \det(A \cdot (n_1 \vec{w}^1 + n_2 \vec{w}^2))$. In addition, we assume that

(3.24) $$\nabla_n Q(n_1, n_2)|_{(n_1^j, n_2^j)} \neq 0 \quad \text{for} \quad j = 1,\ldots,\ell.$$

This is the generic geometric assumption which we mentioned earlier in (3.19b). We note that when the condition in (3.24) is violated, a resonant caustic appears. The $\vec{w}^j$ are obviously the resonant directions since they are all the linear combinations of the two fundamental directions, $\vec{w}^1$ and $\vec{w}^2$ which are also characteristic directions. In 1-D, the coefficients Γ_{pq}^j in one space dimension in (2.46), (2.55) agree to leading order with the celebrated wave interaction coefficients of Glimm when $F_0(u) = I$ - the coefficients Γ_{pq}^j from (3.21) are a multi-D generalization of Glimm's interaction coefficients. This completes our description of the solution of the problem of multi-D oblique planar wave interactions which is given in full detail in [7].

We end this section with a brief discussion of the derivation of (3.21). First, we used the augmented ansatz in (3.20) beyond the collision time, X_0, in an attempt to include all conceivable oblique resonant wave directions as determined by the geometric conditions. With such an ansatz, the strategy of the argument follows in outline the one presented in section 2.3 for resonant periodic wave interactions in one space dimension. By the method of multiple scales, simultaneously we need to obtain a solution $u_2(x,\vec{\theta})$ of an auxilliary system of linear P.D.E.'s with sublinear growth in $\vec{\theta}$ and also a closed system of leading order asymptotic equations. As in (2.43), we need to find necessary and sufficient conditions on an almost periodic function $\tilde{g}(\theta_p, \theta_q)$, so that the auxilliary constant coefficient P.D.E.'s,

$$(3.25) \qquad a_p \frac{\partial u}{\partial \theta_p} + a_q \frac{\partial u}{\partial \theta_q} = \tilde{g}(\theta_p, \theta_q), \quad 1 \leq p, q \leq \ell, \quad p \neq q$$

have a solution $u(\theta_p, \theta_q)$ with sublinear growth. Here a_j is given by the formula,

$$(3.26) \qquad a_j = A \cdot \vec{w}^j, \quad 1 \leq j \leq \ell$$

with $\vec{w}^j$ defined in (3.23). One important difference between 1-D and multi-D is expressed by the following easily proved

LEMMA: The <u>auxilliary system of equations</u> in (3.25) is <u>hyperbolic</u> for every $p \neq q$, $1 \leq p, q \leq \ell$ <u>if and only if</u>

$$(3.27) \qquad \begin{array}{c} \underline{\text{the plane spanned by } \vec{w}^1 \text{ and } \vec{w}^2} \\ \underline{\text{contains a time-like direction}} \\ \underline{\text{for the operator}} \sum_{\ell=0}^{N} A_\ell \partial_{x_\ell} \, . \end{array}$$

Furthermore, $\ell < M$ unless (3.27) is satisfied and then $\ell = M$.

Thus, in 1-D, the auxilliary system is always hyperbolic while in multi-D, for $M \geq 4$ this system is often <u>not</u> hyperbolic; we invite the reader to construct simple examples using the slowness surfaces of a system with $M = 4$ with two circular characteristic cones. For isentropic compressible flow in 2-D, the system in (3.25) is always hyperbolic provided that the condition in (3.24) is satisfied. In general, the geometric condition in (3.24) guarantees that the _operator_ in (3.25) has _real principal type_ - with this condition the auxilliary problem posed in (3.25) can be solved (see [7]). With this information, one way to complete the asymptotic derivation and to obtain the closed system of equations in (3.21) is to use the principle of "exchange of phase functions" in a similar argument as we already presented in detail for 1-D in (2.47)-(2.54).

3.3: The Asymptotic Equations for Resonant Oblique Waves for 2-D Compressible Flow

Here we record the equations from (3.21) which describe resonant oblique plane wave interactions for the equations of 2-D isentropic fluid flow. One might argue with some justification that the asymptotic solution described in section 3.2 has replaced one difficult system of equations by another - the general resonant wave equations in (3.21). Here we report on algebraic calculations presented in detail in [7] which yield the equations in (3.21) for resonant plane wave interactions for 2-D isentropic compressible flow. We will show that after several reductions, essentially the _same simplified resonant system_ as described in section 3.4 in (2.67) occurs in describing oblique interacting waves in a 2-D isentropic fluid - this is a remarkable fact!

The 3×3 hyperbolic system describing isentropic compressible fluid flow in two-space variables is given by

(3.29)
$$\rho_t + \text{div } \vec{m} = 0$$
$$(m_i)_t + \text{div}(\frac{m_i}{\rho} \vec{m}) + p(\rho)_{x_i} = 0, \quad i = 1, 2.$$

Here ρ is the density, $\vec{m} = {}^t(m_1, m_2)$ is the momentum vector, and $p(\rho)$ defines the pressure as a specified function of ρ through an isentropic equation of state; $p = A\rho^\gamma$, $\gamma > 1$ for an ideal gas. In this subsection we will use conventional physical notation with the variable t denoting the time variable rather than x_0 from the previous sections.

We set $u = {}^t(\rho, \vec{m})$ and consider small amplitude perturbations around the constant state

$$u_0 = {}^t(\rho_0, 0, 0) \tag{3.30}$$

with ρ_0 a constant reference density and $c_0 = (p_\rho(\rho_0))^{1/2}$, the corresponding sound speed - of course, we assume $p_\rho(\rho_0) > 0$ to guarantee hyperbolicity. General planar sound waves move at velocity c_0 at the background state ${}^t(\rho_0, 0, 0)$ and have an associated linear phase function and right eigenvector defined by

$$\phi = x \cdot \vec{w} - c_0 t, \quad R(\vec{w}) = {}^t(1, c_0 \vec{w}) \tag{3.31}$$

for any $\vec{w} = (w_1, w_2)$ with $|\vec{w}| = 1$. Any direction $\vec{w}^0 = (w_1^0, w_2^0)$ defines a <u>vorticity wave</u> (steady shear layer solution) at the background state from (3.30) with phase function and right eigenvector defined by

$$\phi^0 = x \cdot w^0, \quad R^0 = {}^t(0, -w_2^0, w_1^0) \tag{3.32}$$

(for convenience, we don't normalize w^0 with $|\vec{w}^0| = 1$). It is not difficult to check that a given sound wave direction and a given vorticity wave direction always satisfy the nondegeneracy condition in (3.24) unless

$$\vec{w}^0 \cdot \vec{w} = 0$$

i.e. the propagating sound wave and vorticity wave are at exactly right angles. On the other hand, two distinct vorticity waves provide an example where the geometric condition in (3.24) is always violated since their linear span is a characteristic hyperplane.

It is well-known that to leading order two oblique sound waves do not resonate and the simplified asymptotic solution with the form in (3.17) continues the solution to leading order beyond the interaction time, X_0 (see [7]). Thus, here we treat the oblique wave interaction of a planar sound wave and a vorticity wave. The equations in (3.29) have obvious rotational invariance so without loss of generality, we consider the interaction of a sound wave with phase function,

$$\phi^1 = x_1 - c_0 t \tag{3.33}$$

and a vorticity wave with phase function

(3.34) $$\phi^2 = x_1(\cos \tilde{\theta} - 1) + x_2 \sin \tilde{\theta}.$$

When the angle, $\tilde{\theta}$, varies with $0 < \tilde{\theta} < 2\pi$, we generate all oblique wave patterns which satisfy the condition in (3.24). According to the recipe in (3.23), new oblique sound waves are generated through the interaction of the vorticity and sound waves associated with (3.34) and (3.33) along the new direction with phase function, ϕ^3, given by

(3.35) $$\phi^3 = \cos \tilde{\theta} \, x_1 + \sin \tilde{\theta} \, x_2 - c_0 t.$$

the equations in (3.20) and (3.21) for the resonant oblique interactions in this special case have the following form: For (3.20),

$$u(x,t,\varepsilon) = \begin{pmatrix} \rho_0 \\ 0 \\ 0 \end{pmatrix} + \varepsilon\sigma_1(x,t,\frac{\phi_1}{\varepsilon}) \begin{pmatrix} 1 \\ c_0 \\ 0 \end{pmatrix} +$$

$$+ \varepsilon\sigma_2(x,t,\frac{\phi^2}{\varepsilon}) \begin{pmatrix} 0 \\ -\sin \tilde{\theta} \\ \cos \tilde{\theta} - 1 \end{pmatrix} + \varepsilon\sigma_3(x,t,\frac{\phi^3}{\varepsilon}) \begin{pmatrix} 1 \\ c_0 \cos \tilde{\theta} \\ c_0 \sin \tilde{\theta} \end{pmatrix}$$

$$+ O(\varepsilon).$$

For (3.21), the vorticity wave is linearly degenerate and the amplitude satisfies the trivial equation,

(3.37a) $$\frac{\partial}{\partial t} \sigma_2(x,t,\theta) = 0$$

i.e., $\sigma_2(x,t,\theta) \equiv \sigma_0(x,\theta)$. The amplitude of the incident sound wave satisfies the resonant equation

(3.37b) $$(\sigma_1)_t + c_0(\sigma_1)_{x_1} + b(\tfrac{1}{2}\sigma^2)_\theta + D_1(\tilde{\theta}) \lim_{T \to \infty} \frac{1}{2T} \int_{-T}^{T} \sigma_0'(x,s-\theta)\sigma_3(x,t,s)ds = 0.$$

The amplitude of the generated sound wave satisfies the resonant equation,

$$(\sigma_3)_t + c_0 \cos\theta (\sigma_3)_{x_1} + c_0 \sin\theta (\sigma_3)_{x_2} + b(\tfrac{1}{2}\sigma^2)_\theta +$$
$$+ D_2(\tilde{\theta}) \lim_{T\to\infty} \frac{1}{2T} \int_{-T}^{T} \sigma_0'(x,-s-\theta)\sigma_1(x,t,s)ds = 0 \ .$$

Here $\sigma_0' = \frac{\partial}{\partial s}\sigma_0(x,s)$ has the physical interpretation as the <u>vortex strength</u> in the perturbed initial <u>shear layer</u> while the coefficients b, $D_1(\tilde{\theta})$, $D_2(\tilde{\theta})$ are given by

(3.38)
$$b = \frac{1}{2}\frac{P_{\rho\rho}}{c_0} + \frac{c_0}{\rho_0}$$
$$D_1(\tilde{\theta}) = \frac{\cos\tilde{\theta}\sin\tilde{\theta}}{\rho_0}$$
$$D_2(\tilde{\theta}) = \frac{-(\cos^2\tilde{\theta}+1)\sin\theta}{2\rho_0} \ .$$

Of course, the requirement, $b \neq 0$, is the genuine nonlinearity condition at ρ_0. Thus, once again we arrive at a coupled system of multi-D scalar conservation laws coupled through convolutions with a known kernel defined essentially by the vortex strength.

We assume that the interaction time is at $x_0 = 0$ and also that the initial conditions for $\vec{\sigma} = (\sigma_1,\sigma_2,\sigma_3)$ are given as periodic functions of θ alone of mean zero and period one - in this case, σ_1 and σ_3 can be chosen as functions of (t,θ) alone and the equations in (3.37) specialize to the

<u>Resonant Oblique Plane Waves for 2-D Isentropic Compressible Flow</u>

(3.39)
$$(\sigma_1)_t + b(\sigma_1^2)_\theta + \int_0^1 K_1(s-\theta)\sigma_3(t,s)ds = 0$$
$$(\sigma_3)_t + b(\sigma_3^2)_\theta + \int_0^1 K_2(-s-\theta)\sigma_1(t,s)ds = 0$$

with the kernels, $K_1(s)$, $K_2(s)$ given by

(3.40)
$$K_1(s) = D_1(\tilde{\theta})\sigma_0'(s)$$
$$K_2(s) = D_2(\tilde{\theta})\sigma_0'(s)$$

with σ_0^1 the vortex strength. This system has essentially the same structure as the system we discussed earlier in section 2.4. One difference is that the integral operator is linear but not skew-symmetric in the Euclidean inner product - however, the key estimate #1 needed for existence of weak solutions is still valid. The numerical experiments which are in progress for the simplified system in (3.39) should indicate any essential differences in the wave behavior for the system in (3.39) as compared to the corresponding behavior for the system in (2.67).

IV. The Rigorous Theory of Nonlinear Geometric Optics for Weak Solutions of Conservation Laws in 1-D

As we have mentioned earlier in the introduction, the formal methods of weakly nonlinear geometric optics, as described in sections II and III, have been used in a wide variety of applied contexts. In particular, the simplified asymptotic methods involving exact solutions of the inviscid Burgers equation have been used <u>after shock waves have formed in these solutions</u> in a variety of applied contexts; qualitatively reliable predictions for weak solutions have been made through this approach. As described in detail in section II of this paper, the tacit assumptions used in constructing the uniformly valid approximation require that these approximate solutions remain smooth for $M \geq 2$ (especially see (2.33c) where the norm $|\sigma_j^1(t,\theta)|_\infty$ is used). Thus, the following theoretical problem is important for understanding these formal asymptotic methods:

PROBLEM: Assess the validity of the approximation of weakly nonlinear geometric
(4.1)   optics after shocks have formed in weak solutions.

Recently, DiPerna and the author [3] have provided a rigorous solution of the basic problem in (4.1) for general genuinely nonlinear systems of conservation laws in one space variable and initial data which is generally of compact support - other results are given in [3] for periodic initial data with the restriction, $M \leq 2$. Section 4.1 contains a description of these results while a sketch of the proof for a scalar convex conservation law is presented in section 4.2. This work indicates in a striking fashion that the methods of weakly nonlinear geometric optics are even better than the predictions of the formal asymptotic theory. These results may surprise some readers who are experts in perturbation theory. Nevertheless, several accessible open problems remain in studying the problem from (4.1) in 1-D and some of these problems are given in the final subsection.

Below, we consider the general weak solution (given by Glimm's method) of the M × M system of conservation laws with small amplitude initial data,

(4.2)
$$F_0(u^\varepsilon)_t + F_1(u^\varepsilon)_x = 0, \quad x \in R^1, \quad t > 0$$
$$u^\varepsilon(x,0) = u_0 + \varepsilon u_1^0(x)$$

When $u_1^0(x)$ has compact support or $u_1^0(x)$ is periodic with $M \leq 2$, in Proposition 1 from section 2.2 we have constructed a formal uniformly valid weakly nonlinear approximation to u^ε in (4.2) in regions of smoothness. This approximation is given by

(4.3a)
$$u_w^\varepsilon = u_0 + \varepsilon \sum_{j=1}^M \sigma_j(\varepsilon t, \phi_j) r_j$$

where $\tau = \varepsilon t$, $\phi_j = x - \lambda_j(u_0)t$, and $\sigma_j(\tau, \theta)$ satisfies the inviscid Burgers equation,

(4.3b)
$$(\sigma_j)_\tau + b_j(\tfrac{1}{2}\sigma_j^2)_\theta = 0, \quad \tau > 0, \quad \theta \in R^1$$
$$\sigma_j(\tau, \theta)|_{\tau=0} = \ell_j \cdot u_1^0(\theta)$$

for $1 \leq j \leq M$ with b_j given in (2.16). In this context, the problem in (4.1) becomes the question,

(4.4) How close are the weak solutions, u^ε, of (4.2) and the asymptotic solutions, u_w^ε, from (4.3) after shock waves have formed?

We remark that in presenting the results from section 2 in the equations from (4.2) and (4.3), we have used the rescaling $x = \frac{x'}{\varepsilon}$, $t = \frac{t'}{\varepsilon}$ this should not cause confusion for the reader.

4.1: The Validity of Nonlinear Geometric Optics for Weak Solutions of 1-D Conservation Laws

Why should the simplified asymptotic solutions from (4.3) continue to approximate the weak solutions of (4.2) after shock waves have formed? First, a wide variety of explicit simple wave solutions of (4.2) and (4.3) were analyzed rigorously in Chapter 1 of [12] with explicit error estimates independent of the first derivatives; second, in the case where the M×M system in (4.2) is genuinely nonlinear so that $b_j \neq 0$, $1 \leq j \leq M$, the earlier work of DiPerna [2] and Liu [11] on the detailed large time asymptotic behavior of solutions of (4.2) for initial data, $u_1^0(x)$, with compact support gave a rigorous proof that the solution of (4.2) asymptotically approaches decoupled N-waves as $T \to \infty$. These decoupled N-waves are not exact solutions of the equations in (4.2) but instead are weak solutions of the decoupled formal asymptotic Burgers equations in (4.3b). Finally, as long as the solutions of (4.2) and (4.3) remain smooth, i.e. until times, $T = o(\varepsilon^{-1})$, it is not difficult to justify the asymptotics via the classical method of characteristics. Since both the short-time and large-time behavior of solutions of (4.2) is well-approximated by solutions of the simplified asymptotic approximation in (4.3), it is reasonable to conjecture that u_w^ε approximates u^ε for weak solutions and perhaps for all time!

The behavior conjectured above has been rigorously proved recently in [3]. In describing this theorem, we use the L^1-norm, $|u|_1 = \int_{-\infty}^{\infty} |u| dx$, and also the L^1-norm of a periodic function with period p, $|u|_1 = \frac{1}{p} \int_0^p |u| dx$. The following result is proved in [3].

Theorem: (Justification of Nonlinear Geometric Optics)

Assume that the general M × M hyperbolic system in (4.2) is <u>genuinely nonlinear</u> at u_0 in <u>all wave fields</u>, i.e. $b_j \neq 0$ in (4.3) for all j with $1 \leq j \leq M$.

A) Assume that $u_1^0(x)$ is an arbitrary function of bounded variation with compact support. Consider the (any) weak solution, $u^\varepsilon(x,t)$, constructed by Glimm's method for the initial value problem in (4.2) and the corresponding weakly nonlinear geometric optics approximation, $u_w^\varepsilon(x,t)$ defined in (4.3), then we have the estimate,

(4.5)
$$\max_{0 < t < +\infty} |u^\varepsilon(\cdot,t) - u^\varepsilon(\cdot,t)|_1 < C\varepsilon^2$$

<u>uniformly for all times</u> where C depends only on the support of $u_1^0(x)$ and the derivatives of $F_j(u)$, $j = 0,1$.

B) Assume that $u_1^0(x)$ is a periodic function with bounded variation per period. For a scalar convex conservation law so that $M = 1$

(4.6)
$$\max_{0 < t < +\infty} |u^\varepsilon(\cdot,t) - u_w^\varepsilon(\cdot,t)|_1 < C\varepsilon^3 .$$

On the other hand, for a pair of conservation laws, i.e. $M = 2$ and periodic initial data, we have the weaker estimate,

(4.7)
$$|u^\varepsilon(\cdot,t) - u_w^\varepsilon(\cdot,t)|_1 < tC\varepsilon^2$$

In B), the constants, C, in (4.6) and (4.7) depend on derivatives of $F_j(u)$, $j = 0,1$, the L^∞ norm of $u_1^0(x)$, and the period, p.

<u>The estimates in (4.5) and (4.6) are much stronger than anticipated by the formal theory</u> and provide very strong supporting evidence for the use of the formal approximations from sections II and III for discontinuous weak solution. Even the somewhat weaker estimate in (4.7) for $M = 2$ and periodic waves is still sufficient to justify u_w^ε as the leading order asymptotic term for <u>discontinuous weak</u> solutions until times of order, $O(\varepsilon^{-1})$.

Why do we use the L^1 norm? Next, we give an elementary example which illustrates this point. We consider small amplitude solutions about the zero background state for the scalar convex conservation law,

(4.8a)
$$u_t + (f(u))_x = 0$$

with $f'(u) = a(u)$, $f''(u) > 0$ and conveniently normalized with $f'(0) = 0$, $f''(0) = 1$. For discontinuous Riemann initial data with the form

$$(4.8b) \qquad u^\varepsilon(x,0) = \begin{cases} \varepsilon, & x < 0 \\ -\varepsilon, & x > 0 \end{cases}$$

we have the explicit solution of (4.8) given by

$$u^\varepsilon(x,t) = \begin{cases} \varepsilon, & x < s^\varepsilon t \\ -\varepsilon, & x > s^\varepsilon t \end{cases}$$

with $s^\varepsilon = (2\varepsilon)^{-1}(f(\varepsilon) - f(-\varepsilon))$. Provided that $f'''(0) \neq 0$, s^ε satisfies $C^{-1}\varepsilon^2 < |s^\varepsilon| < C\varepsilon^2$ with some $C > 0$. For this example, u_w^ε is the solution of the inviscid Burgers equation with the initial data in (4.8b), i.e.

$$u_w^\varepsilon(x,t) = \begin{cases} \varepsilon, & x < 0, \ t > 0 \\ -\varepsilon, & x > 0, \ t > 0. \end{cases}$$

Let's compute the error, $u_w^\varepsilon - u^\varepsilon$, in various norms. In the maximum norm, the deviation of u_w^ε from u^ε is always only $O(\varepsilon)$ due to the difference in shock location. On the other hand, if we compute the deviation in the L^1 norm, we have

$$C^{-1} t \varepsilon^3 < |u^\varepsilon(\cdot,t) - u_w^\varepsilon(\cdot,t)|_1 < Ct\varepsilon^3.$$

This example explains both why we need the L^1-norm to prove the above Theorem for weak solutions rather than the L^∞ norm and also why estimates as presented in (4.7) grow in time due to phase shifts in the shock location. For a more detailed explanation of the factor of t in the estimate from (4.7) in the periodic case with $M = 2$ which we conjecture in [3] is due to the appearance of different phase shifts in the different wave speeds, we refer the reader to section 6 of [3]. The remarkable fact that the estimates in (4.5) and (4.6) are uniform for all time is a consequence of the large time cancellation of shocks and rarefactions of the same family.

We end this subsection with a brief discussion of the proof - the proof does not mimic the derivation of the asymptotic approximations in a straightforward fashion. There are four main aspects to the proof:

1) The use of structural symmetries of the Burgers equation solutions from (4.3)

2) The use of L^1-stability of the total variation to control the error up to intermediate asymptotic times of order ε^{-1}

(4.9)
3) The large-time asymptotic decoupling of general weak solutions of (4.2) into solutions of scalar conservation laws with forcing terms that are suitably small Borel measures following earlier ideas of DiPerna and Liu.

4) The rapid decay of total variation for large time for solutions of scalar convex conservation laws.

In the next subsection, we sketch the proof of the theorem for scalar convex conservation laws and periodic initial data. This gives us the opportunity to illustrate the main ingredients in 1), 2), and 4) of the general proof of the theorem.

We remark that the theorem in [3] is actually only formulated and proved for conservation laws with $F_0(u) = u$; however, since t is a time-like direction for (4.2), we use (2.2a), introduce $w = F_0(u)$ as a new dependent variable, and $F(w) = F_1(F_0^{-1}(w))$ as a new flux function and apply the theorem above to (4.2) in the form

$$w_t + F(w)_x = 0.$$

We leave the elementary details to the reader.

4.2 The Proof of the Theorem for Scalar Convex Conservation Laws with Periodic Initial Data

Here we sketch the proof of the estimate in (4.6) for scalar convex conservation laws and periodic initial data. We will describe the main steps 1), 2), and 4) in this special case and their use in the proof of this estimate.

Our first step is based on the following considerations: Consider the solution, $\tilde{u}^\varepsilon$, of the initial value problem for the inviscid Burgers equation,

(4.10)
$$\tilde{u}^\varepsilon_t + b(\tfrac{1}{2}(\tilde{u}^\varepsilon)^2)_x = 0$$

$$\tilde{u}^\varepsilon(x,0) = u_0 + \varepsilon u_1^0(x).$$

This equation arises as the leading order asymptotic equation from (4.3) for any general genuinely nonlinear system of conservation laws - thus, we might anticipate that the weakly nonlinear approximation from (4.3) is exact without any errors for weak solutions of the special initial value problem in (4.10). How can we justify weakly nonlinear geometric optics for weak solutions of the inviscid Burgers equation in (4.10)? We observe by direct calculation that if $\sigma(\tau,\theta)$ satisfies the inviscid Burgers equation,

(4.11a)
$$\sigma_\tau + b(\tfrac{1}{2}\sigma^2)_\theta = 0$$

then for any $\varepsilon > 0$ and constant u_0,

(4.11b)
$$\tilde{u}^\varepsilon(x,t) = u_0 + \varepsilon\sigma(\varepsilon t, x - bu_0 t)$$

is also a solution of the inviscid Burgers equation

(4.11c)
$$\tilde{u}^\varepsilon_t + b(\tfrac{1}{2}(\tilde{u}^\varepsilon)^2)_x = 0.$$

Thus, solutions of the inviscid Burgers equation are invariant under an additional two-parameter family of structural symmetries besides space-time dilations. If we choose as initial data for σ in (4.11a), the initial value $u_1^0(\theta)$, we observe that the facts in (4.11) justify weakly nonlinear geometric optics exactly for weak solutions of the Burgers equation - this is step 1) from (4.9). Actually, below we use the following mild extension of (4.11):

If $\sigma(\tau,\theta)$ satisfies (4.11a), then $\tilde{u}^\varepsilon = u_0 + \varepsilon\sigma(\varepsilon t, x - (c + bu_0)t)$ is a solution of

(4.12)
$$\tilde{u}^\varepsilon_t + c\tilde{u}^\varepsilon_x + b(\tfrac{1}{2}(\tilde{u}^\varepsilon)^2)_x = 0$$

so that weakly nonlinear geometric optics is justified exactly for weak solution of the conservation law in (4.12).

The facts in the preceding paragraph reduce the proof of the estimate in (4.6) of the Theorem to a direct L^1 estimate for the difference, $u^\varepsilon - \tilde{u}^\varepsilon$, where u^ε, $\tilde{u}^\varepsilon$ satisfy

(4.13a) $$u_t^\varepsilon + (f(u^\varepsilon))_x = 0, \quad u^\varepsilon(x,0) = u_0 + \varepsilon u_1^0(x)$$

and

(4.13b) $$\tilde{u}_t^\varepsilon + c\tilde{u}_x^\varepsilon + b(\tfrac{1}{2}(\tilde{u}^\varepsilon)^2)_x = 0, \quad \tilde{u}^\varepsilon(x,0) = u_0 + \varepsilon u_1^0(x)$$

with $b = f''(u_0)$ and $c = f'(u_0) - f''(u_0)u_0$. We expand $f(u)$ by

$$f(u) = Qf(u,u_0) + R(u,u_0)$$

with $Qf(u,u_0)$, the quadratic part at u_0 given by

(4.14a) $$QF(u,u_0) = f(u_0) + f'(u_0)(u - u_0) + \frac{f''(u_0)}{2}(u - u_0)^2$$

and the remainder $R(u,u_0)$ satisfying

(4.14b) $$R(u,u_0) = O(u - u_0)^3.$$

Thus, $\tilde{u}^\varepsilon$ from (4.13b) satisfies the scalar law,

(4.15a) $$\tilde{u}_t^\varepsilon + (Qf(\tilde{u}^\varepsilon,u_0))_x = 0$$

while, by using the chain rule for BV functions (see section 2 and 4 of [3]), u^ε from (4.13a) satisfies

(4.15b) $$u_t^\varepsilon + (Qf(u^\varepsilon,u_0))_x = \nu$$

where ν is a finite Borel measure on the period strip, $S_{0,T} = \{(x,t) \mid 0 < x < p, 0 < x < T\}$ with total mass on this strip, $|\nu|(S_{0,T})$ estimated by

(4.16) $$|\nu|(S_{0,T}) \leq C \int_0^T \int_0^p |u^\varepsilon - u_0|^2 |\partial_x u^\varepsilon| dx\, dt$$

$$\leq C \int_0^T (\text{Var}_p(u^\varepsilon(\cdot,s)))^3 ds.$$

Here $\text{Var}_p(t(x))$ is the total variation on a period strip; the first estimate in (4.16) is given formally by differentiating (4.14b) and is justified rigorously by using the calculus for BV functions (see sections 2 and 4 of [31]) - the second estimate in (4.16) follows from the fact that the oscillation is dominated by the total variation.

We observe that $\tilde{u}^\varepsilon$ and u^ε are two solutions of the same inhomogeneous scalar conservation law with right hand side given by different finite Borel measures but the same initial data. The second key ingredient, 2) of (4.9) is the general L^1 stability estimate with inhomogeneous terms defined by finite Borel measures proved in section 3 of [3] for scalar convex conservation laws. Since this L^1-stability estimate generalizes the well-known earlier L^1-stability estimates for scalar laws of Volpert-Kruzkhov-Keyfitz, we won't state this stability estimate in detail here. However, a direct consequence is the estimate,

$$(4.17) \qquad |u^\varepsilon(\cdot,t) - \tilde{u}^\varepsilon(\cdot,t)|_1 \leq 2|v|(S_{0,T})$$

for the two weak solutions with the same initial data satisfying (4.15). By combining (4.16) and (4.17), we obtain

$$(4.18) \qquad |u^\varepsilon(\cdot,t) - \tilde{u}^\varepsilon((\cdot,t)|_1 \leq C \int_0^T (\text{Var}_p(u^\varepsilon(\cdot,s)))^3 ds.$$

The second ingredient in the stability estimates for 2) in (4.9) is the well-known fact that

$$(4.19) \qquad \text{Var}_p u^\varepsilon(\cdot,t) \leq \text{Var}_p u_0^\varepsilon = \varepsilon.$$

A direct substitution of the estimate from (4.19) into (4.18) yields

$$(4.20) \qquad |u^\varepsilon(\cdot,t) - u_w^\varepsilon(\cdot,t)|_1 \leq ct\varepsilon^3$$

since $u_w^\varepsilon = \tilde{u}^\varepsilon$. This estimate is strong enough to justify the weakly nonlinear approximation for large times of order $o(\varepsilon^{-2})$. However, we can do better if we exploit the ideas mentioned in 4) of (4.9). It is well-known that for periodic waves, the total variation decays rapidly in time, in fact,

$$(4.21) \qquad \text{Var}_p u^\varepsilon(\cdot,t) \leq \frac{C}{1+t} \text{Var}_p u_0^\varepsilon = \frac{C}{1+t}\varepsilon$$

and substitution of (4.21) into (4.18) improves the estimate in (4.20) to

$$|u^\varepsilon(\cdot,t) - u^\varepsilon_w(\cdot,t)|_1 \leq C\varepsilon^3 \int_0^t (1 + s)^{-3} ds \leq C_1 \varepsilon^3.$$

This completes our sketch of the proof of the estimate in (4.6).

4.3: Some Accessible Open Problems in the Rigorous Theory of Weakly Nonlinear Geometric Optics

Problem #1: As described in section 4.1, the main theorem in [3] is proved under the hypothesis that <u>every</u> wave field for the hyperbolic system in (4.2) is genuinely nonlinear. <u>Extend the main theorem in [3] for initial data of compact support to systems with eigenvalues</u> that are <u>either genuinely nonlinear</u> or <u>linearly degenerate</u>. The equations of compressible fluid flow from section 2.4 are the prototypical example. In particular, assess whether the estimate in (4.5) remains valid in this case.

Problem #2: It is not difficult to extend the formal theory of weakly nonlinear approximations as developed in section 2 to general mixed initial boundary value problems for the hyperbolic system in (4.2) on the space-time quarter-space, $x > 0$, $t > 0$ with suitable general boundary conditions on $x = 0$ and initial data of compact support. Also, the use of Glimm's method to prove existence of global small amplitude weak solutions has been carried out in detail in [23]. Formulate and prove a suitable analogue of the main theorem in [3] for data of compact support for the general mixed problem described above.

Problem #3: Extend and generalize the main theorem of [3] to data with compact support and suitable systems like those occurring in nonlinear elasticity with eigenvalues that are neither genuinely nonlinear nor linearly degenerate. Here the formal asymptotic solutions from section 2 needs to be modified suitably - see section 1.5 and 1.6 of [12] for the modifications of single wave expansions in the formal asymptotics. The system in 1-D with $M = 2$ describing nonlinear hyperelasticity is the simplest starting point for systems. The scalar case under the assumptions, $f^{(1)}(u_0) = a(u_0)$, $f^{(j)}(u_0) = 0$, $2 \leq j \leq p - 1$, $f^{(p)}(u_0) \neq 0$ can be handled by the same proof as we gave in section 4.2 with minor modifications - in fact, uniform decay estimates as in (4.21) are also known in these situations.

Problem 4: Study the validity of the resonant asymptotic approximations discussed in sections 2.3 and 2.4 for general periodic weak solutions of conservation laws. This is the most difficult problem of this section.

Problem 5: Study the large time asymptotic behavior of periodic waves for general pairs of conservation laws with genuinely nonlinear wave fields - i.e. M = 2. The asymptotics from section 4.2 yield a conjectured behavior involving explicit periodic solutions of the Burgers equation. Can one prove that the estimate in (4.6) is sharp for periodic waves? More importantly, if we regard the weakly nonlinear approximation u_w^ε as an approximation to solutions u^ε, then the examples in (4.8) and in [3] indicate that the resolution of these methods, according to the ideas of P.D. Lax (see [9]), is even better than the integral estimates from (4.6). Can this be established rigorously?

BIBLIOGRAPHY

[1] Choquet-Bruhat, V., "Ondes asmptotiques et approchées pour systèmes d'équations aux dérivées partielles nonlinéaires, J. Math. Pures et Appl., 48, 117-158 (1969).

[2] DiPerna, R., "Decay and asymptotic behavior of solutions to nonlinear hyperbolic systems of conservation laws", Indiana Univ. Math. J., 24, 1047-1071 (1975).

[3] DiPerna, R. and Majda, A., "The validity of nonlienar geometric optics for weak solutions of conservation laws", Commun. Math. Physics, April 1985, 35 pp.

[4] Hunter, J., "A ray method for slowly modulated nonlinear waves" (preprint, March 1984).

[5] Hunter, J. and Keller, J.B., "Weakly nonlinear high frequency waves", Commun. Pure Appl. Math.

[6] Hunter, J. and Keller, J.B., "Weak shock diffraction" (in press, Wave Motion).

[7] Hunter, J., Majda, A., and Rosales, R., "Resonantly interacting weakly nonlinear hyperbolic waves, II: several space variables" (in preparation).

[8] Landau, L.D., "On shock waves at large distances from their place of origin", J. Phys. USSR 9, 495-500 (1945).

[9] Lax, P.D., "Accuracy and resolution in the computation of solutions of linear and nonlinear equations", in Recent Advances in Numerical Analysis, pp. 107-118, DeBoor, C. and Golub, G. ed. New York Academic Press (1978).

[10] Lighthill, M.J., "A method for rendering approximate solutions to physical problems uniformly valid", Phil. Mag. 40, 1179-1201 (1949).

[11] Liu, T.P., "Decay to N-waves of solutions of general systems of nonlinear hyperbolic conservation laws", Comm. Pure Appl. Math. 30, 585-610 (1977).

[12] Majda, A., Compressible Fluid Flow and Systems of Conservation Laws in Several Space Variables, Appl. Math. Science Series #53, Springer-Verlag, New York, 1984.

[13] Majda, A. and Rosales, R., "A theory for spontaneous Mach system formation in reacting shock fronts, I: the basic perturbation analysis", S.I.A.M. J. Appl. Math. 43, 1310-1334 (1983).

[14] Majda, A. and Rosales, R., "A theory for spontaneous Mach stem formation in reacting shock fronts, II: steady wave bifurcations and the evidence for breakdown", Stud. Appl. Math. 71, 117-148 (1984).

[15] Majda, A. and Rosales, R., "Resonantly interacting weakly nonlinear hyperbolic waves, I: a single space variable", Stud. Appl. Math. 71, 149-179 (1984).

[16] Nayfeh, A., "A comparison of perturbation methods for nonlinear hyperbolic waves", in Singular Perturbations and Asymptotics, pp. 223-276, Meyer, R. and Parter, S. editors, Academic Press, New York, 1980.

[17] Rosales, R., and Majda, A., "Weakly nonlinear detonation waves", SIAM J. Appl. Math. 43, 1086-1118 (1983).

[18] Whitham, G., "The flow pattern of a supersonic projectile", Commun. Pure Appl. Math. 5, 301-348 (1952).

[19] Woodward, P., "Piecewise-parabolic methods for astrophysical fluid dynamics", L.L.L. preprint November 1983.

[20] Woodward, P., "Simulation of the Kelvin-Helmholtz instability of a supersonic slip surface with the piecewise parabolic method", L.L.L. preprint, March 1984.

[21] Majda, A., Rosales, R., and Schonbek, M., "A system of integro-differential equations arising in resonant wave asymptotics", (to appear).

[22] Katznelson, Y., An introduction to Harmonic Analysis, Dover, New York, 1976.

[23] Goodman, J., "The Mixed Initial Boundary Value Problem for Conservation Laws", Ph.D. Thesis Stanford, August, 1982.

ON THE CONSTRUCTION OF A MODULATING MULTIPHASE WAVETRAIN FOR A PERTURBED KdV EQUATION

By

David W. McLaughlin*

Department of Mathematics and
Program in Applied Mathematics
University of Arizona
Tucson, Arizona 95721

Abstract

This paper summarizes the status of a direct construction of an asymptotic representation of a modulating multiphase wavetrain for a class of perturbed kdV equations. This class includes the kdV-Burgers equation. The calculations apply on a "boundary" between dispersive and dissipative behavior. The construction proceeds by standard asymptotic methods. The result of the construction is an invariant representation of the reduced equations which permits their diagonalization. While mathematically the construction is incomplete, care is taken to identify the mathematical status of each step in the construction. The equivalence of this constructive approach with postulated averages of conservation laws is established for two phase waves. Finally, the Young measure for this program is constructed explicitly.

I. Introduction

In this paper we study the initial value problem for a perturbed Korteweg de Vries (kdV) equation:

$$U_t - 6UU_x + \epsilon^2 U_{xxx} + \beta\, f(U, \epsilon U_x, \epsilon^2 U_{xx}, \ldots) = 0$$
$$U(x, t=0) = U_{in}\left(\frac{x}{\epsilon}\,;\, x\right).$$

---

*Supported in part by N.S.F. Grant # MCS 79-3533, in part by the U.S. Air Force Office of Scientific Research under grant AFOSR-81-0253, and in part by the U.S. Army, while on leave 1980-82 at Courant Institute, New York University.

Here ε denotes the small parameter ($0 < \varepsilon \ll 1$), while $\beta > 0$ is a constant of order unity. The initial data U_{in} depends upon two spatial scales, a fast scale x/ε and a slower scale x. The reason we tie the fast scale in the initial data to the small parameter ε in the equation is that the kdV equation ($\beta=0$) in the "small dispersion limit" ($0 < \varepsilon << 1$) is known to develop oscillations whose spatial wavelength is $O(\varepsilon)$. Actually, we will restrict our attention to a special class of initial data - a slowly varying N-phase waveform - which we will describe in detail later.

Our goal is to construct a representation of the solution $U = U^{(\varepsilon)}(x,t)$ which is valid for small, but finite ε.

There are two general reasons for our interest in this problem. The first is technical. We desire to develop a method for the construction of $U^{(\varepsilon)}$ which (i) is general enough to include external perturbations (such as dissipation) of a completely integrable system, (ii) shows the validity of our prescription [1,2,3] of "averaged conservation laws" for the description of modulating N phase wavetrains, (iii) is a sufficiently standard mathematical technique that the construction can provide a first step toward a rigorous proof of the validity of the representation. There are cases (the sine-Gordon equation, [4,5] for example) where the wave train is modulationally unstable. To study this instability one must retain additional terms in the modulation equations. The technique we are developing is sufficiently standard that a systematic representation of such higher order terms can be attempted.

The second reason we are interested in this problem concerns the Korteweg-de Vries - Burgers equation itself. Consider the special perturbation $f = -\varepsilon^2 U_{xx}$. Then the equation becomes

$$U_t - 6 U U_x + \varepsilon^2 U_{xxx} - \delta U_{xx} = 0$$

with $\delta = \beta\varepsilon^2$. Relax this relation between δ and ε for the moment and think of the equation as depending upon two independent small parameters, ε and δ. When the dispersive term is absent ($\varepsilon \equiv 0$), the solution U tends, as $\delta \to 0$, to a sequence of shocks whose speeds are fixed by the entropy condition. On the other hand, when the dissipative term is absent ($\delta = 0$), the solution U becomes very oscillatory for small ϵ. In this case the weak limit of $U^{(\epsilon)}$ as $\epsilon \downarrow 0$ is described by a family of nonlinear, hyperbolic equations [6,7,8,9]. When both the dissipative and dispersive terms are present, but small ($0 < \epsilon, \delta << 1$), shocks will describe the weak limit provided the dissipation dominates. Recently it has been established [10] that shocks will describe the weak limit as $\epsilon \downarrow 0$, $\delta \downarrow 0$, provided the dissipation parameter $\delta \geqslant \beta\epsilon$. Our calculations ($\delta = \beta \epsilon^2$) describe the boundary between dissipative and dispersive behavior. For stronger dissipation, shocks apply; on this boundary, the behavior for small ϵ is described by the equations derived in this paper; for still weaker dissipation, the weak limit is described by [6,7,8,9].

The method that we are developing has several main advantages: (i) It constructs a representation of the wave which, since it is valid for finite, but small ϵ, retains the oscillatory structure of the wave. (ii) It extends naturally to small perturbations of integrable systems.

This work has two shortcomings. The first concerns a limitation of the ansatz itself. It demands a very restricted class of initial data-slowly varying N-phase waves. This means that, for general data, it is not valid uniformly in space; rather, it applies only in the vicinity of the "shock front". Some sort of matching must be developed to make the representations uniformly valid. The second shortcoming concerns the work reported herein. This work, although systematic, is not rigorous; we view it as the first step toward a rigorous derivation. In the text we will clearly identify those points where mathematical rigor is absent.

This work continues our study of the modulation of completely integrable wave trains [1,2,3,4,5]. The procedure of averaging conservation laws to obtain modulation equations, as well as placing these equations in Riemann invariant form, was initiated in [11,12] for the single phase case. The constructive method employed here follows single phase work in [13,14] and the multi-phase procedure of [15]. In the single phase case the results are not new, although our arguments are somewhat more systematic than those in the literature; in the multiphase case, our results are new because of our use of the completely integrable exact theory. In the single phase case, dissipative perturbations of nonlinear, dispersive wave trains have been analysed previously [12,16,17]. The (weak) zero dispersion limit of the kdV equation is studied in [8,9,10].

The last section of this paper is motivated by the rigorous work of [10,19,20].

Finally, I acknowledge many conversations with R. DiPerna, H. Flaschka, M.G. Forest, C.D. Levermore, H. McKean, G. Papanicolaou and S. Venakidas. My work has certainly benefited from each of these interactions.

II. Definition of the Problem

Consider the initial value problem

$$U_t - 6 U U_x + \varepsilon^2 U_{xxx} + \beta f(U, \varepsilon U_x, \varepsilon^2 U_{xx}, \ldots) = 0, \qquad \text{(II.1a,b)}$$

$$U(x,t=0) = W_{in}^{(N)} \left(\frac{x}{\varepsilon} ; \vec{\lambda}(x) \right).$$

Here $0 < \varepsilon << 1$ and $\beta = O(1)$. The initial data is a slowly modulating N-phase waveform for the kdV equation which we now describe.

The kdV equation has a family of exact solutions of the form

$$U(x,t) = W_N \left(\frac{\Theta_1(x,t)}{\varepsilon}, \ldots, \frac{\Theta_N(x,t)}{\varepsilon} ; \lambda_0, \ldots, \lambda_{2N} \right) \qquad \text{(II.2a)}$$

where the N "phases" $\Theta_1, \ldots, \Theta_N$ depend linearly upon x and t,

$$\Theta_j(x,t) \equiv \kappa_j x - \omega_j t . \qquad \text{(II.2b)}$$

The waveform W_N is 2π periodic in each argument $\theta_j \equiv \Theta_j/\varepsilon$. The family of "N phase waves" (II.2a) is indexed by 2N+1 parameters $\lambda = (\lambda_0, \lambda_1, \ldots, \lambda_{2N})$, which fix the N spatial wavenumbers $\kappa = (\kappa_1, \ldots, \kappa_N)$ and the N temporal frequencies $\omega = (\omega_1, \ldots, \omega_N)$. The explicit formulas for $\kappa = \vec{\kappa}(\lambda)$ and $\omega = \vec{\omega}(\lambda)$, as well as more details about these N phase waves, will be given later. For now, we only remark that in the single phase (N=1) case, (II.2) is a 2π periodic traveling wave solution of the kdV equation.

The initial data (II.1b) is described in terms of an N phase wave W_N as follows: Prescribe smooth functions $\vec{\lambda} = \vec{\lambda}(x)$ in terms of which

$$W_{in}^{(N)}(\tfrac{x}{\varepsilon}; \vec{\lambda}(x)) \equiv W_N(\tfrac{\Theta_1(x)}{\varepsilon}, \ldots, \tfrac{\Theta_N(x)}{\varepsilon}; \vec{\lambda}(x)) \qquad (II.3a,b)$$

where

$$\partial_x \Theta_j \equiv \kappa_j(\vec{\lambda}(x)), \quad \vec{\lambda}(x) \text{ prescribed.}$$

Our problem is to construct a representation of $U = U^\varepsilon$ which is valid for small ε. We use a standard asymptotic method which begins with an ansatz:

$$U \sim U^\varepsilon [\tfrac{\Theta_1^\varepsilon(x,t)}{\varepsilon}, \ldots, \tfrac{\Theta_N^\varepsilon(x,t)}{\varepsilon}; x, t]. \qquad (II.4)$$

Here the (x,t) dependence of $\Theta^\varepsilon = (\Theta_1^\varepsilon, \ldots, \Theta_N^\varepsilon)$ is to be determined. We define

$$\kappa^\varepsilon(x,t) \equiv \partial_x \Theta^\varepsilon(x,t)$$
$$\omega^\varepsilon(x,t) \equiv \partial_t \Theta^\varepsilon(x,t), \qquad (II.5)$$

and note that

$$\partial_t U^\varepsilon \to (\tfrac{\omega^\varepsilon \cdot \nabla}{\varepsilon} + \partial_t) U^\varepsilon$$

$$\partial_x U^\varepsilon \to (\tfrac{\kappa^\varepsilon \cdot \nabla}{\varepsilon} + \partial_x) U^\varepsilon,$$

where $\nabla U^\varepsilon \equiv (\tfrac{\partial}{\partial \Theta_1} U^\varepsilon, \ldots, \tfrac{\partial}{\partial \Theta_N} U^\varepsilon) = \varepsilon(\tfrac{\partial}{\partial \Theta_1} U^\varepsilon, \ldots, \tfrac{\partial}{\partial \Theta_N} U^\varepsilon)$. Of course, definition (II.5) implies N consistency (or integrability) conditions for Θ_j:

$$\partial_t \kappa^\varepsilon = \partial_x \omega^\varepsilon. \qquad (II.6)$$

In terms of this ansatz, equation (II.1a) becomes

$$\frac{1}{\varepsilon}[\omega^\varepsilon \cdot \mathcal{W}^\varepsilon - 6U^\varepsilon \kappa^\varepsilon \cdot \mathcal{W}^\varepsilon + (\kappa^\varepsilon \cdot \nabla)^3 U^\varepsilon]$$

$$+ [U_t^\varepsilon + 6U^\varepsilon U_x^\varepsilon + 3((\kappa^\varepsilon \cdot \nabla)^2 U_x^\varepsilon + (\kappa^\varepsilon \cdot \nabla)(\kappa_x^\varepsilon \cdot \nabla) U^\varepsilon) + \beta f(U^\varepsilon, (\kappa^\varepsilon \cdot \nabla + \varepsilon \partial_x) U^\varepsilon \dots)]$$

$$+ \varepsilon [3\kappa^\varepsilon \cdot \mathcal{W}_{xx}^\varepsilon + 3\kappa_x^\varepsilon \cdot \mathcal{W}_x^\varepsilon + \kappa_{xx}^\varepsilon \cdot \mathcal{W}^\varepsilon] + \varepsilon^2 [U_{xxx}^\varepsilon] = 0 \ .$$

In this equation, expansions of the form

$$\begin{aligned} U^\varepsilon &\sim W + \varepsilon\, U^{(1)} + \varepsilon^2\, U^{(2)} + \dots \\ \kappa^\varepsilon &\sim \kappa + \varepsilon\, \kappa^{(1)} + \varepsilon^2\, \kappa^{(2)} + \dots \\ \omega^\varepsilon &\sim \omega + \varepsilon\, \omega^{(1)} + \varepsilon^2\, \omega^{(2)} + \dots \end{aligned} \qquad (II.7)$$

lead to the following sequence of problems:

$$O(\varepsilon^{-1}): \quad (\omega \cdot \nabla)\, W - 6 W (\kappa \cdot \nabla) W + (\kappa \cdot \nabla)^3 W = 0 \qquad (II.8a)$$

$$O(\varepsilon^j): \qquad L\, U^{(j)} + F^{(j)} = 0\ ,\ j = 0, 1, 2, \dots \qquad (II.8b)$$

where the linear operator L is given by

$$L \equiv \omega \cdot \nabla - 6 (\kappa \cdot \nabla) W + (\kappa \cdot \nabla)^3 \ . \qquad (II.9)$$

The inhomogeneity $F^{(0)}$ is given by

$$F^{(0)} = [W_t - 6WW_x + 3((\kappa \cdot \nabla)^2 W_x + (\kappa \cdot \nabla)(\kappa_x \cdot \nabla) W) + \beta\, f(W, (\kappa \cdot \nabla) W, \dots)]$$

$$+ [\omega^{(1)} \cdot \mathcal{W} - 6 W\, \kappa^{(1)} \cdot \mathcal{W} + 3(\kappa \cdot \nabla)^2 (\kappa^{(1)} \cdot \nabla)\, W] \ , \qquad (II.10)$$

with similar, but more complicated, formulas for $F^{(j)}$, $j \geq 1$. The sequence of problems (II.8) must now be studied; however, we first summarize some background material from the theory of the inverse spectral representation of kdV. The background material of section III will be used throughout the remainder of this paper.

III. Inverse Spectral Theory for the kdV Equation

The background material in this section may be found in [1,21] which contain references to the original literature. Let $q = q(y, \tau)$ satisfy the kdV equation,

$$q_\tau = 6 q q_y - q_{yyy} . \tag{III.1}$$

By considering the "Lax representation" of this equation, one realizes that it arises as the integrability condition for the linear system

$$[- \partial_{yyy} + 2(q \partial_y + \partial_y q)] \psi = 4\lambda \partial_y \psi$$

$$\partial_\tau \psi = -2 q_y \psi + 2(q+2\lambda) \partial_y \psi . \tag{III.2a,b}$$

The function ψ is known as a "squared eigenfunction". Equation (III.2a) acts as an eigenvalue problem for the squared eigenfunction ψ; equation (III.2b) defines its time flow. The pair (III.2a,b) is compatible since q satisfies kdV.

The pair of equations (III.2) is fundamental in the theory of the kdV equation. Here we use the pair (i) to generate an infinite family of conservation laws, (ii) to provide a representation of the N-phase wave solutions of (II.8a), (iii) to provide formulas for the fluxes and densities of the conservation laws in terms of the N-phase waves, and (iv) to compute necessary averages.

III.A. Fundamental Conservation Law

Let q solve kdV and ψ solve the pair (III.2). Then it immediately follows that ψ satisfies the conservation law

$$\partial_\tau [\psi] + \partial_y [6(q-2\lambda)\psi - 2\partial_{yy} \psi] = 0 . \tag{III.3}$$

This fundamental conservation law generates an infinite family as follows: One seeks a solution of (III.2a) which has the asymptotic behavior

$$\psi(y,\tau;\lambda) \simeq \frac{1}{\sqrt{2\lambda}} + \sum_{j=1}^{\infty} \psi_j(y,\tau)(2\lambda)^{-j-1/2} \quad \text{as} \quad \lambda \to \infty . \tag{III.4}$$

This ansatz in (III.2a) leads to a recursion relation for the coefficients $\psi_j(y,\tau)$,

$$\partial_y \psi_{j+1} = [-\frac{1}{2} \partial_{yyy} + q \partial_y + \partial_y q] \psi_j , \quad i = 1,2,... \tag{III.5}$$

$$\psi_0 = 1.$$

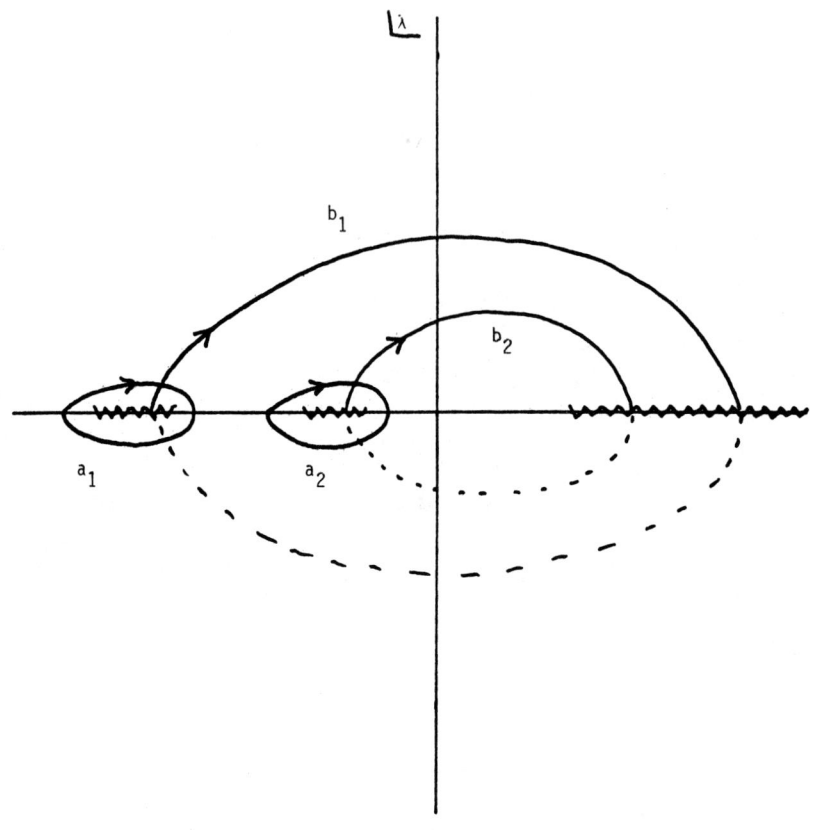

Figure 1

Recursion relation (III.5), together with certain choices of integration constants, provides formulas for ψ_j in terms of q and its derviatives. We list the first few:

$$\psi_0 = 1$$
$$\psi_1 = q$$
$$\psi_2 = \tfrac{1}{2}(3q^2 - q_{yy})$$
$$\psi_3 = \tfrac{5}{2} q^3 + \tfrac{5}{4} q_y^2 - \partial_y(\tfrac{5}{2} q q_y - \tfrac{1}{4} q_{yyy})$$
$$\psi_4 = \tfrac{35}{8} q^4 + \tfrac{35}{4} q q_y^2 + \tfrac{7}{8} q_{yy}^2 + \partial_y [- 5q^2 q_y$$
$$+ \tfrac{1}{2} q q_{yyy} - \tfrac{3}{4} q_y q_{yy} - \tfrac{1}{2} \partial_y \psi_3]$$

By inserting the asymptotic behavior (III.4) into the fundamental conservation law (III.3), we obtain a family of conservation laws for the kdV equation:

$$\partial_\tau [\psi_j] + \partial_y [6(q \psi_j - \psi_{j+1}) - 2\partial_{yy} \psi_j] = 0, \quad j = 1,2,\ldots \qquad (III.7)$$

III.B. Solutions of the Adjoint Linearized kdV Equation

The squared eigenfunction ψ generates solutions of the adjoint linearized kdV equation. Let q and $Q = q + \delta q$ denote two solutions of the kdV equation. Then, as their difference goes to zero, δq satisfies the linear equation

$$\partial_\tau \delta q - 6 \partial_y q \delta q + \partial_{yyy} \delta q = 0, \qquad (III.8)$$

with formal adjoint

$$\partial_\tau \psi - 6q \partial_y \psi + \partial_{yyy} \psi = 0 \qquad (III.9)$$

Let q denote a solution of kdV, and ψ denote a solution of the pair (III.2). Then it immediately follows that ψ satisfies the adjoint equation (III.9); thus, the coefficients ψ_j are also solutions of the adjoint equation.

III. C. The μ Representation of kdV Waves

The squared eigenfunction system (III.2a,b) can be used to generate representations of N-phase wavetrains. This construction begins with the observation that system (III.2a,b) admits a first integral [22]:

$$\frac{1}{2} \psi \psi_{yy} - \frac{1}{4} \psi_y^2 - (q-\lambda) \psi^2 = R^2(\lambda) \qquad (III.10)$$

To construct N-phase waves, we fix (2N+1) real constants $\vec{\lambda} = (\lambda_0, \lambda_1, \ldots, \lambda_{2N})$,

$$\lambda_0 < \lambda_1 < \ldots < \lambda_{2N} , \qquad (III.11)$$

and demand that the first integral $R^2(\lambda)$ be the polynomial

$$R^2(\lambda) = \prod_{k=0}^{2N} (\lambda - \lambda_k) . \qquad (III.12)$$

This situation is achieved by seeking a solution $\psi^{(N)}$ of the squared eigenfunction system which is polynomial in λ of degree N:

$$\psi^{(N)}(y,\tau;\lambda) = \prod_{j=1}^{N} (\lambda - \mu_j(y,\tau)) . \qquad (III.13)$$

Insertion of this ansatz into (III.2) leads to the "μ-representation of kdV waves with N degrees of freedom":

$$q_N(y,\tau) = \Lambda - 2 \sum_{j=1}^{N} \mu_j(y,\tau) , \qquad (III.14a)$$

$$\psi^{(N)}(y,\tau;\lambda) = \prod_{j=1}^{N} (\lambda - \mu_j(y,\tau)) , \qquad (III.14b)$$

where $\Lambda = \sum_{j=0}^{2N} \lambda_j$, and where the $\mu_j(y,\tau)$ are constrained by

$$\lambda_0 < \lambda_1 < \mu_1(y,\tau) < \lambda_2 < \lambda_3 < \mu_2(y,\tau) < \lambda_4 < \ldots < \lambda_{2N-1} < \mu_N(y,\tau) < \lambda_{2N}, \qquad (III.14c)$$

and satisfy the ordinary differential equations

$$\partial_y \mu_j = -2i \frac{R(\mu_j)}{\prod_{i \neq j}(\mu_j - \mu_i)} \qquad (III.14d)$$

$$\partial_\tau \mu_j = -2i \left[2\left(\Lambda - 2\sum_{i \neq j} \mu_i\right)\right] \frac{R(\mu_j)}{\prod_{i \neq j}(\mu_j - \mu_i)} . \qquad (III.14e)$$

When evaluated on q_N, the fundamental conservation law (III.3) has the µ-representation

$$\partial_\tau T + \partial_y \chi = 0 , \qquad (III.14f)$$

$$T = \psi = \frac{\psi^{(N)}(y,\tau;\lambda)}{R(\lambda)} = \frac{\prod_{j=1}^{N}(\lambda - \mu_j(y,\tau))}{\sqrt{\prod_{k=0}^{2N}(\lambda - \lambda_k)}} \qquad (III.14g)$$

$$\chi = \left[6\Lambda - 12\left(\lambda + \sum_{j=1}^{N} \mu_j(y,\tau)\right)\right] \frac{\prod_{j=1}^{N}(\lambda - \mu_j(y,\tau))}{\sqrt{\prod_{k=0}^{2N}(\lambda - \lambda_k)}} - 2T_{yy} . \qquad (III.14h)$$

Formulas (III.14) summarize the "µ-representation of kdV waves with N-degrees of freedom". We now show that these exact solutions kdV are "N phase wavetrains"; that is, they are solutions which are quasi-periodic in space and time which depend upon N phases.

III.D. The θ Representation of kdV Waves

The wave q_N admits an equivalent representation which results once an Abel transformation is used to integrate the µ equations (III.14d,e). On the Riemann surface

$\mathcal{R} = [\lambda, R(\lambda) \equiv \sqrt{\prod_{0}^{2N}(\lambda - \lambda_k)}]$, one fixes a canonical set of a-b cycles. On this Riemann surface, we introduce the following objects: (i) a basis of holomorphic differentials,

$$\psi_i \equiv \sum_{j=1}^{N} c_{ij} \lambda^{j-1} \frac{d\lambda}{R(\lambda)}, \qquad (III.15a)$$

normalized by the condition

$$\oint_{a_i} \psi_j = \delta_{ij} ; \qquad (III.15b)$$

(ii) A symmetric matrix with positive definite imaginary part,

$$B_{ij} \equiv \oint_{b_i} \psi_j ; \qquad (III.15c)$$

(iii) Two differentials of the second kind

$$\Omega_1 \equiv [-\frac{1}{2}\lambda + \sum_{j=1}^{N} D_j \lambda^{j-1}] \frac{d\lambda}{R(\lambda)} , \qquad (II.16a,b)$$

$$\Omega_2 \equiv [-\frac{1}{2}\lambda^{N+1} + \frac{1}{4} \Lambda \lambda^N + \sum_{j=1}^{N} E_j \lambda^{j-1}] \frac{d\lambda}{R(\lambda)} ,$$

where the coefficient $\Lambda \equiv \sum_{k=0}^{2N} \lambda_k$, and the coefficients D_j and E_j are uniquely determined by the normalization conditions

$$\oint_{b_i} \Omega_j = 0$$

These quantities form the ingredients for a change of variables from $(\mu_1,\ldots,\mu_N) \to (\Theta_1,\ldots,\Theta_N)$:

$$\Theta_j = (B^{-1})_{ij} \sum_{k=1}^{N} \int_{\mu_k^0}^{\mu_k} \psi_j . \qquad (III.17)$$

If $\mu_j(y,\tau)$ satisfies the differential equations (III.14d,e), then the new variables $\Theta_j(y,\tau)$ satisfy

$$\partial_y \Theta_j = \kappa_j \qquad (II.18a,b)$$

$$\partial_\tau \Theta_j = \omega_j , \qquad (II.18a,b)$$

where the constants κ_j and ω_j are defined in terms of $\vec{\lambda}$:

$$\kappa_j = \kappa_j(\vec{\lambda}) = -\oint_{a_j} \Omega_1$$
$$\omega_j = \omega_j(\vec{\lambda}) = -12 \oint_{a_j} \Omega_2 . \qquad (II.19a,b)$$

Using theta functions, one can invert transformation (III.17) and give formulas for μ_j (and therefore q_N) in terms of $(\Theta_j,\ldots,\Theta_N)$:

$$q_N = \Lambda + \Gamma - 2\partial_{yy} \log [\Theta(z(\Theta);B)] , \qquad (III.20a)$$

where the theta function is defined (for $z \in \mathbb{C}^N$) by

$$\Theta(z;B) \equiv \sum_{m \in \mathbb{Z}^N} \exp\{\pi i [2(m,z) + (m,Bm)]\} \qquad (\text{III.20b})$$

and $z(\Theta)$ denotes the linear map

$$z_j(\Theta) = \sum_{j=1}^{N} \varepsilon \frac{B^{(j)}}{2\pi} \Theta_j + d_j , \qquad (\text{III.20c})$$

where $B^{(j)}$ is the jth column of the period matrix B. Finally, the constant Γ is given by

$$\Gamma \equiv -2 \sum_{j=1}^{N} \oint_{a_j} \lambda \, \psi_j \qquad (\text{III.20d})$$

and $\vec{d}$ denotes a real constant which plays no role in the following.

In this manner the theory of the exact kdV equation has generated a solution (III.20a) in the form

$$q_N = q_N(\Theta_1, \ldots, \Theta_N; \vec{\lambda})$$

where the (y,τ) dependence enters only through the phases,

$$\Theta_j(y,t) = \kappa_j y + \omega_j \tau + \overset{\circ}{\Theta}_j ,$$

and where the wave form is 2π periodic in each individual phase. Thus, the constants κ_j and ω_j are enterpreted physically as spatial wave numbers and temporal frequencies.

This completes the summary of that material from the inverse spectral representation of exact kdV waves which we need for this paper. Now we return to the analysis of the sequence of problems (II.8).

IV. The Leading Order ($O(\varepsilon^{-1})$) Problem

In this section we construct solutions of (II.8a). Fix 2N+1 real constants $\vec{\lambda} = (\lambda_0 < \lambda_1 < \ldots < \lambda_{2N})$, and construct the B matrix and the differentials Ω_j as in (III.15) and (III.16). Define the wave vector $\vec{\kappa} = \vec{\kappa}(\vec{\lambda})$ and the frequency vector $\vec{\omega} = \vec{\omega}(\vec{\lambda})$ by

$$\vec{\kappa}_j(\lambda) = -\oint_{a_j} \Omega_1$$

$$\vec{\omega}_j(\lambda) = -12 \oint_{a_j} \Omega_2 \ .$$
(IV.1a,b)

For this $\vec{\kappa} = \vec{\kappa}(\lambda)$ and $\vec{\omega} = \vec{\omega}(\lambda)$, we seek $W: T^N$(N-Torus) $\to R$ which satisfies the $O(\varepsilon^{-1})$ problem:

$$(\omega \cdot \nabla)W - 6W(\kappa \cdot \nabla)W + (\kappa \cdot \nabla)^3 W = 0 \ . \tag{IV.2}$$

Using material in Section III, we find solutions of (IV.2):

$$W = W(\overset{o}{0}; \vec{\theta}, \lambda) = q_N(\vec{\theta} - \overset{o}{\vec{\theta}}; \lambda) \ , \tag{IV.3}$$

Equivalently, we may use the μ-representation:

$$W = W(\overset{o}{0}; \vec{\theta}, \lambda) = \Lambda - 2 \sum_{j=1}^{N} \mu_j(\vec{\theta} - \overset{o}{\vec{\theta}}; \lambda) \ , \tag{IV.4}$$

where the μ variables satisfy

$$(\kappa \cdot \nabla)\mu_j = -2i \frac{R(\mu_j)}{\prod_{i \neq j} (\mu_i - \mu_j)} \tag{IV.5a}$$

$$(\omega \cdot \nabla)\mu_j = -2i \left[2\Lambda - 2\sum_{i \neq j} \mu_i\right] \frac{R(\mu_j)}{\prod_{i \neq j} (\mu_j - \mu_i)} \ . \tag{IV.5b}$$

In this manner we generate $3N+1$ real parameter family of solutions $W: T^N \to R$ of (IV.2). N of the parameters, $\overset{o}{\Theta} = (\overset{o}{\theta}_1, \ldots, \overset{o}{\theta}_N)$, are trivial in that they merely center the N phases. The remaining $2N+1$ parameters $\lambda = (\lambda_0, \lambda_1, \ldots, \lambda_{2N})$ carry qualitative information about the wave. For example, they determine the wave numbers $\kappa = (\kappa_1, \ldots, \kappa_N)$ and the frequencies $\omega = (\omega_1, \ldots, \omega_N)$ by (IV.1). In addition, they determine the mean of the wave W:

$$<W> = \frac{1}{(2\pi)^N} \int_{T^N} W(\Theta) \, d^N\Theta \ . \tag{IV.6}$$

Remark (i) Physically, it seems more natural to co-ordinatize the N phase waves by the N spatial wave numbers $\vec{\kappa}$, N temporal frequencies $\vec{\omega}$, and mean $<W>$ rather than by the $2N+1$ parameters λ. However, the λ co-ordinates are better understood mathematically.

Remark (ii) For the single phase (N=1) case, equation (IV.2) is an ordinary differential equation of third order which is easy to analyse. One quickly shows that all of its 2π periodic solutions belong to the 4 parameter family (IV.3) with parameters $(\lambda_0, \lambda_1, \lambda_2, \overset{o}{\Theta})$.

Remark (iii) For N>1, I suspect that all solutions of (IV.2) which genuinely depend upon all N phases ($\frac{\partial W}{\partial \theta_j} \neq 0$ for any j) belong to family (IV.3). For, assume there exists $W: T^N \to R$ which solves (IV.2), depends genuinely upon all N phases, and does not belong to family (IV.3). Then $W(y,\tau) = W(\kappa_1 y + \omega_1 \tau, \ldots, \kappa_N y + \omega_N \tau)$ is a solution of the kdV equation which is quasi-periodic in y and τ with exactly N frequencies, and yet is not an "N-gap potential" for inverse spectral theory. I think that no such solution exists.

Remark (iv) Notice that in this theory it is easy to shut off a phase, say θ_j. One seeks a W which is independent of θ_j. If κ_j and ω_j go to infinity, this situation is forced upon us. On the other hand, if κ_j and ω_j vanish (a soliton limit), $\partial/\partial\theta_j$ is removed from equation (IV.2). In any case, such situations must be understood before a modulation theory sufficiently general to create and destroy phases can be developed. Such generality is needed for a uniformly valid approximation.

V. Solvability Theory for the Linear Problems

Fix W, a solution of (IV.2) in family (IV.3). All of the $O(\epsilon^j)$, $j \geq 0$, problems (II.8) are of the form

$$LU + F = 0, \qquad (V.1)$$

with a prescribed inhomogeneity $F: T^N \to \mathbb{R}$. Here the linear operator L is defined in terms of W by

$$LU \equiv (\omega \cdot \nabla)U - 6(\kappa \cdot \nabla) W U + (\kappa \cdot \nabla)^3 U. \qquad (V.2)$$

(We work in the Hilbert space of functions $U: T^N \to R$ which are square integrable over the torus T^N.)

For the solvability theory of (V.1), we need to understand $n(L)$, the null space of L, and $n(L^t)$, the null space of the adjoint of L. Here the formal adjoint L^t is given by

$$L^t V = -(\omega \cdot \nabla)V + 6W(\kappa \cdot \nabla)V - (\kappa \cdot \nabla)^3 V. \tag{V.3}$$

We have the following fact concerning the null space $n(L^t)$:

Theorem V.1: (a) $\psi^{(N)} \equiv \prod_{j=1}^{N}(\lambda - \mu_j) \in n(L^t) \;\forall\; \lambda$

(b) $\psi \equiv \dfrac{\psi^{(N)}}{R(\lambda)} \simeq \sum_{j=0}^{\infty} \psi_j (2\lambda)^{-j-1/2} \in n(L^t) \;\forall\; \lambda$

(c) $\psi_j \in n(L^t) \;\forall\; j = 0,1,\ldots$. (V.4a,b,c)

The proof of this theorem follows immediately from the material around (III.14), together with the fact that $\mu_j(y,t)$ depends upon (y,t) only through the phases $\theta_j(y,t) = \kappa_j y + \omega_j t$.

Theorem V.2: Formulas (V.4) generate only $(N+1)$ linearly independent members of $n(L^t)$. These may be represented as

(a) $\{\sigma_0, \sigma_1, \ldots, \sigma_N\}$, where $\prod_{j=1}^{N}(\lambda - \mu_j) = \sum_{j=0}^{N} \sigma_j \lambda^j$,

(V.5a,b)

(b) $\{\psi_0, \psi_1, \ldots, \psi_N\}$.

Theorem V.3: (a) $\dfrac{\partial W}{\partial \theta_j} \in n(L) \;\forall\; j = 1, 2, \ldots, N$

(b) $1 - 6(\kappa \cdot \nabla_\omega) W \in n(L)$.

Using formulas (V.5), we have $(N+1)$ solvability conditions which are necesary if (V.1) is to have a solution. For $N=1$, simple analysis of the ordinary differential equation shows that these are actually necessary and sufficient. We have

Theorem V.4: Let $N = 1$. Then

(a) $n(L^t) = \text{span}\,\{1, W\}$ (V.7)

(b) $n(L) = \text{span}\,\{W_\theta,\; 1 - 6\kappa W_\omega\}$

Thus, in the N=1 case, both $\eta(L)$ and $\eta(L^\dagger)$ have dimension 2.

Remark   For $N > 1$, we suspect that $\eta(L)$ and $\eta(L^\dagger)$ have dimension N+1. If so, $\eta(L^\dagger) = \text{span}\{\psi_0,\ldots,\psi_N\}$. We have not succeeded in proving this. For now, the solvability conditions

$$(\psi_j, F) = 0, \text{ for } j = 0,1,\ldots, N \tag{V.8}$$

are necessary for $N>1$; necessary and sufficient for $N=1$.

VI. The $O(\epsilon^0)$ Problem

Armed with this solvability theory, we return to the sequence of linear problems (II.8b). Explicitly, the $O(\epsilon^0)$ problem is

$$LU^{(0)} + \tilde{F}^{(0)} + \tilde{\tilde{F}}^{(0)} = 0, \tag{VI,1a}$$

$$\tilde{F}^{(0)} = [W_t - 6WW_x + 3((\kappa\cdot\nabla)^2 W_x + (\kappa\cdot\nabla)(\kappa_x\cdot\nabla)W)$$

$$+ \beta f(W, \kappa\cdot\nabla W, \ldots)] \tag{VI,1b}$$

$$\tilde{\tilde{F}}^{(0)} = [(\omega^{(1)}\cdot\nabla)W - 6W\,\kappa^{(1)}\cdot\nabla W + 3(\kappa\cdot\nabla)^2(\kappa^{(1)}\cdot\nabla)W].$$

In the source, one does not have to worry about $\tilde{\tilde{F}}^{(0)}$ because for this part the source, we have an explicit solution. Recall that W satisfies

$$(\omega\cdot\nabla)W - 6W(\kappa\cdot\nabla)W + (\kappa\cdot\nabla)^3 W = 0.$$

Define 2N functions on the torus T^N by

$$X^{(\omega_j)} \equiv \frac{\partial W}{\partial \omega_j}, \tag{VI,2}$$

$$X^{(\kappa_j)} \equiv \frac{\partial W}{\partial \kappa_j}, \quad j = 1,2,\ldots N.$$

Then, by differentiating the W equation, one finds

$$LX^{(\omega_j)} + \frac{\partial}{\partial\theta_j} W = 0$$

$$LX^{(\kappa_j)} - 6W\frac{\partial}{\partial\theta_j} W + 3(\kappa\cdot\nabla)^2 \frac{\partial}{\partial\theta_j} W = 0. \tag{VI,3}$$

Thus, using (VI,3) we may write $U^{(0)}$ as follows:

$$U^{(0)} = \sum_{j=1}^{N} [\omega_j^{(1)} \chi^{(\omega_j)} + \kappa_j^{(1)} \chi^{(\kappa_j)}] + \tilde{U}^{(0)} \tag{VI,4a}$$

where

$$L \tilde{U}^{(0)} + \tilde{F}^{(0)} = 0 \tag{VI,4b}$$

Thus, the solvability theory need only treat $\tilde{F}^{(0)}$.

Theorem (V,2) provides $N+1$ necessary conditions for solvability,

$$(\psi_j, \tilde{F}^{(0)}) = 0 \quad \text{for} \quad j = (0,1,\ldots,N) . \tag{VI,5}$$

In this manner, we arrive at the modulation equations which must be satisfied:

$$\partial_t \kappa_j = \partial_x \omega_j \quad , \quad j = 1,2,\ldots,N \tag{VI,6}$$

$$(\psi_j, W_t - 6WW_x + 3[(\kappa \cdot \nabla)^2 W_x + (\kappa \cdot \nabla)(\kappa_x \cdot \nabla)W] + \beta f(W, \kappa \cdot \nabla W, ..)) = 0,$$
$$j = 0,1,\ldots,N .$$

Equations (VI,6) are the main result of this section. They provide a system of $2N+1$ first order partial differential equations for the $2N+1$ parameters $\vec{\lambda} = (\lambda_0, \ldots, \lambda_{2N})$. The first N equations result from consistency of the ansatz; the last $N+1$ from necessary solvability conditions. Equations (VI,6) provide a closed system which depends only upon $\vec{\lambda}, \vec{\lambda}_x, \vec{\lambda}_t$. In the next section, we place this system in manageable form.

If the null space is exactly $N+1$ dimensional (as we know for $N=1$ and suspect for $N>1$), these solvability conditions are necessary and sufficient to ensure the existence of $U^{(1)} : T^N \to R$. At this stage, one could proceed in two different directions. (i) One can continue to generate higher order terms $U^{(j)}$ in the expansion of U^ε. The corrections to the frequencies ω_j^ε, the wave numbers κ_j^ε, and the mean $\langle U^\varepsilon \rangle$ will provide sufficient freedom to ensure solvability at each stage. In this manner, a formal asymptotic expansion of the form $U^\varepsilon \sim W_N + \varepsilon U^{(1)} + \varepsilon^2 U^{(2)} + \ldots$ could be constructed.
(ii) Alternatively, one could truncate the expansion at $W_N(\frac{\Theta(x,t)}{\varepsilon}); \vec{\lambda}(x,t))$ where the $\vec{\lambda}(x,t)$ satisfy the modulation equations (VI,6) and attempt to prove th

$u^\epsilon - W_N$ is $O(\epsilon)$ with some uniformity. The second direction is the most important.

VII. Connection Between the Modulation Equation and Averaged Conservation Laws

The modulation equations

$$\partial_t \kappa_j = \partial_x \omega_j, \qquad j = 1,2,\ldots,N$$

$$(\psi_j, W_t - 6WW_x + 3[(\kappa\cdot\nabla)^2 W_x + (\kappa\cdot\nabla)(\kappa_x\cdot\nabla)W] + \beta f(W, \kappa\cdot\nabla W,\ldots)) = 0, \qquad \text{(VII.1a,b)}$$
$$j = 0,1,2,\ldots,N,$$

although a closed system for $\vec{\lambda}(x,t)$, appear to be a complicated system of nonlinear partial differential equations. In [3] we show that, even in the presence of an external perturbation f, these modulation equations are actually very tractable provided (VII.1b) can be replaced by $N+1$ averaged conservation laws. In this section we derive the validity of this replacement for $N = 1,2$. (This is sufficient to treat 2 phase waves.)

VII.A. Averaged Conservation Laws

One approach to deriving modulation equations is to average conservation laws, a procedure which we can now describe. One <u>assumes</u> that the exact equation,

$$U_t - 6UU_x + \epsilon^2 U_{xxx} + \beta f(U, \epsilon U_x, \epsilon^2 U_{xx}, \ldots) = 0, \qquad \text{(VII.1)}$$

has a solution of the form

$$U \sim W_N\left(\frac{\Theta_1(x,t)}{\epsilon}, \ldots, \frac{\Theta_N(x,t)}{\epsilon}\right) \vec{\lambda}(x,t)) + O(\epsilon), \qquad \text{(VII.2)}$$

which is 2π periodic in each phase. In addition, it has conservation laws of the form

$$\partial_t T(U, \epsilon U_x, \ldots) + \partial_x X(U, \epsilon U_x, \ldots) + \beta G(U, \epsilon U_x, \ldots) = 0 \qquad \text{(VII.3)}$$

for any solution of (VII.1). In particular, evaluating on solution (VII.2), conservation law (VII.3) takes the form ($\partial_t \to \frac{1}{\epsilon}\omega\cdot\nabla + \partial_t$, etc.)

$$\frac{1}{\epsilon} [(\omega \cdot \nabla) T + (\kappa \cdot \nabla) X] \tag{VII.4}$$

$$+ \left[\frac{\partial}{\partial t} T(W_N, (\kappa \cdot \nabla) W_N, \ldots) + \frac{\partial}{\partial x} X(W_N, (\kappa \cdot \nabla) W_N, \ldots) + \beta G(W_N, (\kappa \cdot \nabla) W_N, \ldots) \right]$$

$$+ \ldots = 0 .$$

When averaged over T^N, the $O(\epsilon^{-1})$ term averages to zero, and one is left with the averaged conservation law

$$\partial_t < T(W_N, (\kappa \cdot \nabla) W_N, \ldots) > + \partial_x < X(W_N, (\kappa \cdot \nabla) W_N, \ldots) > \tag{VII.5}$$

$$+ \beta < G(W_N, (\kappa \cdot \nabla) W_N, \ldots) > = 0 .$$

This is one equation among $2N+1$ unknowns.

Each kdV conservation law will lead to an averaged conservation law of the form (VII.5). Indeed, consider j^{th} kdV density as generated by the recursion relation (III.5), $\psi_j = \psi_j(U, \epsilon \partial_x U, \ldots, (\epsilon \partial_x)^{k_j} U)$. Here k_j is the order of the highest derivative of U in ψ_j. We compute:

$$\partial_t \psi_j = \sum_{\ell=0}^{k_j} \psi_{j,\ell} \, \partial_t (\epsilon \partial_x)^\ell U$$

$$= \sum_{\ell=0}^{k_j} \psi_{j,\ell} (\epsilon \partial_x)^\ell [6UU_x - \epsilon^2 U_{xxx} - \beta f(U, \epsilon U_x, \ldots)]$$

$$= - \partial_x X_j - \beta \sum_{\ell=0}^{k_j} \psi_{j,\ell} (\epsilon \partial_x)^\ell f(U, \epsilon U_x, \ldots),$$

where $\psi_{j,\ell}$ denotes a partial derivative. Thus, we obtain the pertured conservation law

$$\partial_t \psi_j + \partial_x X_j = - \beta \sum_{\ell=0}^{k_j} \psi_{j,\ell} (\epsilon \partial_x)^\ell f(U, \epsilon U_x, \ldots),$$

with the right hand side explicitly given in terms of the jth density and the external perturbation f. Evaluating this conservation law on the wave form (VII.2), and averaging over the torus T^N we derive the perturbed avaraged conservation law

$$\partial_t \langle \psi_j \rangle + \partial_x \langle 6(W\psi_j - \psi_{j+1}) \rangle + \beta \langle G_j \rangle, \qquad (\text{VII.6a})$$

where

$$\langle G_j \rangle \equiv \sum_{\ell=0}^{k_j} (\psi_{j,\ell}, (\kappa \cdot \nabla)^\ell f(W, \kappa \cdot \nabla W, \ldots)), \qquad (\text{VII.6b})$$

and where the densities ψ_j are evaluated on the N phase wave form W. For later use, we list the first three:

$$\partial_t \langle W \rangle - \partial_x \langle 3W^2 \rangle + \beta \langle f \rangle = 0$$

$$\partial_t \langle \frac{W^2}{2} \rangle - \partial_x \langle 2W^2 + \frac{3}{2}(\kappa \cdot \nabla W)^2 \rangle + \beta \langle wf \rangle = 0 \qquad (\text{VII.7a,b,c})$$

$$\partial_t \langle W^3 + \frac{1}{2}(\kappa \cdot \nabla W)^2 \rangle - \partial_x \langle \frac{9}{2} W^4 + 12W(\kappa \cdot \nabla W)^2 + \frac{3}{2}((\kappa \cdot \nabla)^2 W)^2$$

$$+ \beta \langle 3W^2 f + (\kappa \cdot \nabla) W (\kappa \cdot \nabla) f \rangle = 0 \quad .$$

VII.B. Connection between Averaged Conservation Laws and Null Space Modulation Equations

Consider the null space equations (VI.5),

$$(\psi_j, \hat{F}^{(0)}) = 0, \qquad j = 0,1,2,\ldots, N,$$

$$\hat{F}^{(0)} = W_t - 6WW_x + 3((\kappa \cdot \nabla)^2 W_x + (\kappa \cdot \nabla)(\kappa_x \cdot \nabla)W) + \beta f(W, (\kappa \cdot \nabla)W, \ldots) \quad . \qquad (\text{VII.8a,b})$$

We show the first three of these are equivalent to averaged conservation laws; indeed, they become (VII.6) for j = 1,2,3. Since $\psi_0 = 1$, the first is immediate:

$$0 = (\psi_0, \hat{F}^{(0)}) = (1, \hat{F}^{(0)})$$

$$= \partial_t \langle W \rangle - \partial_x \langle 3W^2 \rangle + \beta \langle f \rangle$$

The next is almost as simple:

$$0 = (\psi_1, \hat{F}^{(0)}) = (W, \hat{F}^{(0)})$$

$$= (W, W_t) - (W, 6WW_x) + 3(W, (\kappa \cdot \nabla)^2 W_x + (\kappa \cdot \nabla)(\kappa_x \cdot \nabla)W) + \beta(W,f)$$

$$= \frac{1}{2} \partial_t \langle W^2 \rangle - 2 \partial_x \langle W^3 \rangle - 3(\kappa \cdot \nabla W, \partial_x (\kappa \cdot \nabla)W) + \beta(W,f)$$

$$= \partial_t \langle \frac{W^2}{2} \rangle - \partial_x \langle 2W^3 + \frac{3}{2}(\kappa \cdot \nabla W)^2 \rangle + \beta \langle W f \rangle.$$

The verification that the third null space equation yields the third averaged conservation law is more tedious. It uses an extra ingredient, $\partial_t \kappa_j = \partial_x \omega_j$. This calculation begins as follows:

$$
\begin{aligned}
0 &= 2(\psi_2, F^{(0)}) = (3W^2 - (\kappa\cdot\nabla)^2 W, F^{(0)}) \\
&= (3W^2 - (\kappa\cdot\nabla)^2 W, W_t - 3(W^2)_x + 3(\kappa\cdot\nabla)^2 W_x + 3(\kappa\cdot\nabla)(\kappa_x\cdot\nabla)W + \beta f) \\
&= (3W^2, W_t) - ((\kappa\cdot\nabla)^2 W, W_t) - 9(W^2, (W^2)_x) + 3((\kappa\cdot\nabla)^2 W, (W^2)_x) \\
&\quad + 9(W^2, (\kappa\cdot\nabla)^2 W_x) - 3((\kappa\cdot\nabla)^2 W, (\kappa\cdot\nabla)^2 W_x) \\
&\quad + 9(W^2, (\kappa\cdot\nabla)(\kappa_x\cdot\nabla)W) - 3((\kappa\cdot\nabla)^2 W, (\kappa\cdot\nabla)(\kappa_x\cdot\nabla)W) \\
&\quad + 3\beta(W^2, f) - \beta((\kappa\cdot\nabla)^2 W, f) \\
&= \partial_t \langle W^3 \rangle + ((\kappa\cdot\nabla)W, (\kappa\cdot\nabla)W_t) - \tfrac{9}{2}\partial_x\langle W^4 \rangle - 3((\kappa\cdot\nabla)W, (\kappa\cdot\nabla)(W^2)_x) \\
&\quad - 9((\kappa\cdot\nabla)W^2, (\kappa\cdot\nabla)W_x) - 3((\kappa\cdot\nabla)^2 W, (\kappa\cdot\nabla)^2 W_x) - 9((\kappa\cdot\nabla)W^2, (\kappa_x\cdot\nabla)W) \\
&\quad - 3((\kappa\cdot\nabla)^2 W, (\kappa\cdot\nabla)(\kappa_x\cdot\nabla)W) + \beta\langle f[3W^2 - (\kappa\cdot\nabla)^2 W]\rangle \\
&= \partial_t \langle W^3 \rangle + \partial_t \left\langle \tfrac{[(\kappa\cdot\nabla)W]^2}{2} \right\rangle - ((\kappa\cdot\nabla)W, (\kappa_t\cdot\nabla)W) - \partial_x \langle \tfrac{9}{2} W^4 \rangle \\
&\quad - 3((\kappa\cdot\nabla)^2 W, (\kappa\cdot\nabla)^2 W_x) - 3((\kappa\cdot\nabla)^2 W, (\kappa\cdot\nabla)(\kappa_x\cdot\nabla)W) \\
&\quad - 3((\kappa\cdot\nabla)W, (\kappa\cdot\nabla)(W^2)_x) - 9((\kappa\cdot\nabla)W_x, (\kappa\cdot\nabla)W^2) \\
&\quad - 9((\kappa\cdot\nabla)W^2, (\kappa_x\cdot\nabla)W) + \beta\langle f[3W^2 - (\kappa\cdot\nabla)^2 W]\rangle
\end{aligned}
$$

Continuing,

$$
\begin{aligned}
0 &= \partial_t \langle W^3 + \tfrac{1}{2}(\kappa\cdot\nabla W)^2 \rangle - \partial_x \langle \tfrac{9}{2} W^4 \rangle + \beta\langle f[3W^2 - (\kappa\cdot\nabla)^2 W]\rangle \\
&\quad - ((\kappa\cdot\nabla)W, (\kappa_t\cdot\nabla)W) - \partial_x \langle \tfrac{3}{2}[(\kappa\cdot\nabla)^2 W]^2 \rangle \\
&\quad + 3((\kappa\cdot\nabla)^2 W, (\kappa\cdot\nabla)(\kappa_x\cdot\nabla)W) \\
&\quad - \partial_x \langle 12W(\kappa\cdot\nabla W)^2 \rangle + 12(W_x, (\kappa\cdot\nabla W)^2) + 24(W, [(\kappa\cdot\nabla)W](\kappa\cdot\nabla)W_x) \\
&\quad + 24(W, [(\kappa\cdot\nabla)W](\kappa_x\cdot\nabla)W) - 6((\kappa\cdot\nabla)W, W(\kappa\cdot\nabla)W_x) - 6([(\kappa\cdot\nabla)W]^2, W_x) \\
&\quad - 18((\kappa\cdot\nabla)W_x, W(\kappa\cdot\nabla)W) - 18(W(\kappa\cdot\nabla)W, (\kappa_x\cdot\nabla)W).
\end{aligned}
$$

Therefore,

$$
\begin{aligned}
\partial_t \langle W^3 &+ \tfrac{1}{2}(\kappa\cdot\nabla W)^2 \rangle - \partial_x \langle \tfrac{9}{2} W^4 + 12W(\kappa\cdot\nabla W)^2 \\
&+ \tfrac{3}{2}[(\kappa\cdot\nabla)^2 W]^2 \rangle + \beta\langle f[3W^2 - (\kappa\cdot\nabla)^2 W]\rangle \\
&= ((\kappa\cdot\nabla)W, (\kappa_t\cdot\nabla)W) - 3((\kappa\cdot\nabla)^2 W, (\kappa\cdot\nabla)(\kappa_x\cdot\nabla)W) \\
&\quad - ((\kappa\cdot\nabla)W, 6W_x(\kappa\cdot\nabla)W + 6(\kappa_x\cdot\nabla)W) \quad\quad\quad\quad\quad\quad\quad\quad\text{(VII.9)} \\
&= ((\kappa\cdot\nabla)W, (\kappa_t\cdot\nabla)W + 3(\kappa\cdot\nabla)^2(\kappa_x\cdot\nabla)W \\
&\quad - 6W_x(\kappa\cdot\nabla)W - 6W(\kappa_x\cdot\nabla)W) .
\end{aligned}
$$

Thus, the third null space condition will yield the third averaged conservation law provided the expression

$$((\kappa\cdot\nabla)W, (\kappa_t\cdot\nabla)W + 3((\kappa\cdot\nabla)^2(\kappa_x\cdot\nabla)W - 6W_x(\kappa\cdot\nabla)W - 6W(\kappa_x\cdot\nabla)W) \quad (VII.10)$$

vanishes. To reach this point, we merely integrated the null space condition by parts several times. Now we employ the first N modulation equations

$$\kappa_t = \omega_x$$

to replace expression (VII.10) by

$$((\kappa\cdot\nabla)W, (\omega_x\cdot\nabla)W + 3(\kappa\cdot\nabla)^2(\kappa_x\cdot\nabla)W - 6W_x(\kappa\cdot\nabla)W - 6W(\kappa_x\cdot\nabla)W). \quad (VII.10')$$

To show this expression vanishes, we use the equation satisfied by the N phase wave:

$$(\omega\cdot\nabla)W - 6W(\kappa\cdot\nabla)W + (\kappa\cdot\nabla)^3W = 0.$$

Differentiating this equation with respect to x yields

$$(\omega\cdot\nabla)W_x - 6W(\kappa\cdot\nabla)W_x - 6W_x(\kappa\cdot\nabla)W + (\kappa\cdot\nabla)^3W$$
$$+ (\omega_x\cdot\nabla)W - 6W(\kappa_x\cdot\nabla)W + 3(\kappa\cdot\nabla)^2(\kappa_x\cdot\nabla)W = 0.$$

Finally, take the inner product with $(\kappa\cdot\nabla)W$:

$$((\kappa\cdot\nabla)W, (\omega\cdot\nabla)W_x - 6W(\kappa\cdot\nabla)W_x - 6W_x(\kappa\cdot\nabla)W + (\kappa\cdot\nabla)^3W_x$$
$$+ (\omega_x\cdot\nabla)W - 6W(\kappa_x\cdot\nabla)W + 3(\kappa\cdot\nabla)^2(\kappa_x\cdot\nabla)W) = 0$$

That is,

$$((\kappa\cdot\nabla)W_x, (\omega\cdot\nabla)W - 6W(\kappa\cdot\nabla)W + (\kappa\cdot\nabla)^3W)$$
$$+ ((\kappa\cdot\nabla)W, (\omega_x\cdot\nabla)W + 3(\kappa\cdot\nabla)^2(\kappa_x\cdot\nabla)W - 6W_x(\kappa\cdot\nabla)W - 6W(\kappa_x\cdot\nabla)W) = 0.$$

The first term vanishes by the fact that W solves the N phase equation; thus, we arrive at the vanishing of (VII.10')!

In summary, we have shown that (i) the first two null space equations imply the first two averaged conservation laws, and that (ii) the third null space equation, together with the consistency conditions $\kappa_t = \omega_x$, imply the third averaged conservation law.

Remark (i). Presumably, the jth null space equation, together with the consistency conditions, implies the jth averaged conservation law. To prove this statement, one needs a more abstract argument than the explicit calculation described above. As yet, we have not succeeded.

VIII. An Invariant Form of the Modulation Equations

In this section, we assume the correct modulation equations are of the form

$$\partial_t \kappa_j = \partial_x \omega_j \quad j = 1,2,\ldots,N$$

(VIII.1a,b)

$$\partial_t <\psi_j> + \partial_x <\chi_j> + \beta <G_j> \quad j = 1,2,\ldots, N+1 ,$$

and we summarize some results of [3] in order to emphasize that these modulation equations are indeed tractable.

In [3], we establish the following

<u>Theorem</u> The modulation equations (VIII.1) admit an equivalent representation in terms of the differentials Ω_1, Ω_2 :

$$\partial_t \Omega_1 - 12 \partial_x \Omega_2 + \beta \, dF = 0, \quad \text{(VIII.2)}$$

where F is a meromorphic function on the Riemann surface R of the form

$$F(\lambda) = \frac{A(\lambda)}{R(\lambda)}, \quad A(\lambda) = \sum_{j=0}^{N} \alpha_j \lambda^j . \quad \text{(VIII.3a)}$$

The coefficients α_j are fixed in terms of $<G_j>$ by the linear system

$$\sum_{k=0}^{j} \rho_{j-k} \left[(N-k) \alpha_{N-k} - \frac{1}{2} \sum_{m=0}^{k} \left(\sum_{\ell=0}^{2N} \lambda_\ell^{k-m} \right) \alpha_{N-m} \right] = \frac{1}{2^{j+1}} <G_{j+1}>, \quad j = 0,1,2,\ldots N.$$

(VIII.3b)

Here

$$\frac{-2}{\sqrt{\prod_{\ell=0}^{2N} (1-\lambda_\ell \xi^2)}} = \sum_{k=0}^{\infty} \rho_k \xi^{2k} .$$

Representation (VIII.2) is fundamental. It contains alternative representations and quickly shows they are equivalent. For example, the most useful mathematical form is an immediate

Corollary (Riemann Invariant Form of the Modulation Equations).

By evaluating the invariant representation $\partial_t \Omega_1 - \partial_x \Omega_2 + \beta dF = 0$ at the branch points, one obtains

$$\partial_t \lambda_\ell + [\, s^{(\ell)}(\vec{\lambda})\,]\partial_x \lambda_\ell = \beta \frac{A(\lambda_\ell)}{\sum_{j=1}^{N+1} D_j \lambda_\ell^{j-1}} \quad , \quad \ell = 0,1,\ldots,2N, \qquad \text{(VIII.4a)}$$

where the ℓth characteristic speed $s^{(\ell)}(\vec{\lambda})$ is given by

$$s^{(\ell)}(\vec{\lambda}) \equiv \frac{-12 \sum_{j=1}^{N+2} E_j \lambda_\ell^{j-1}}{\sum_{j=1}^{N+1} D_j \lambda_\ell^{j-1}}. \qquad \text{(VIII.4b)}$$

A fully nonlinear modulation theory cannot be simpler than the Riemann invariant form (VIII.4a). We emphasize the "internal perturbations' described in the introduction cause the parameters to modulate with the characteristic speeds $s^{(\ell)}(\vec{\lambda})$. The external perturbation f provides the right hand side of the modulation equations.

IX. The Weak Limit as a Measure

Recently, in some very interesting mathematical work [19,20,10], weak limits of solutions of nonlinear pde's have been described by a measure. When the nonlinear pde is dissipation dominated, as in Burgers' equation, the measure is simply a Dirac measure supported on the shock. When oscillations persist as in the small dispersion limit of KdV, the weak limit is more interesting. As yet, the measure has not been rigorously characterized.

Our purpose in this final section is to calculate, with heuristic reasoning, the measure which describes the weak limit as $\varepsilon \to 0$ of problem (II.1a), in a region of space-time where the solution is described by a modulating N-phase wave.

Since modulation theory constructs the solution for small but finite ε, our representation certainly contains enough information to calculate the measure very explicitly.

Consider the solution of

$$U_t - 6UU_x + \varepsilon^2 U_{xxx} + \beta f(U, \varepsilon U_x, \ldots) = 0, \qquad (IX.1)$$

as constructed in this paper. Namely, in a region S of space-time, let U^ε be the modulating N phase wave

$$U^\varepsilon(x,t) = W_N \left(\frac{\Theta_1(x,t)}{\varepsilon}, \ldots, \frac{\Theta_N(x,t)}{\varepsilon}; \vec{\lambda}(x,t) \right)$$

$$+ \varepsilon U^{(1)} \{\frac{\vec{\Theta}}{\varepsilon}; x,t\} + \ldots, \qquad (IX.2)$$

with the modulation of the parameters $\vec{\lambda}(x,t)$ described by

$$\partial_t \Omega_1 - 12 \partial_x \Omega_2 + \beta dF = 0 \qquad (IX.3)$$

Fix some t and a spatial interval I such that $(t,I) \in S$; further, let $\phi: R \to R$ denote any test function with support within I. Fix $f: R \to R$ [1] and consider

$$(\phi, f(U^{(\varepsilon)}(\cdot,t))) = \int_{-\infty}^{\infty} \phi(x,t) f(U^{(\varepsilon)})(x,t) \, dx,$$

which one must consider in order to describe the weak limit $f(U^\varepsilon(\cdot,t))$ as $\varepsilon \downarrow 0$.

---

1) The function f could be of the form $f[U, \varepsilon U_x, \varepsilon^2 U_{xx}, \ldots]$.

We compute:

$$(\phi, f(U^{(\epsilon)}(\cdot,t))) = \int_{-\infty}^{\infty} \phi(x,t) \, f(U^{(\epsilon)}(x,t)) dx$$

$$= \sum_{j=-\infty}^{\infty} \int_{x_j-\Delta/2}^{x_j+\Delta/2} \phi(x,t) \, f(U^{(\epsilon)}(x,t)) dx$$

$$\simeq \sum_{j=-\infty}^{\infty} \int_{x_j-\Delta/2}^{x_j+\Delta/2} \phi(x,t) \, f[W_N(\frac{\Theta(t,x)}{\epsilon}; \vec{\lambda}(t,x))] dx$$

$$\simeq \sum_{j=-\infty}^{\infty} \phi(x_j,t) \int_{x_j-\Delta/2}^{x_j+\Delta/2} f[W_N(\frac{\Theta(t,x)}{\epsilon}; \vec{\lambda}(t,x_j))] dx$$

$$\quad (\Delta \text{ tiny})$$

$$= \sum_{j=-\infty}^{\infty} \phi(x_j,t) \, \Delta \, \{\frac{1}{\Delta} \int_{x_j-\Delta/2}^{x_j+\Delta/2} f[W_N(\frac{\vec{k}(x_j,t)x - \vec{\beta}(x_j,t)}{\epsilon}); \vec{\lambda}(t,x_j)]\}$$

$$= \sum_{j=-\infty}^{\infty} \phi(x_j,t) \, \Delta \, \{\frac{\epsilon}{\Delta} \int_{-\Delta/2\epsilon}^{\Delta/2\epsilon} f[W_N(\vec{\kappa}y - \vec{\beta}; \vec{\lambda}(t,x_j)) dy\}$$

$$\simeq \sum_{j=-\infty}^{\infty} \phi(x_j,t) \, \Delta \, \{\frac{1}{(2\pi)^N} \int_{T^N} f[W_N(\theta_1,\ldots,\theta_N; \vec{\lambda}(x_j,t)] d^N\theta\}$$

$$(\epsilon \to 0 \text{ and ergodicity})$$

$$\simeq \int_{-\infty}^{\infty} \phi(x,t) < f(W_N; \vec{\lambda}(x,t)) > dx \quad (\Delta \to 0)$$

Thus, we compute

$$\lim_{\epsilon \downarrow 0} (\phi, f(U^{\epsilon}(\cdot,t))) = (\phi, <f>) \quad \forall \phi ;$$

that is,

$$f(U^{\epsilon}(x,t)) \xrightarrow[\epsilon \to 0]{} <f(W_N(\cdot; \vec{\lambda}(x,t))> \quad (IX.4)$$

$$= \frac{1}{(2\pi)^N} \int_{T^N} f[W_N(\theta; \vec{\lambda}(x,t))] \, d^N\theta.$$

The measure itself can be characterized very explicitly by using the μ-coordinates for the torus (see VII.18):

$$f(U^\varepsilon(x,t)) \xrightarrow[\varepsilon \to 0]{} \frac{1}{V} \oint_{b_1} \cdots \oint_{b_N} f[\Lambda(x,t) - \sum_{j=1}^{N} \mu_j]$$

$$\cdot \frac{\prod_{i>j}^{N} (\mu_i - \mu_j)}{\prod_{i=1}^{N} |R(\mu_i; \vec{\lambda}(x,t))|} d\mu_1 \wedge \cdots d\mu_N$$

(By translation, we can remove the function $\Lambda(x,t)$ from the argument of f.)

Formula (IX.5) is the main result of this section. It shows that the weak limits $\varepsilon \to 0$ can be characterized by a measure, and <u>gives an explicit formula for that measure</u>. Notice that the (x,t) dependence of the measure is through $\vec{\lambda}(x,t)$ which satisfies $\Omega_t \Omega_1 - 12 \partial_x \Omega_2 + BdF = 0$. This is precisely the measure used in [1].

<u>Remark</u>. Our calculation is limited to the region S where the solution is an N phase wave. It is not uniformly applicable. However, we believe it indicates that the general weak limit, which would be valid for all space, should be characterized by the quadratic variational problem of [8,9].

X. Conclusion

The work reported in this article was done during my visit to the Courant Institute in 1980-1982. It has not been published prior to these proceedings. Articles written since that time which are of related interest include [23] and [24]. The latter discusses the propagation of nonlnear oscillations in a rather general context.

References

1. H. Flaschka, M.G. Forest, and D.W. McLaughlin, "Multiphase averaging and the inverse spectral solution of the Korteweg de Vries equation," Comm. Pure Appl. Math. <u>33</u>, 1980, pp. 739-784.

2. D.W. McLaughlin, "Modulations of Kdv Wavetrains," Physica D<u>3</u>, 1981, pp. 335-343.

3. M.G. Forest and D.W. McLaughlin, "Modulations of Perturbed KdV Wavetrains," SIAM J. Appl. Math. <u>44</u>, 1984, 287-300.

4. M.G. Forest and D.W. McLaughlin, "Modulations of sinh-Gordon and sine-Gordon wavetrains," Stud. Appl. Math. 68, 1983, pp. 11-59.

5. N. Ercolani, M.G. Forest, and D.W. McLaughlin, "Modulational stability of two phase sine-Gordon wavetrains," Stuc. Appl. Math. 71, 1984, 91-101.

6. A.V. Gurevich and L.P. Pitaevskii, "Nonstationary Structure of a Collionsless Shock Wave," Sov. Phys. JETP 38, 1974.

7. B. Fornberg and G.B. Whitham, "A numerical and theoretical study of certain nonlinear wave phenomena," Phil. Trans. Roy. Soc. Lond. 289, 1978, pp. 373-404.

8. P.D. Lax and C.D. Levermore, "Zero dispersion limit for the KdV equation," Proc. Nat. Acad. Science (U.S.A.), 1979. Also, Comm. Pure. Appl. Math. 36, 1983, 253-290. Comm. Pure Appl. Math. 36, 1983, 571-594. Also Comm. Pure Appl. Math. 36, 1983, 809-829.

9. S. Venekides, Thesis, New York University, 1982.

10. M.E. Schonbek, "Convergence of solutions to nonlinear dispersive equations," preprint, U. Rhode Island, 1981.

11. G.B. Whitham, "Nonlinear dispersive waves," Proc. Roy. Soc. A 283, 1965, pp. 238-261.

12. G.B. Whitham, *Linear and Nonlinear Dispersive Waves*, Wiley-Interscience, New York, 1974.

13. R. Miura adn M. Kruskal, "Application of a nonlinear WKB method to the Korteweg de Vries equation," SIAM J. Appl. Math. 26, 1974, pp. 376-395.

14. J.C. Luke, "A perturbation method for nonlinear dispersive wave problems," Proc. Roy. Soc. A 292, 1966 (403-412).

15. M.J. Albowitz and D.J. Benney, "The evolution of multi-phase modes for nonlinear dispersive waves," Stud. Appl. Math. 49, 1970, pp. 225-238.

16. Jiminez, Thesis, Cal Tech, 1973.

17. E.N. Pelinovsky and S. Kh. Sharvatsky, "Breaking of stationary waves in nonlinear dispersive media," Physica D 3, 1980, pp. 317-328.

18. R. Rosales, private notes.

19. L. Tartar, "Compensated compactness and applications to partial differential equations," *Nonlinear Analysis and Mechanics*: Heriot-Watt Symposium, Vol IV, Research Notes in Mathematics 39, R.S. Knops, Ed., Pitman Publishing, 1979.

20. R. Diperna, "Measure valued solutions to conservation laws," Duke Univ. Preprint (1984).

21. H.P. McKean and P. van Moerbeke, "The spectrum of Hill's equation," Invent. Math. 30, 1975, 11, 217-274.

22. I.M. Gel'fand and L.A. Dikii, "Integrable nonlinear equations and the Liouville theorem," Funkt. Analiz. Egr. Prilozheniya 13, 1979, pp. 8-20.

23. S. Venekides, these proceedings.

24. N. Ercolani, M. Forest, and D.W. McLaughlin, "Oscillations and Instabilities in Near Integrable PDE's" proc. of Sante Fe Conference on Evolution Equation, 1985 (to appear).

EVIDENCE OF NONUNIQUENESS AND OSCILLATORY SOLUTIONS IN COMPUTATIONAL FLUID MECHANICS

J. W. Nunziato, D. K. Gartling, and M. E. Kipp
Sandia National Laboratories
Albuquerque, New Mexico 87185

1 Introduction

In the study of partial differential equations, analysis has traditionally focused on mathematical problems in one space dimension which are well-posed; that is, problems for which solutions exist, are unique, and depend continuously on the data (*i.e.*, stable problems). However, as we venture into multidimensions and begin to investigate the consequences of the nonlinear behavior of fluids, we find that the governing partial differential equations may change type at some point in the flow field. Furthermore, the solutions may be ill-behaved in the sense that they may not exist or they may have jump discontinuities such as shocks; the solutions may have turning points or bifurcate, ultimately leading to multiple solutions; or the solutions may be oscillatory and may become unstable in a given region. These types of solutions are currently receiving considerable attention and new, more sophisticated analysis techniques are required in order to discover the properties of these solutions.

In the meantime, due to the press of trying to solve technologically important problems, numerical analysts are attempting to solve a wide range of nonlinear partial differential equations using standard techniques. In this process they, too, are discovering a wide variety of ill-behaved solutions, and as a result, are confronted with answering the question: Is it the lack of robustness of the numerical method that has lead to an ill-behaved solution, or is it in fact a consequence of the mathematical description of the problem?

In this paper, we will review some of our recent experiences in computing solutions for nonlinear fluids in relatively simple, two-dimensional geometries. The purpose of this discussion will be to display by example some of the interesting but difficult questions that arise when ill-behaved solutions are obtained numerically. In particular, we will consider two examples. As the first example, we will consider a nonlinear elastic (compressible) fluid with chemical reactions and discuss solutions for detonation and detonation failure in a two-dimensional cylinder. In this case, the numerical algorithm utilizes a finite-difference method with artificial viscosity (von Neumann–Richtmyer method) and leads to two, distinctly different, stable solutions depending on the time step criterion used. Physically, this problem is characterized by two disparate time scales; one based on the acoustic transit time across a fluid element and the other associated with the rate of chemical decomposition of the element. The correct solution is obtained using a numerical stability criterion based on the smallest time scale. The second example to be considered involves the convection of a viscous fluid in a rectangular container as a result of an exothermic polymerization reaction. A solidification front develops near the top of the container and propagates down through the fluid, changing the aspect ratio of the region ahead of the front. Using a Galerkin-based finite element method, a numerical solution of the partial differential equations is obtained which tracks the front and correctly predicts the fluid temperatures near the walls. However, the solution also exhibits oscillatory behavior with regard to the number of cells in the fluid ahead of the front and in the strength of the cells. More definitive experiments and analysis are required to determine whether this oscillatory phenomena is a numerical artifact or a physical reality.

There are numerous other examples in computational fluid mechanics in which the solution is ill-behaved or may not even exist (lack of convergence). We want to emphasize that the two problems described above were selected for illustrative purposes because they involve very simple two-dimensional geometries and thus provide a very clear picture of the role of material nonlinearities and the numerical solution procedure. These particular problems were also selected since they provided examples of two distinctly different numerical procedures which have become fairly standard.

2 Detonation Failure in a Chemically Reacting Elastic Fluid

2.1 Physics of the Problem

The problem of detonation of a chemically reacting, elastic (compressible) fluid has been one of considerable interest for the past forty years. There is ample experimental evidence that, depending on the boundary conditions, it is possible to propagate a detonation wave without change of shape and structure for large distances and yet, in other cases, again depending on boundary conditions, there is evidence that the detonation will fail and the shock wave decay. For example, shown in Figure 1 is the result of a series of four experiments for the explosive fluid nitromethane in a cylindrical geometry [1]. Notice that when the cylinder is virtually unconfined, say with mylar (Figure 1d), the detonation extinguishes before propagating very far. Detonation failure can also occur in the glass tubes if the tube diameter is sufficiently small. Clearly, in the glass tube with the 15.5 mm diameter, there appears to be a beginning of failure on the left side of the cylinder, but then reignition occurs and detonation continues. For larger cylinder diameters, the detonation wave proceeds along the column with little difficulty. In the remainder of this section we will describe the mathematical model that corresponds to the physical picture of detonation of a chemically reacting, elastic fluid and discuss the numerical method we employed to study two-dimensional detonation phenomena (for more details, see Kipp and Nunziato [2]).

2.2 Mathematical Description

In studying detonation phenomena in chemically reacting fluids, we seek solutions in some region Ω of the conservation equations of mass, momentum, and energy which contain shock waves; *i.e.*, weak solutions. Assuming the fluid is a nonconductor of heat and neglecting body forces, the conservation equations can be expressed as

$$\dot{v} = \nabla \cdot \mathbf{u} , \tag{2.1}$$

$$\dot{\mathbf{u}} = -\nabla p , \tag{2.2}$$

$$\dot{e} = -p\dot{v} + w, \tag{2.3}$$

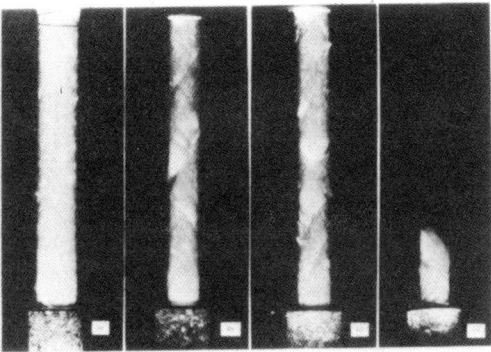

Figure 1. Open camera light records of detonation in pure nitromethane in 150 mm long tubes (a) glass confinement (i.d. 19.5 mm/o.d. 22 mm), (b) and (c) glass confinement (i.d. 15.5 mm/o.d. 17.9 mm), (d) Mylar confinement (i.d. 16.0 mm/o.d. 16.07 mm) [1].

respectively, where ∇ is the gradient operator with respect to the Lagrangian space coordinates which have been scaled by the specific volume of the fluid in the reference state. Also, in Equation (2.1)-(2.3), p is the pressure, e is the specific internal energy, **u** is the velocity vector, v is the specific volume, and w is the external heat supply. The superimposed dot indicates a material time derivative.

To account for the presence of chemical reactions, conservation of mass can be written for each constituent of the fluid mixture. Assuming a single uni-directional reaction

$$A \rightarrow B ,$$

the reaction can be characterized by the extent of reaction a, $0 \leq a \leq 1$, and conservation of mass yields the evolutionary equation

$$\dot{a} = r = \hat{r}(v, T, a) \tag{2.4}$$

where r is the reaction rate. Clearly, (2.4) describes the chemical kinetics of the system in a global sense. Finally, for consistency with thermodynamics, we require every solution of the conservation equations to satisfy the entropy inequality

$$\dot{S} \geq \frac{w}{T} \tag{2.5}$$

where S is the specific entropy and T is the absolute temperature.

For an elastic fluid with chemical reactions, we assume there exists a free energy function of the form

$$\psi = \hat{\psi}(v, T, a) . \tag{2.6}$$

Then the specific internal energy is given by

$$e = \psi + TS ,$$

and using standard thermodynamic arguments, it is not difficult to show that

$$p = -\frac{\partial \hat{\psi}}{\partial v}, \quad S = -\frac{\partial \hat{\psi}}{\partial T}. \tag{2.7}$$

The entropy inequality is satisfied if and only if

$$\mu r \leq 0 \tag{2.8}$$

where μ is the chemical potential,

$$\mu = \frac{\partial \hat{\psi}}{\partial a}.$$

It is important to note that by virtue of the thermodynamic formulas (2.7), it is possible to rewrite conservation of energy in terms of an evolutionary heat equation of the form

$$c_v \dot{T} = -TG\dot{v} - Qr + w \tag{2.9}$$

where c_v is the specific heat at constant volume, G is the pressure-temperature modulus, and Q is the heat of thermal decomposition:

$$c_v = \frac{\partial \hat{e}}{\partial T}, \quad G = \frac{\partial \hat{p}}{\partial T}, \quad Q = \frac{\partial \hat{e}}{\partial a}.$$

Also, by virtue of these results, the entropy growth in any problem must satisfy

$$T\dot{S} = -\mu r + w \tag{2.10}$$

Clearly, if the external heat supply w vanishes, then (2.8) and (2.10) assert that the entropy must be a monotone increasing function of time and will have a stationary point (usually a maximum) when the reaction rate vanishes.

It remains only to specify the form of the reaction rate r for the fluid. For example, the thermal decomposition of nitromethane is characterized by two parallel reactions of second order. Thus we assume that the reaction is given by

$$\dot{a} = r = (1-a)\{(1-a)k_1 + ak_2\} \tag{2.11}$$

where k_1 and k_2 are Arrhenius rates of the form

$$k_i = A_i \exp(-\theta_i/T), \qquad (i = 1, 2), \tag{2.12}$$

A is the frequency factor and θ is the activation temperature. Nitromethane kinetics and the precise values of these parameters are discussed fully in [2]. Note that if k_1 and k_2 are positive and $k_1 > k_2$, then the extent of reaction is strictly monotone and asymptotically stable; that is, for fixed temperature T^* and every value of a in the interval $[0,1]$, the initial-value problem

$$\dot{a} = \hat{r}(T^*, a), \tag{2.13}$$

$$a(0) = a_0,$$

has a unique smooth solution $a(t)$, $t \geq 0$ such that

$$a(t) \to 1 \text{ as } t \to \infty . \tag{2.14}$$

The equilibrium state $a = 1$ is referred to as chemical completion. The entropy inequality (2.8) implies that for fixed volume v^* and temperature T^* the free energy is a minimum at chemical completion; i.e.,

$$\hat{\psi}(v^*, T^*, a) \geq \hat{\psi}(v^*, T^*, 1) . \tag{2.15}$$

Clearly, the free energy function $\hat{\psi}(v^*, T^*, a)$ is a Lyapunov function for the extent of reaction a [3].

In summary, we are interested in weak solutions to the conservation equations

$$\dot{v} = \nabla \cdot \mathbf{u} , \tag{2.16}$$

$$\dot{\mathbf{u}} = -\nabla p , \tag{2.17}$$

and the evolutionary equations (with $w = 0$)

$$\dot{T} = -\frac{1}{c_v}\{TG\dot{v} + Q\dot{a}\} , \tag{2.18}$$

$$\dot{a} = (1-a)\{(1-a)k_1 + ak_2\} . \tag{2.19}$$

For most fluids,

$$c_v > 0 , \ G > 0$$

and for exothermic reactions which are uni-directional ($\dot{a} > 0$),

$$Q < 0 .$$

It is important to note that the coefficients of (2.18) and the pressure p are determined by the free energy ψ which depends on the state variables v, T, a. For the present study, we employed a linear mixing rule for the free energy ψ in which case

$$\psi = (1-a)\hat{\psi}_f(v, T) + a\hat{\psi}_g(v, T)$$

where ψ_f represents the frozen response of the fluid and ψ_g is the free energy of the gas products at chemical completion. Typically, $\hat{\psi}_f$ and $\hat{\psi}_g$ are highly nonlinear functions; ψ_f is evaluated from Hugoniot and thermophysical data, while ψ_g is given as non-ideal gas with a single co-volume (Abel's equation of state) [2].

Before turning to the numerical solution of the system (2.16)–(2.19), it is appropriate to point out that the inclusion of chemical reactions could possibly alter the hyperbolic character of the conservation equations. To address this issue, we restrict our attention to two dimensions in which case the system of equations (2.16)–(2.19) is equivalent to

$$\mathbf{A}\dot{\mathbf{U}} + \mathbf{B} \cdot \nabla \mathbf{U} + \hat{\mathbf{f}}(\mathbf{U}) = 0 \tag{2.20}$$

where

$$\mathbf{U} = (v, u_1, u_2, T, a)$$

is the solution vector. **A** and **B** are 5×5 matrices of the form

$$[\mathbf{A}] = \begin{bmatrix} 1 & 0 & 0 & 0 & 0 \\ 0 & 1 & 0 & 0 & 0 \\ 0 & 0 & 1 & 0 & 0 \\ 0 & 0 & 0 & 1 & 0 \\ 0 & 0 & 0 & 0 & 1 \end{bmatrix}, \qquad (2.21)$$

$$[\mathbf{B}] = - \begin{bmatrix} 0 & e_1 & e_2 & 0 & 0 \\ (C^2 - \frac{TG^2}{c_v})e_1 & 0 & 0 & -Ge_1 & (h-Q)\frac{G}{c_v}e_1 \\ (C^2 - \frac{TG^2}{c_v})e_2 & 0 & 0 & -Ge_2 & (h-Q)\frac{G}{c_v}e_2 \\ 0 & -\frac{TG}{c_v}e_1 & -\frac{TG}{c_v}e_2 & 0 & 0 \\ 0 & 0 & 0 & 0 & 0 \end{bmatrix}, \qquad (2.22)$$

and $\mathbf{e}_i (i = 1,2)$ are the unit vectors in the Lagrangian coordinate frame. The forcing vector $\hat{\mathbf{f}}(\mathbf{U})$ is

$$[\hat{\mathbf{f}}(\mathbf{U})] = (0, 0, 0, \frac{Qr}{c_v}, -r) . \qquad (2.23)$$

In obtaining this form, we have used the notation

$$C^2 = \frac{\partial \hat{p}}{\partial v} + \frac{TG^2}{c_v} , \quad h = Q - \frac{c_v}{G}\frac{\partial \hat{p}}{\partial a} , \qquad (2.24)$$

where C is the sound speed and h is the heat of reaction.

A surface M defined by

$$\hat{\phi}(\mathbf{X}, t) = 0 , \quad \mathbf{X} = (X_1, X_2) , \qquad (2.25)$$

is **characteristic** with respect to (2.20) at the point $(\mathbf{X}, t)$ if

$$\det(\mathbf{B} \cdot \mathbf{n} - \lambda \mathbf{A}) = 0 \qquad (2.26)$$

where

$$\lambda = -\frac{\dot{\phi}}{|\nabla \phi|} , \qquad (2.27)$$

is the velocity of the surface and

$$\mathbf{n} = \frac{\nabla \phi}{|\nabla \phi|} \qquad (2.28)$$

is the unit normal to the surface. The quasilinear system (2.20) is said to be **hyperbolic** if all numbers λ are real, **elliptic** if all numbers λ are complex, and **mixed** if there are both real and complex values of λ.

It is clear from (2.26) that since **A** is a unit matrix, the characteristics of (2.20) are the eigenvalues of the matrix $\mathbf{B} \cdot \mathbf{n}$. Using (2.22)–(2.24) and tedious algebra, (2.26) yields the characteristics

$$\lambda = 0 , \quad \text{and} \; \pm C . \qquad (2.29)$$

At first glance, these characteristics would appear to be the usual ones pertaining to a nonlinear compressible fluid; that is, representing a contact surface ($\lambda = 0$) and a backward ($\lambda = -C$) and forward ($\lambda = +C$) facing characteristic along which signals can propagate with speed C. This is indeed the case if C is real. However, if the free energy is such that, as the material reacts, the parameter C^2 defined by (2.25) becomes negative (i.e., $\partial \hat{p}/\partial v < 0$) somewhere in the flow field, then the governing equations undergo a change in type. This can seriously influence the choice of numerical methods employed to obtain solutions of (2.16)–(2.19). [1]

Finally, we note in passing that Von Neumann [6] studied the one-dimensional counterpart of the system (2.16)-(2.19) and demonstrated the existence of unique steady wave solutions containing a shock front. These solutions correspond to the current notion of a detonation wave where the point of chemical completion in the wave profile is identified as the Chapman-Jouguet (CJ) state [7].

2.3 An Explicit Finite Difference Method

Our interest here is in numerical solutions to the partial differential equations (2.16) and (2.17) and the ordinary differential equations (2.18) and (2.19), subject to the appropriate initial and boundary conditions, for two-dimensional axisymmetric geometries.

The solution method utilizes an explicit finite difference algorithm developed by Wilkins [8] and implemented in the Lagrangian wave propagation code, TOODY [9]. This algorithm is based on an averaging method to arrive at finite difference approximations for the partial derivatives of the pressure p. To be more specific, let Ω be a regular region in two dimensions and consider a point (X_1, X_2) in the region (Figure 2). Then, the partial derivatives $\partial p/\partial X_1$, $\partial p/\partial X_2$ at

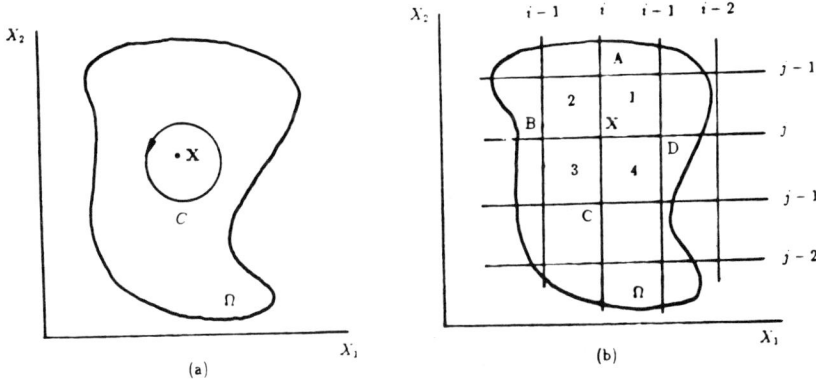

Figure 2. An arbitrary region Ω in two-dimensional space (a) and the corresponding finite difference mesh (b).

---

[1] This fact was first pointed out to one of the authors (JWN) by D. B. Hayes (Sandia National Labs) in February 1980 in the context of viscoelastic (Maxwell) solids. Joseph, et. al. [4] have recently shown that the equations for viscoelastic (Maxwell) fluids can undergo a change of type in the flow field and made a similar observation about numerical solution methods. In studying sink flow of a Maxwell fluid, Gartling [5] has found that finite element solutions failed to converge for certain types of higher-order elements.

$(\mathbf{X} = X_1, X_2)$ are defined by

$$\frac{\partial p}{\partial X_1} = \lim_{A \to 0} \left[\frac{1}{A} \int_C p \, dX_2 \right],$$

$$\frac{\partial p}{\partial X_2} = \lim_{A \to 0} \left[-\frac{1}{A} \int_C p \, dX_1 \right],$$

where C is a counterclockwise closed contour containing (X_1, X_2) and A is the area enclosed by C (Figure 2a). If the region Ω is discretized into rectangular zones, then the contour C is the smallest possible quadrilateral (ABCD in Figure 2b) containing (X_1, X_2) and the limit is achieved by linear interpolation along the sides of the quadrilateral. Thus, the partial derivatives of p at (X_1, X_2) are approximated by

$$\frac{\partial p}{\partial X_1}\Big|_{i,j} = \frac{1}{A_{i,j}} \{p_1 [(X_2)_{i,j+1} - (X_2)_{i+1,j}]$$

$$+ p_2 [(X_2)_{i-1,j} - (X_2)_{i,j+1}]$$

$$+ p_3 [(X_2)_{i,j-1} - (X_2)_{i-1,j}]$$

$$+ p_4 [(X_2)_{i+1,j} - (X_2)_{i,j-1}]\},$$

$$\frac{\partial p}{\partial X_2}\Big|_{i,j} = -\frac{1}{A_{i,j}} \{p_1 [(X_1)_{i,j+1} - (X_1)_{i+1,j}]$$

$$+ p_2 [(X_1)_{i-1,j} - (X_1)_{i,j+1}]$$

$$+ p_3 [(X_1)_{i,j-1} - (X_1)_{i-1,j}]$$

$$+ p_4 [(X_1)_{i+1,j} - (X_1)_{i,j-1}]\},$$

where the subscripts i, j refer to the location of the nodal points in the coordinate frame and p_k ($k = 1, 2, 3, 4$) refers to the value of p in zone k. In the present scheme, the value of p in zone 3 is assigned to the nodal point $\mathbf{X}_{i,j}$. The area $A_{i,j}$ is simply the area of the quadrilateral $ABCD$:

$$A_{i,j} = \frac{1}{2} \{[(X_2)_{i,j-1} - (X_2)_{i,j+1}][(X_1)_{i+1,j} - (X_1)_{i-1,j}]$$

$$+ [(X_2)_{i+1,j} - (X_2)_{i-1,j}][(X_1)_{i,j+1} - (X_1)_{i,j-1}]\}.$$

Using these finite difference analogs, the integration of the conservation equations proceeds incrementally in time through space. At time $t = 0$, all quantities are defined at all grid points by the initial data. The computations required to advance all values by a step in time Δt^n are performed at all nodes so that no nodal point is advanced to time $t^{n+1} = t^n + \Delta t^n$ until all nodal points have been advanced to t^n. The velocity $\mathbf{u} = (u_1, u_2)$ and position $\mathbf{X} = (X_1, X_2)$ are advanced by

$$\mathbf{u}^{n+\frac{1}{2}} = \mathbf{u}^{n-\frac{1}{2}} + \frac{1}{2}(\Delta t^n + \Delta t^{n-1}) \cdot \dot{\mathbf{u}}^n,$$

$$\mathbf{X}^{n+1} = \mathbf{X}^n + (\Delta t^n)\mathbf{u}^{n+\frac{1}{2}}.$$

For axisymmetry geometries, the new specific volume v_k^n of the kth zone can be computed by assuming that the mesh is of unit angle thickness and that X_2 is the radial coordinate. For example,

$$v_3^n = \tfrac{1}{3} A_3^n \left[(X_2)_{i,j}^n + (X_2)_{i-1,j}^n + (X_2)_{i,j-1}^n + (X_2)_{i-1,j-1}^n \right]$$
$$+ \tfrac{1}{6} \left[(X_1)_{i-1,j-1}^n (X_2)_{i,j}^n - (X_1)_{i,j}^n (X_2)_{i-1,j-1}^n \right] \cdot \left[(X_2)_{i-1,j}^n - (X_2)_{i,j-1}^n \right]$$
$$+ \tfrac{1}{6} \left[(X_1)_{i-1,j}^n (X_2)_{i,j-1}^n - (X_1)_{i,j-1}^n (X_2)_{i-1,j}^n \right] \cdot \left[(X_2)_{i,j}^n - (X_2)_{i-1,j-1}^n \right]$$

where

$$A_3^n = \tfrac{1}{2} \left[(X_1)_{i,j-1}^n - (X_1)_{i-1,j}^n \right] \left[(X_2)_{i,j}^n - (X_2)_{i-1,j-1}^n \right]$$
$$+ \tfrac{1}{2} \left[(X_1)_{i,j}^n - (X_1)_{i-1,j-1}^n \right] \left[(X_2)_{i-1,j}^n - (X_2)_{i,j-1}^n \right] .$$

Of course, the specified volume rate $\dot{v}^{n+1}$ can then be evaluated according to

$$\dot{v}^{n+1} = \frac{v^{n+1} - v^n}{\Delta t^n} .$$

The equation of state is called each zone cycle with a new specific volume v^n and is expected to return a pressure, temperature, and sound speed consistent with this specific volume. Pressures are used in the subsequent integration of the conservations equations and the sound speed is used to calculate a stable timestep for the calculations to advance in time. In traditional equations of state, the constitutive models are sufficiently simple that straightforward differencing techniques can be implemented to solve energy conservation and the constitutive model of the material for the required variables, while maintaining second-order accuracy consistent with the remainder of the code (see Swegle [9]). It is important to note that this algorithm employs the method of artificial viscosity to maintain smooth solutions to shock wave problems. During each timestep, this involves the addition of a viscosity to the pressure calculated in the equation of state which is quadratic in the specific volume rate:

$$q = - \left\{ \frac{1}{v^2} b_1 \sqrt{A_3} \, |\dot{v}| + C b_2 \right\} \sqrt{A_3} \frac{\dot{v}}{v^2}$$

where b_1 and b_2 are constant viscosity coefficients.

In the case of a reactive fluid, the equation of state routine must also provide the temperature and the reaction coordinate. With the inclusion of artificial viscosity, (2.18) must be replaced with an ordinary differential equation of temperature, T, of the form

$$\dot{T} = -\frac{1}{c_v} \{(q + TG)\dot{v} + Q\dot{a}\} . \quad (2.30)$$

The coupling of equations (2.16)–(2.17), (2.19) and (2.30) is such that differencing of (2.19) and (2.30) is not practical. Instead, we have made use of a technique successfully applied in other problems in which (2.19) and (2.30) were integrated using an external ordinary differential equation (ODE) solver. The solver [10], which uses an Adams' predictor-corrective method, integrates the differential equations across the timestep Δt^n used in the algorithm. That is, for each zone cycle, the information at the end of the previous timestep (T^n, a^n) is used as initial data for equations (2.19) and (2.30), which are then integrated to the end of the timestep to give the new values T^{n+1}, a^{n+1}. However, in this integration the updated specific volume v^{n+1} and artificial viscosity q^{n+1} are used. Then in order to ensure smooth integration by the ODE solver, the specific volume and viscosity are assumed to vary linearly over the current timestep. That is, for $t^n \le t \le t^{n+1}$,

$$v(t) = \frac{v^{n+1} - v^n}{\Delta t^n}(t - t^n) + v^n ,$$

$$q(t) = \frac{q^{n+1} - q^n}{\Delta t^n}(t - t^n) + q^n .$$

The principal advantage of this method is that accurate solutions of (2.19) and (2.30) are possible in cases when single step differencing is not expected to yield good results (if it is even possible). A second benefit is the freedom to change the chemical kinetics at will without having to redrive a new set of difference equations. This is extremely useful since often times the kinetics are not well-defined and several models need to be studied.

In order to carry out numerical calculations, it remains only to identify the correct, stepwise stable, timestep corresponding to the finite difference analogs which will ensure convergence of the numerical solution to the correct solution of the nonlinear partial differential equations describing the problem. In the case of nonlinear problems such as those posed here, it is extremely difficult to establish the correct timestep criterion which will guarantee convergence. As a result, it has been necessary to rely on stability analyses with the assumption that the conditions for stability also ensure convergence. This is common practice and almost all the numerical algorithms for shock wave problems make use of the stability analysis of Richtmyer and Morton [11] which showed that, in one-dimension, the finite difference equations for an ideal polytropic gas are stepwise stable if the timestep satisfies the CFL condition

$$\Delta t \le \frac{\Delta X}{C^*} \qquad (2.31)$$

where C^* is interpreted as the maximum sound speed in the problem at each instant of time and ΔX is the mesh size.

It is important to note, however, that in the present model for a reactive fluid, there exists another characteristic time, that of the rate of reaction

$$\Delta t_R = \frac{\Delta a}{\dot{a}} .$$

This time is considerably shorter than the hydrodynamic timestep and, as a result, can have significant influence on the condition for stability of the finite difference analogs. To determine the correct timestep in this situation, we appeal to the work of Hicks [12] who analyzed the stability of the one-dimensional finite difference analogs for a viscoelastic (Maxwell) material described by the constitutive equation

$$\dot{p} = -\frac{C^2}{v_0^2}\dot{v} - \frac{1}{\tau_r}(p - p_e(v)) .$$

Here τ is a characteristic relaxation time which is much less than the characteristic acoustic time for a problem. Hicks showed that for these types of materials the stability condition is much more restrictive than the CFL condition and, in fact, the timestep must satisfy

$$\Delta t \leq \frac{\Delta X}{C^*} \left\{ -\frac{(\Delta X)}{4C^*\tau} + \sqrt{\left(\frac{\Delta X}{4C^*\tau}\right)^2 + 1} \right\} \qquad (2.32)$$

It is not difficult to show (Nunziato and Kipp [2]) that the current model for a chemically reacting fluid can be cast in the form of the rate-dependent model considered by Hicks, where the relaxation time is characterized by the kinetics of the chemical reaction involved. For a fluid such as nitromethane, the characteristic time τ is 0.26 (10^{-9})s. This value is used in (2.32) to generate the timestep/zone size curve shown in Figure 3. Notice that for extremely small zone sizes, the timestep calculated from (2.32) converges to the CFL condition; whereas for large zone sizes the timestep converges to a value of 2 τ. Clearly, this timestep can be orders of magnitude less than the CFL time step one would compute on the basis of the detonation velocity for nitromethane (6.25 km/s).

2.4 Numerical Results and Discussion

As the inner diameter of a thick-walled tube decreases, the cylinder of a chemically reacting fluid contained therein will eventually be of such a diameter that detonation is not self-sustaining. For a fluid such as nitromethane, confined in a thick-walled brass tube, this critical diameter occurs at 2.84 ± 0.42 mm [13]. The geometry employed to simulate this property is shown in Figure 4, in which the left boundary of the cylinder is a reflection plane. A no-slip condition is imposed at the fluid/brass interface. The booster consists of a 2 mm length of an explosive PBX 9404 whose pressure is computed from the Chapman-Jouguet state. In these numerical simulations, the initial zone size was 0.1 mm square with a fixed timestep of 0.2 (10^{-9}) seconds. This is consistent with (2.32) and results in a timestep which is about one order of magnitude smaller than the CFL timestep.

The sequence of Figure 5 for a nitromethane radius of 1.1 mm illustrates the progression from ignition of the nitromethane to eventual collapse of the detonation. Initially, the wave front unloads into the side wall and a release wave propagates from the wall through the reaction zone. This release wave cools the reaction zone and subsequently extends its length in the vicinity of the wall. The resulting decrease in the reaction zone is accompanied with increased wave front curvature which promotes further slowing of the reaction. This failure process continues with the detonation converging onto the axis of the tube and separation of the shock front from the reacted fluid is observed to occur. Under these initial conditions, the length of nitromethane that has reacted is approximately 6 mm. With the same initial conditions, a nitromethane radius of 1.2 mm reacts for 13 mm before detonation failure occurs. At a radius of 1.3 mm, the simulation of the reaction becomes stable and propagates as a steady detonation wave (Figure 6).

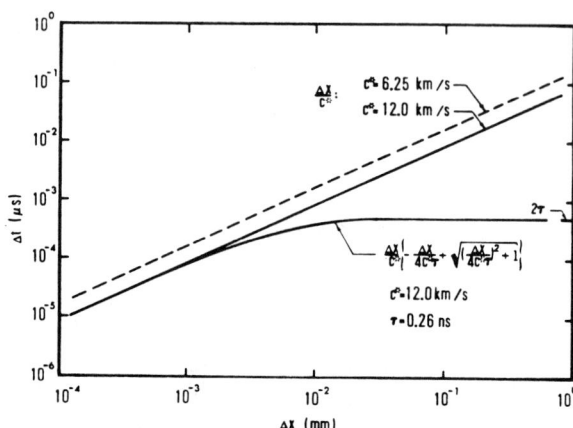

Figure 3. Mesh size dependence of CFL and resolved reaction zone timestep restrictions for stability of explicit finite difference method.

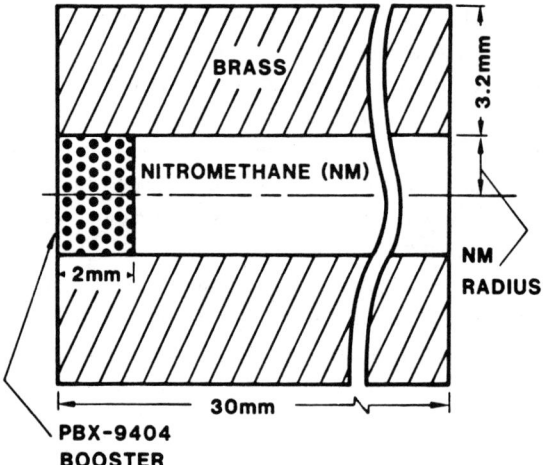

Figure 4. Geometry for studies of detonation and detonation failure in a column of nitromethane with brass confinement.

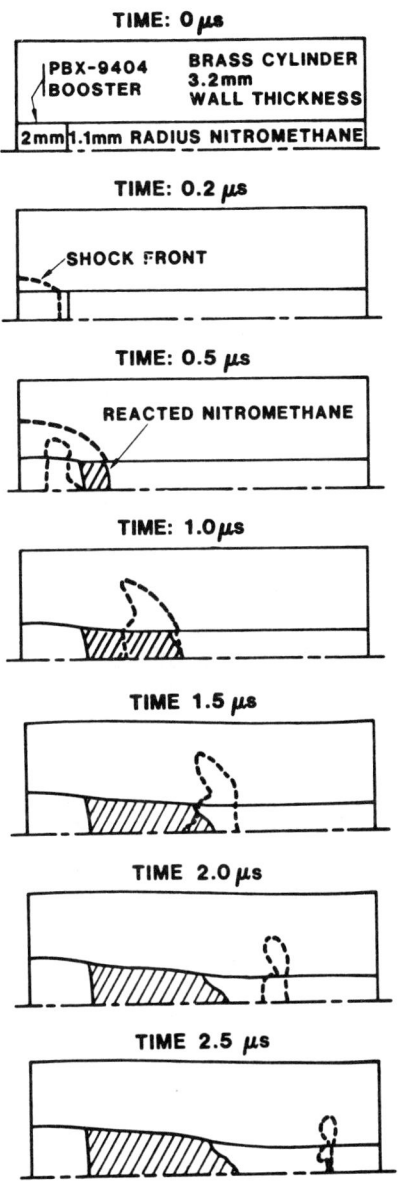

Figure 5. Detonation failure in a 1.1 mm radius column of nitromethane.

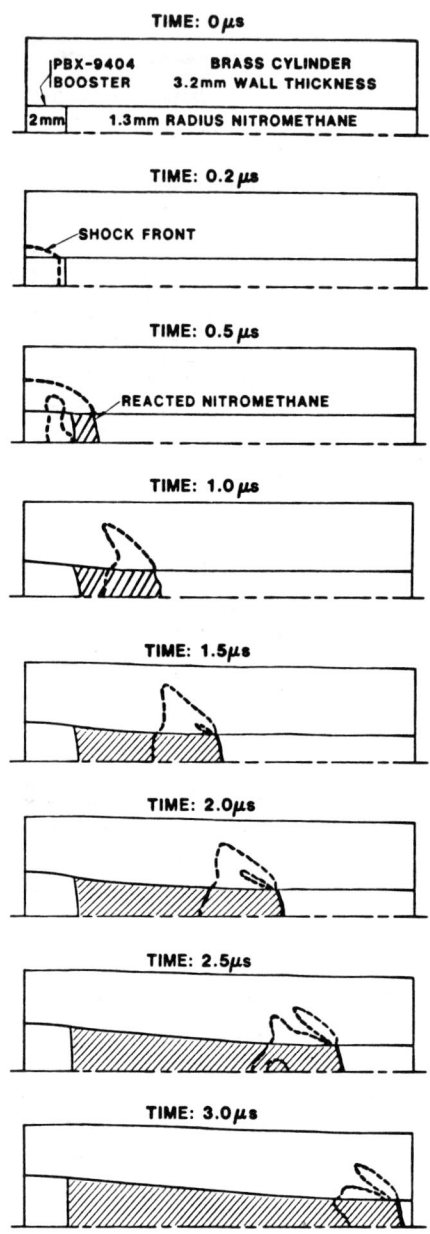

Figure 6. Detonation wave propagation in a 1.3 mm radius column of nitromethane.

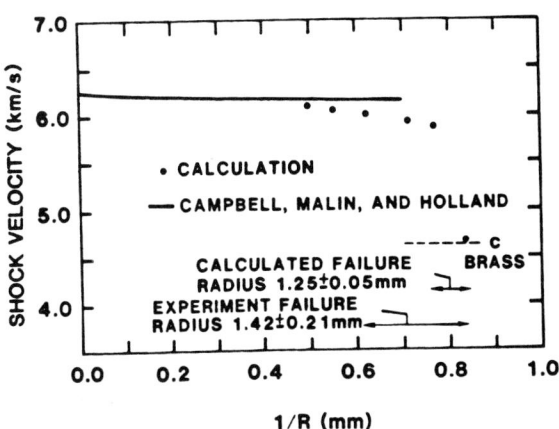

Figure 7. Velocity of the detonation wave as a function of confined diameter.

The radius of nitromethane which we compute as marking the boundary between stable detonation and failure falls at 1.2 to 1.3 mm. This radius is within the experimental error found by Malin [13] for brass confinement. The calculations made for nitromethane contained in brass tubes may be summarized in terms of the shock velocity of the detonation front. Figure 7 is a plot of shock velocity versus inverse radius of the nitromethane. The line is data from Campbell, et. al [14] and indicates that the failure diameter is approached with a gradual drop in detonation velocity. Our computed velocities are included in Figure 7 and show very good agreement with the experimental data. On the basis of this comparison one would conclude that indeed we have a numerical technique which does converge to the correct physical solution.

Having demonstrated that we can simulate detonation failure in a thick-walled brass tube, an important question immediately arises. These computations are extremely expensive. In this case we have a problem involving roughly 3,000 zones in the nitromethane alone in which we have to make approximately 10^4 timesteps in order for the wave to reach the end of the tube. Computations on a CDC 7600 machine run at approximately 10^6 zone cycles per hour and take roughly 1 to 2 hours. Computations on the CRAY-1 offer improvements of a factor of 2 and complete vectorization would at least provide a factor of 4. Nevertheless, these are very simple geometries. To do practical problems will require more complex geometries and significantly longer running times. To this end, it is natural to ask if we can run at a much coarser timestep such as the CFL condition. Normally, shock wave computations run at 95 percent of the CFL condition. To address this question, we ran computations using the CFL timestep and we were surprised to find another stable solution. However, it is not the correct solution as indicated by the experimental data. A comparison of the numerical results using both the CFL criterion and the reaction-controlled criterion are shown in Figure 8 for a radius of 1.2 mm. In this case, we see that the solution using the CFL condition is perfectly stable and results in a burn front that

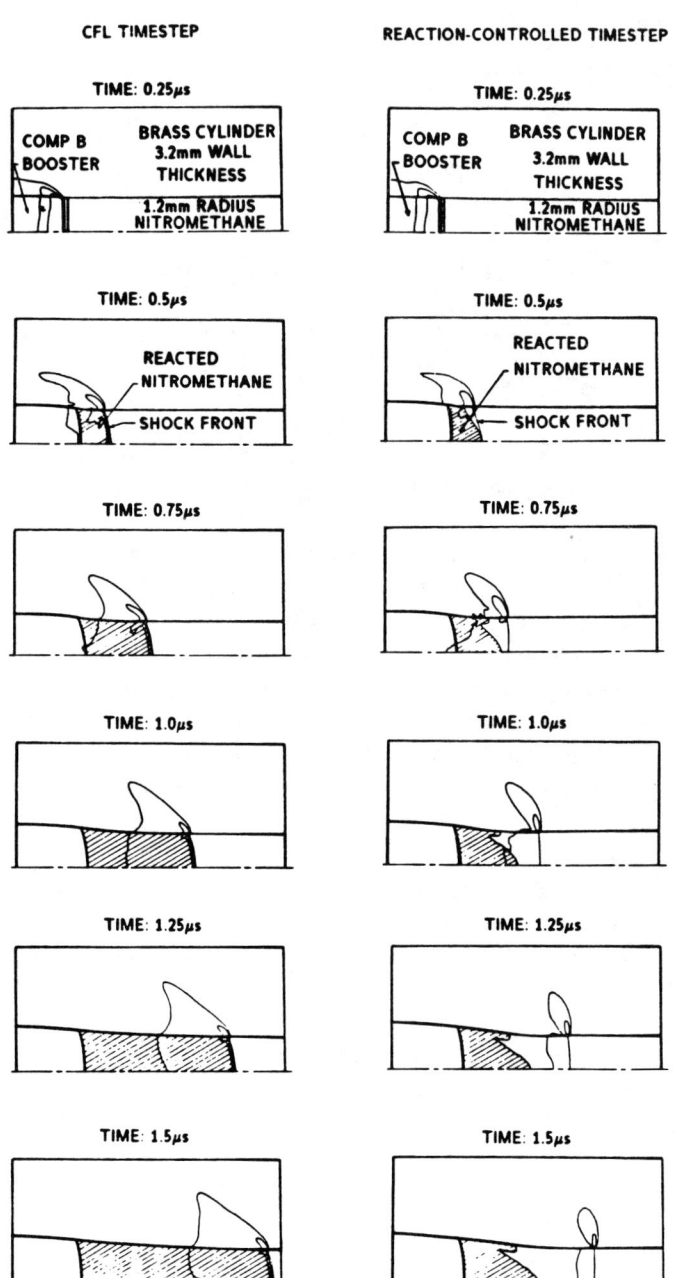

Figure 8. Comparison of detonation wave solutions using the normal CFL timestep and the reaction-controlled timestep.

propagates completely down the length of a tube. On the other hand, the results obtained using the reaction-controlled timestep results in detonation failure. We believe that this evidence of multiple stable solutions is a consequence of the numerical method; not a consequence of the physical model. To make this point, we have examined the pressure history and the reaction coordinate history at the brass wall for both the CFL timestep and the reaction-controlled timestep. Differences in behavior at the wall in the nitromethane are shown in Figure 9. The reaction-controlled calculation has a slightly longer rise time and a lower amplitude in pressure than the CFL timestep calculation. Nevertheless, we believe these differences are not significant and the overall results are comparable. This implies that we have not introduced any excess numerical diffusion in the computation as a result of the large number of time cycles using the reaction-controlled timestep. What is particularly important here is the reaction coordinate history. In the case of the CFL condition, the reaction quickly goes to completion; whereas in the case of the reaction-controlled timestep, the reaction starts slowly at first and then goes to completion over a finite time. It is clear from these results that with the CFL condition, the algorithm cannot discriminate the effect of the chemical reaction and the wave propagates at the detonation velocity. This corresponds to what is usually referred to as a "CJ burn". It is only in the case where we are able to resolve the reaction using the reaction-controlled timestep that we can properly account for the physical effects associated with the chemical decomposition of the material and permit the lateral release waves to influence the reaction zone as the detonation wave propagates along the cylinder. It is this wave interaction which leads to detonation failure.

We find these results less than satisfactory in view of the current interest in developing more complex kinetic models for chemically reacting flows. It is clear that in situations where there are a large number of multiple reactions including radicals, whose life times may be extremely short, it will be next to impossible numerically to resolve their effects since the timestep restriction

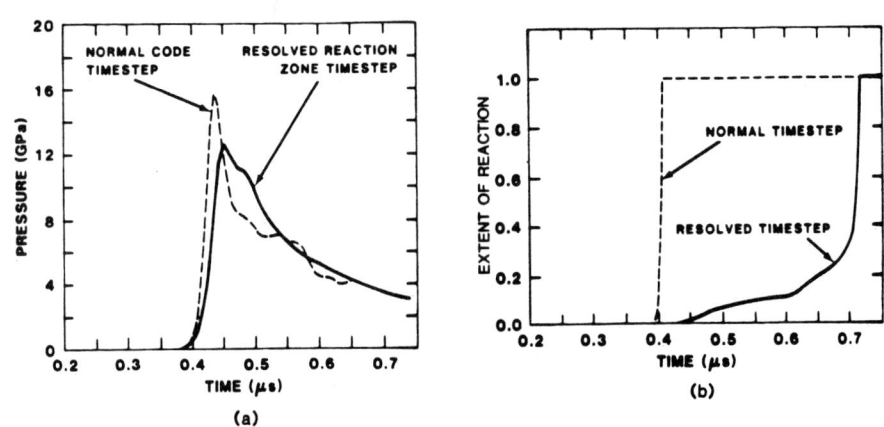

Figure 9. Comparison of the pressure (a) and extent of reaction (b) solutions at the brass wall using the normal CFL timestep and the resolved reaction zone timestep.

will be determined by the shortest reaction time. This supports the notion that, in flow problems where the fluid mechanics and the chemical reactions are highly coupled, one should develop what might be termed homogenized or effective reaction kinetic models which can be characterized by single characteristic time. At least in that case one is no worse off than we have been in this study with regard to timestep restrictions.

3 Solidification of a Chemically Reacting Viscous Fluid

3.1 Physics of the Problem

Epoxies are chemically complex materials that are formed by the combination of a resin and a curing agent. When these components are mixed, the result is a viscous incompressible liquid that undergoes an exothermic chemical reaction. As a consequence of this chemical reaction, the liquid resin gels to form a solid material. In Figure 10 a schematic is shown of a rectangular container containing the liquid resin at the time when the upper part of the container has begun to solidify [15]. As solidification occurs, the affected region increases due to a solidification front which propagates down through the container. Ahead of the front, there is convection in the liquid phase due to the volumetric heating associated with the polymerization reaction. As a result of the propagating solidification front, the aspect ratio of the fluid phase changes and it would be natural to expect some changes in the number of cells observed. The experiments of this process also show what are called "marbles" that develop right in front of the solidification front. These are believed due to non-Newtonian effects; effects which presently are not included in the physical model to be described here.

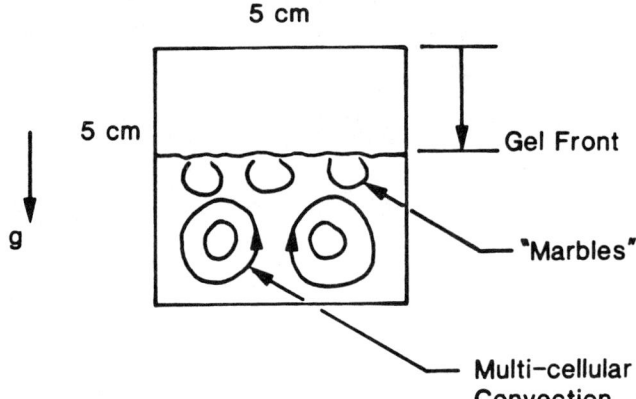

Figure 10. Schematic of an epoxy curing experiment using an epoxy resin (DGEBA) and a primary amine curing agent (MPDA). Solidification starts at 75 min. at a temperature of 344 K.

3.2 Mathematical Description

In considering the curing of an epoxy resin we are interested in the heat and mass transfer in the liquid phase as well as the heat transfer in the solid phase. In the liquid phase, the epoxy is assumed to be incompressible within the limits of the Boussinesq approximation and to behave as a Newtonian fluid. The flow is assumed to be laminar and sufficiently slow to allow the viscous dissipation of the fluid to be ignored. Furthermore, changes in density as a result of solidification are assumed to be negligible. In this case, conservation of mass, momentum, and energy can be written as

$$\nabla \cdot \mathbf{u} = 0 , \tag{3.1}$$

$$\rho_0 \left[\frac{\partial \mathbf{u}}{\partial t} + (\nabla \mathbf{u})^T \mathbf{u} \right] = -\nabla p + \nabla \cdot \mathbf{S} - \rho \mathbf{g} , \tag{3.2}$$

$$\rho_0 \left[\frac{\partial e}{\partial t} + \mathbf{u} \cdot \nabla e \right] = -\nabla \cdot \mathbf{q} + \rho_0 w , \tag{3.3}$$

where the deviatoric stress $\mathbf{S}$ is

$$\mathbf{S} = \mu(\nabla \mathbf{u} + \nabla \mathbf{u}^T) , \tag{3.4}$$

the density ρ is given by the Boussinesq relation

$$\rho = \rho_0 + \beta(T - T_0) , \tag{3.5}$$

and the heat flux $\mathbf{q}$ is given by Fourier's law

$$\mathbf{q} = -K\nabla T . \tag{3.6}$$

In (3.1)–(3.6), ∇ is the gradient operator in an Eulerian coordinate system, $\mathbf{u}$ is the velocity, p is the pressure, e is the specific internal energy, T is the absolute temperature, $\mathbf{g}$ is the gravitational body force, w is the external heat supply, μ is the viscosity, β is the thermal expansion coefficient, and K is the thermal conductivity. The subscript 0 indicates the reference state.

We characterize the solidification process by the extent of reaction a, $0 \leq a \leq 1$, which in physical terms is interpreted as the ratio of the mass of solidified material (per unit volume) to the total mass of material (per unit volume). Conservation of mass and momentum (due to diffusion) for the solidification phase results in

$$\frac{\partial a}{\partial t} + \mathbf{u} \cdot \nabla a = -\nabla \cdot \mathbf{f} + \hat{r}(a, T) \tag{3.7}$$

where the diffusive flux $\mathbf{f}$ is given by

$$\mathbf{f} = -D\nabla a , \tag{3.8}$$

D is the binary diffusion coefficient, and r is the reaction rate. The reaction kinetics for the epoxy are taken to be second-order with Arrhenius rate constants [15]:

$$r = \hat{r}(a, T) = (1-a)\{k_1 + ak_2\} \tag{3.9}$$

where

$$k_i = A_i \exp(-\theta_i/T) > 0 \quad (i = 1, 2) . \tag{3.10}$$

Note that with the fluid at rest and no diffusion, the extent of reaction a has all the stability properties described in Section 2.

It remains to specify the constitutive equation for the specific internal energy e and how the material properties μ, β, D, and K depend on the extent of reaction a. To characterize the energy in both the mobile liquid phase and in the solid phase, we employ a linear mixing rule and write

$$e = (1-a)\hat{e}_f(T) + a\hat{e}_s(T) . \tag{3.11}$$

Then conservation of energy (3.3), along with (3.6), becomes the heat equation

$$\rho_0 c_p \left[\frac{\partial T}{\partial t} + \mathbf{u} \cdot \nabla T \right] = \nabla \cdot (K \nabla T) - Q \left[\frac{\partial a}{\partial t} + \mathbf{u} \cdot \nabla a \right] + \rho_0 w \tag{3.12}$$

where c_p is the specific heat of the mixture:

$$c_p = (1-a)(c_p)_f + a(c_p)_s = \frac{\partial \hat{e}}{\partial T} , \tag{3.13}$$

and Q is the heat of thermal decomposition:

$$Q = e_s - e_f = \frac{\partial \hat{e}}{\partial a} . \tag{3.14}$$

Note that the energy in the fluid is determined by conduction, convection, and the volumetric heating due to the exothermic polymerization reaction ($Q < 0$).

For the present analysis, we considered a simplified model in which all the material properties $(\mu, \beta, K, D, c_p, Q)$ were taken to be constant except the viscosity μ. In order to characterize the solidification, the viscosity is taken be a strongly increasing function of the extent of reaction a once it reaches the gel point of 0.6. A large viscosity results in the fluid velocities going to zero and, in this limit, the above description reduces to the analysis of heat transfer in the solid phase. In summary, we are seeking smooth solutions of the partial differential equations (3.1)–(3.6) (with $w = 0$):

$$\nabla \cdot \mathbf{u} = 0 , \tag{3.15}$$

$$\rho_0 \left[\frac{\partial \mathbf{u}}{\partial t} + (\nabla \mathbf{u})^T \mathbf{u} \right] = -\nabla p + \nabla \cdot \left[\mu(a)(\nabla \mathbf{u} + \nabla \mathbf{u}^T) \right] - [\rho_0 + \beta(T - T_0)] \mathbf{g} , \tag{3.16}$$

$$\rho_0 c_p \left[\frac{\partial T}{\partial t} + \mathbf{u} \cdot \nabla T \right] = \nabla \cdot (K \nabla T) - Q \left[\nabla \cdot (D \nabla a) + (1-a)\{k_1 + ak_2\} \right] , \tag{3.17}$$

$$\frac{\partial a}{\partial t} + \mathbf{u} \cdot \nabla a = \nabla \cdot (D \nabla a) + (1 - a)\{k_1 + ak_2\} \ . \tag{3.18}$$

To complete the description of the problem, boundary and initial data are required for the dependent variables $\mathbf{u}, T$, and a. The liquid epoxy is assumed to be initially at rest with a uniform temperature. At the boundary of the fluid, the no-slip condition is applied to the fluid, a convection heat transfer coefficient is used to model the heat transfer to the surrounding environment, and there is no flux of reacted material.

We also considered the characteristics of this system of partial differential equations in two-dimensions using the analysis outlined in Section 2. Briefly, the system (3.15) – (3.18) can be written in the form (2.20) where the solution vector $\mathbf{U}$ is now a 12-vector:

$$\mathbf{U} = (u_1,\ u_2,\ p,\ S_{11},\ S_{12},\ S_{22},\ q_1,\ q_2,\ T,\ f_1,\ f_2,\ a)\ .$$

The coefficients $\mathbf{A}$ and $\mathbf{B}$ are 12×12 matrices and, in this case, the characteristic problem

$$\det(\mathbf{B} \cdot \mathbf{n} - \lambda \mathbf{A}) = 0$$

is identically satisfied for any λ. This implies that the presence of chemical reactions does not alter the basic viscous flow problem; that is, every time plane is a characteristic surface for the equations. This is a fundamental feature of parabolic systems and it is evident that we need not concern ourselves with changes in type as in Section 2.

3.3 An Implicit Finite Element Method

The numerical solution of the system of partial differential equations outlined above was undertaken using a Galerkin based, finite element method and implemented in a Navier-Stokes flow code NACHOS (see Gartling, et. al [16]). This method has enjoyed considerable success in solving low Rayleigh number, buoyantly-driven convection problems such as the one considered here, and it offers distinct advantages over finite difference methods for problems with complex geometries. The major complications in the present problem concern the change of phase in the epoxy and the solution of the advection-diffusion equation describing the chemical reaction.

As usual, the numerical procedure requires the two-dimensional region Ω under consideration to be discretized into geometrically simple subdomains with the boundary conditions applied to the boundary Γ (Figure 11). Then, the set of partial differential equations (3.15) – (3.18) are approximated using the Galerkin finite element form of the method of weighted residuals. Let the dependent variables be approximated by expansions of the form

$$u_i(\mathbf{x}, t) = \sum_{n=1}^{N} \Phi_n(\mathbf{x}) u_i^n(t) = \mathbf{\Phi}^T(\mathbf{x}) \mathbf{u}_i(t)\ , \qquad (i = 1, 2)$$

$$p(\mathbf{x}, t) = \sum_{m=1}^{M} \Psi_m(\mathbf{x}) p^m(t) = \mathbf{\Psi}^T(\mathbf{x}) \mathbf{P}(t)\ ,$$

$$\tag{3.19}$$

$$T(\mathbf{x},t) = \sum_{n=1}^{N} \Theta_n(\mathbf{x})T^n(t) = \mathbf{\Theta}^T(\mathbf{x})\mathbf{T}(t) ,$$

$$a(\mathbf{x},t) = \sum_{n=1}^{N} \Xi_n(\mathbf{x})a^n(t) = \mathbf{\Xi}^T(\mathbf{x})\mathbf{a}(t) ,$$

where $\mathbf{\Phi}$, $\mathbf{\Psi}$, $\mathbf{\Theta}$, and $\mathbf{\Xi}$ are vectors of interpolation or shape functions defined on the discretized domain, $\mathbf{u}_i$, $\mathbf{P}$, $\mathbf{T}$ and $\mathbf{a}$ are vectors of nodal point unknowns, and superscript T indicates a vector transpose. The basis functions in (3.19) are of the usual type, being low order polynomials and constructed such that the nth (or mth) function has a unit value at node n (or m) and is zero at all other nodes (i.e., a function with local support). Also, as noted in (3.19) the pressure basis functions are different from the velocity functions with the pressure being defined at M nodes while the velocities are defined at N nodes ($N > M$). The necessity for this form of mixed interpolation procedure is related to certain requirements for the satisfaction of the incompressibility condition (3.15). The complexity of this issue is beyond the scope of the present discussion. A detailed treatise on this topic and the general question of basis functions for incompressible fluid flow applications is available in the paper by Sani, et al. [17]. Very often the temperature, the extent of reaction, and the velocity components are approximated with the same basis functions since their appearance in the governing equations requires the same smoothness.

Substituting the above approximations for u_1, u_2, p, T, and a into the basic field equations yields a set of relations of the following form

$$\begin{aligned}
\text{Momentum:} \quad & f_1(\mathbf{\Phi}, \mathbf{\Psi}, \mathbf{\Theta}, \mathbf{u}_i, \mathbf{P}, \mathbf{T}, \mathbf{a}) = R_1 , \\
\text{Incompressibility:} \quad & f_2(\mathbf{\Phi}, \mathbf{u}_i) = R_2 , \\
\text{Energy:} \quad & f_3(\mathbf{\Phi}, \mathbf{\Theta}, \mathbf{u}_i, \mathbf{T}, \mathbf{a}) = R_3 , \\
\text{Reaction-Diffusion:} \quad & f_4(\mathbf{\Phi}, \mathbf{\Theta}, \mathbf{u}_i, \mathbf{T}, \mathbf{a}) = R_4 ,
\end{aligned} \quad (3.20)$$

where the R_i represent the residuals or errors resulting from the use of the approximations in (3.19). The Galerkin form of the method of weighted residuals reduces the residuals in (3.20) to zero, in a weighted sense, by making the errors orthogonal to the global interpolation functions. The orthogonality condition is expressed by the inner products given by

$$\begin{aligned}
\text{Momentum}: \quad & \int_\Omega \mathbf{\Phi} f_1 d\Omega = \int_\Omega \mathbf{\Phi} R_1 d\Omega = 0 , \\
\text{Incompressibility}: \quad & \int_\Omega \mathbf{\Psi} f_2 d\Omega = \int_\Omega \mathbf{\Psi} R_2 d\Omega = 0 , \\
\text{Energy}: \quad & \int_\Omega \mathbf{\Theta} f_3 d\Omega = \int_\Omega \mathbf{\Theta} R_3 d\Omega = 0 , \\
\text{Reaction-Diffusion}: \quad & \int_\Omega \mathbf{\Xi} f_4 d\Omega = \int_\Omega \mathbf{\Xi} R_4 d\Omega = 0
\end{aligned} \quad (3.21)$$

on the fluid domain Ω.

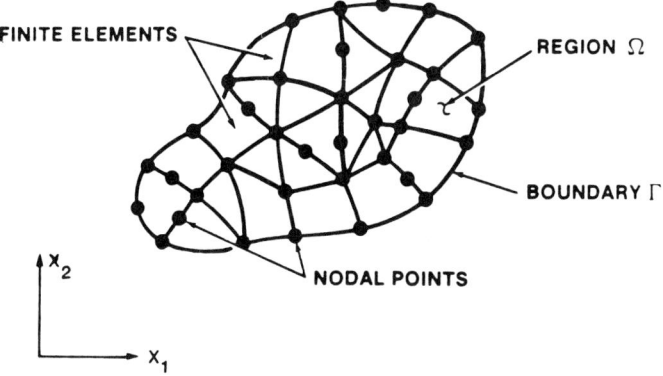

Figure 11. Finite element mesh on an arbitrary region Ω in two-dimensional space.

Omitting the algebraic manipulations, it can be shown that the nonlinear equations reduce to matrix equations:

Momentum:
$$\mathbf{M}\frac{\partial \mathbf{u}_i}{\partial t} + [\mathbf{C}(\mathbf{u}_i) + \mathbf{K}(\mathbf{T},\mathbf{a})]\mathbf{u}_i + \mathbf{QP} = \mathbf{F}(\mathbf{T}) , \qquad (3.22)$$

Incompressibility:
$$\mathbf{Q}^T \mathbf{u}_i = \mathbf{0} , \qquad (3.23)$$

Energy:
$$\mathbf{N}(\mathbf{T},\mathbf{a})\frac{\partial \mathbf{T}}{\partial t} + [\mathbf{D}(\mathbf{u}_i,\mathbf{T},\mathbf{a}) + \mathbf{L}(\mathbf{T},\mathbf{a})]\mathbf{T} = \mathbf{G}(\mathbf{u}_i,\mathbf{T},\mathbf{a}) , \qquad (3.24)$$

Reaction-Diffusion:
$$\frac{\partial \mathbf{a}}{\partial t} + [\mathbf{J}(\mathbf{u}_i) + \mathbf{A}(\mathbf{T},\mathbf{a})]\mathbf{a} = \mathbf{R}(\mathbf{T},\mathbf{a}) , \qquad (3.25)$$

where the most general dependence of each term on the dependent variables is shown explicitly. It is important to recognize some of the basic features of the matrix equations (3.22)–(3.25) since these characteristics will heavily influence the choice of a solution procedure for the various types of convection problems. These matrix equations represent the discrete analogs of the basic conservation equations with each term representing a particular physical process. The **M** and **N** terms are the mass and capacitance terms with **N** being a function of **T** and **a** due to possible variation of the specific heat with temperature and the extent of reaction. The **C**, **D**, and **J** operators represent the velocity-dependent convective transport terms; specific heat variations dictate the dependence of **D** on the temperature and the extent of reaction. The diffusion terms are represented by **K**, **L**, and **A** which may vary with temperature and extent of reaction due to the viscosity, thermal conductivity, and binary diffusivity dependencies. The **Q** term represents the pressure gradient operator and $\mathbf{Q}^T$ the solenoidal constraint. Finally, the **F**, **G**, and **R** vectors contain the body force, viscous dissipation, volumetric heating and chemical reaction terms plus the natural boundary conditions for the problem *i.e.*, applied tractions or heat fluxes. It should be noted that the matrices **K**, **L**, and **A**, as well as the forcing functions **F**, **G**, and

R, will in general be modified by the occurrence of non-zero natural boundary conditions. An inspection of the structure of the individual matrices in (3.22) shows that **M, N, K, L** and **A** are symmetric, while **C, D,** and **J** are unsymmetric. Thus, for general flows, the solution procedure must deal with an unsymmetric system. Also, it is obvious from the general dependencies that the equations can be highly coupled and very nonlinear.

Clearly, the system of equations (3.22)–(3.25) are continuous in time and thus represent a system of nonlinear, ordinary differential equations for the nodal point unknowns. A wide variety of solution methods exist for such systems. However, practical limitations, due to the large size and special structure of the matrices involved, limit the choices for possible integration algorithms. It is also appropriate to observe that (3.22)–(3.25) has a special form which precludes the use of purely explicit integration procedures. The absence of a time derivative for the pressure renders the system singular in the pressure and forces the pressure to be considered as an implicit variable. Because of this limitation on the pressure, the method used here to solve (3.22)–(3.25) is implicit. In particular, we used a trapezoid rule, which is a second-order accurate method from the Adams-Moulton family of integration methods. Omitting the detailed derivation, the trapezoid rule when applied to (3.22)–(3.25) produces

$$\left[\frac{2}{\Delta t^n}\mathbf{M} + \mathbf{C}^{n+1} + \mathbf{K}^{n+1}\right]\mathbf{u}_i^{n+1} + \mathbf{QP}^{n+1} = \left[\frac{2}{\Delta t^n}\mathbf{M} - \mathbf{C}^n - \mathbf{K}^n\right]\mathbf{u}_i^n - \mathbf{QP}^n + \mathbf{F}^n + \mathbf{F}^{n+1} ,$$

$$\mathbf{Q}^T \mathbf{u}_i^{n+1} = \mathbf{0} ,$$

(3.26)

$$\left[\frac{2}{\Delta t^n}\mathbf{N}^{n+1} + \mathbf{D}^{n+1} + \mathbf{L}^{n+1}\right]\mathbf{T}^{n+1} = \left[\frac{2}{\Delta t^n}\mathbf{N}^n - \mathbf{D}^n - \mathbf{L}^n\right]\mathbf{T}^n + \mathbf{G}^n + \mathbf{G}^{n+1} ,$$

$$\left[\frac{2}{\Delta t^n} + \mathbf{J}^{n+1} + \mathbf{A}^{n+1}\right]\mathbf{a}^{n+1} = \left[\frac{2}{\Delta t^n} - \mathbf{J}^n - \mathbf{A}^n\right]\mathbf{a}^n + \mathbf{R}^n + \mathbf{R}^{n+1} ,$$

where the superscript n indicates the time level at which the function is evaluated and Δt^n is the timestep, $\Delta t^n = t^{n+1} - t^n$.

The equations (3.26) are recognized as a highly coupled, nonlinear set of algebraic equations. It is possible to split the system and consider a sequential solution of the momentum/continuity equation followed by the energy equation and the reaction-diffusion equation. In addition to the splitting of the system, a quasilinearization of the system (about the n time level) is needed due to the nonlinearities and dependencies on variables at the advanced time level. Invoking these procedures leads to an algorithm of the following form

Cycle $n + 1$:

Step 1 –

$$\left[\frac{2}{\Delta t^n}\mathbf{M} + \mathbf{C}^n + \mathbf{K}^n\right]\mathbf{u}_i^{n+1} + \mathbf{QP}^{n+1} = \overline{\mathbf{F}}^n ,$$

$$\mathbf{Q}^T \mathbf{u}_i^{n+1} = \mathbf{0} ,$$

(3.27)

Step 2 –
$$\left[\frac{2}{\Delta t^n}\mathbf{N}^n + \mathbf{D}(\mathbf{u}_i^{n+1}, \mathbf{T}^n, \mathbf{a}^n) + \mathbf{L}^n\right]\mathbf{T}^{n+1} = \overline{\mathbf{G}}(\mathbf{u}_i^{n+1}, \mathbf{T}^n, \mathbf{a}^n),$$

Step 3 –
$$\left[\frac{2}{\Delta t^n} + \mathbf{J}\left(\mathbf{u}_i^{n+1}\right) + \mathbf{A}\left(\mathbf{T}^{n+1}, \mathbf{a}^n\right)\right]\mathbf{a}^{n+1} = \overline{\mathbf{R}}\left(\mathbf{u}_i^{n+1}, \mathbf{T}^{n+1}, \mathbf{a}^n\right).$$

The functions $\overline{\mathbf{F}}$, $\overline{\mathbf{G}}$, and $\overline{\mathbf{R}}$ represent the right-hand side of (3.26). The procedure given by (3.27) may be used to advance the $\mathbf{u}_i$, $\mathbf{P}$, $\mathbf{T}$, and $\mathbf{a}$ variables in time without iteration within a timestep. The combination of quasilinearization and sequential processing of the equations leads to an algorithm that is computationally inexpensive per timestep but requires relatively small timesteps to achieve acceptable accuracy. Iteration to converge the solution at the $n+1$ time plane can be used (which permits each equation to be solved on its own time scale), but this substantially increases the computer cost of this simplified algorithm. This procedure, with a slightly different integration method, has been used on similar problems by Gartling [18,19] with good success.

3.4 Numerical Results and Discussion

This numerical procedure has been used to simulate a laboratory experiment involving the curing of a small sample of epoxy [14]. The problem consists of a 5 cm cubicle crucible that is filled with a well-stirred mixture of an epoxy resin and a primary amine curing agent. The crucible is held in a constant temperature environment oven until the gelation process is completed. As part of the experiment, temperature histories were measured at several points within the crucible and a laser Schlieren technique was used to observe the progression of the gel process. The epoxy properties and reaction parameters used in the simulations were determined from the work of Hirschbuehler [14] and Kamal, *et al.* [20].

For purposes of the numerical simulation, a two-dimensional vertical section of crucible was modeled using eight-node quadrilateral elements and various mesh refinements with element densities ranging from 5×10 to 16×25 elements. A schematic of the geometry and boundary conditions is shown in Figure 12 along with the finite element grid. The curing process for this particular epoxy extends over a period of approximately one hour. Using timesteps of 30 seconds, the simulation required approximately 450 CPU seconds on the CDC 7600.

Based on the results of the numerical simulation, the following observations can be made. For the first 45 to 50 minutes, the epoxy undergoes a slow heating as a result of the exothermic chemical reaction. The heating is very uniform over the volume with the fluid motion being similar to that observed in other studies of volumetrically heated fluids [19]. Time dependent secondary flows appear at the midplane of the cavity and interact with the main vortex flow. The continued rise in fluid temperature eventually leads to an increase in the reaction rate. At

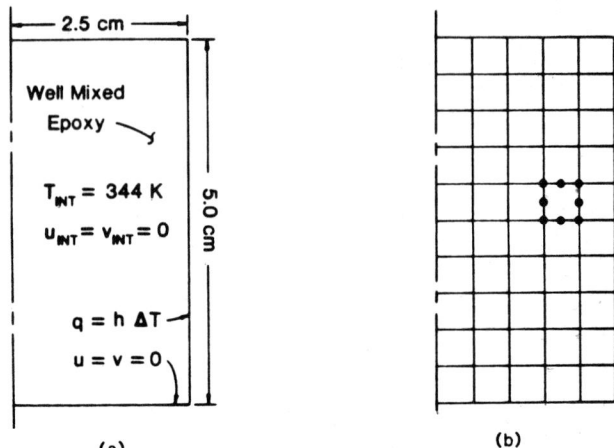

Figure 12. Geometry and boundary conditions (a) for the epoxy curing problem. The finite element mesh (b) used 8-node quadrilaterals.

approximately 50 minutes into the simulation, temperatures near the top of the cavity reach a critical value and the reaction accelerates rapidly. Gelation, which occurs first at the highest temperature points, begins near the top of the cavity and proceeds downward in the form of a solidification front. Gelation continues for approximately 5 minutes with a solidification front moving continually downward at approximately 0.5 cm min. Secondary flows were observed to appear and disappear in the fluid layer at the bottom of the cavity ahead of the gel front. These features appear to be in qualitative agreement with experimental observation and it is conjectured that they are a consequence of the rapidly changing aspect ratio. The apparent instabilities are evident in the streamline profiles shown in Figure 13 where we see significant changes in the number of cells and the direction of flow.

Shown in Figure 14 is a comparison of the computed and measured solidification front trajectory and the temperature histories at two points in the cavity. Considering the complexity of the problem and the uncertainty in the material properties and initial conditions, the agreement is good. However, the lack of agreement of late time in the temperature histories indicates that a boundary condition or the material properties are not known to a sufficient accuracy. Precisely, what influence this has on the oscillatory flow field is unknown.

The numerical solution procedure employed here is not as rigorous as one might like. Time lagging the coefficients raises serious questions about the precise nature of the flow and its oscillatory character. The source of these oscillations will only be determined for certain after more careful experiments have been performed and the equations have been subjected to more detailed analysis. We have focused our own research on more robust solution procedures for the fully-coupled system of equations which use predictor-corrector integration routines. In this instance, the solution procedure is such that the solution to the **entire** set of equations is advanced uniformly in time. Of course, this approach is considerably more expensive because of the large bandwidth of the solution matrix.

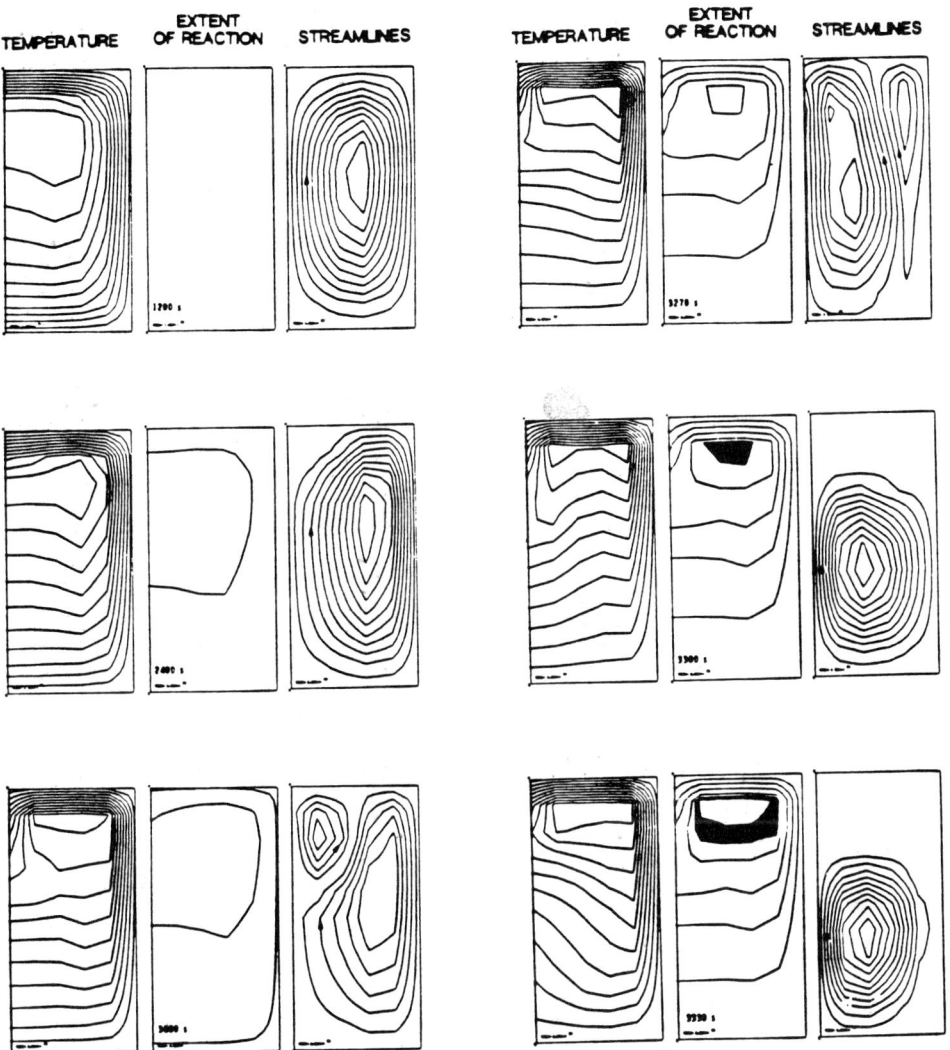

Figure 13. Temperature contours, extent of reaction contours, and streamlines in one-half of the container. The region inside the darkened zone on the extent of reaction contours is solidified material ($a = 1$). The arrow on the streamlines indicates the direction of flow.

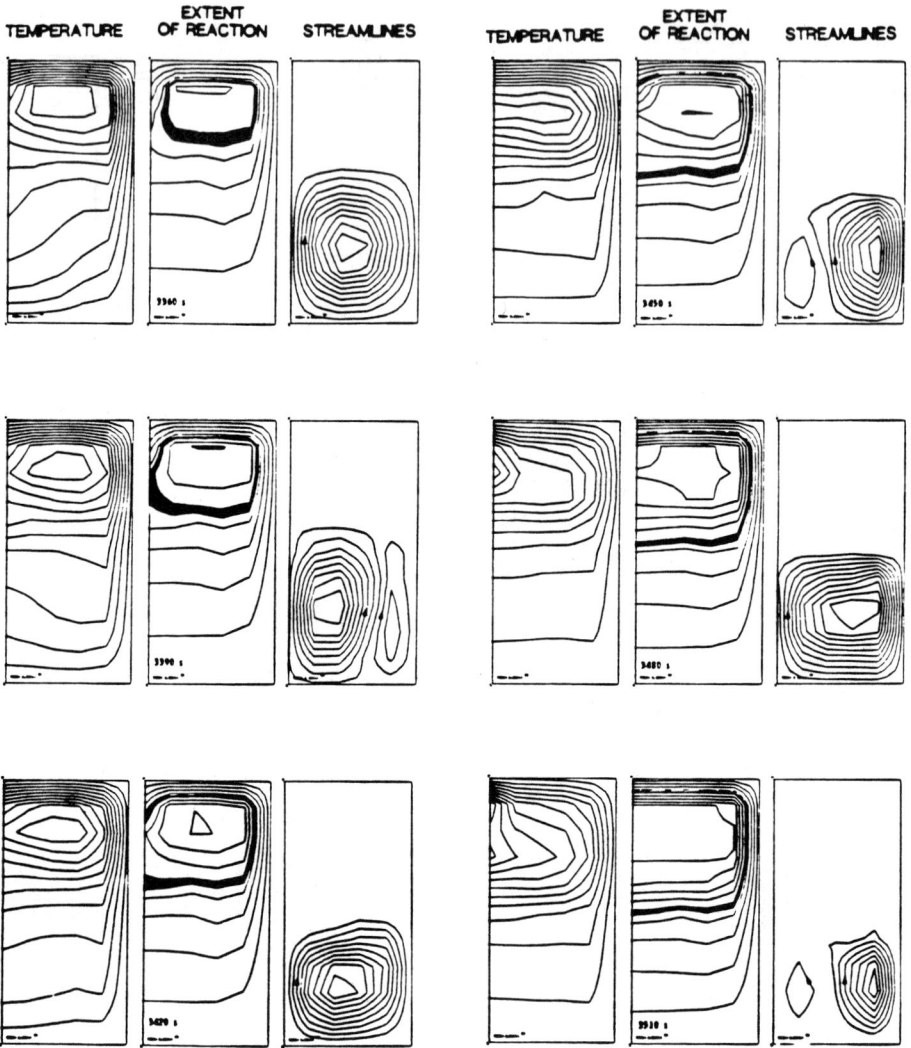

Figure 13. (Continued)

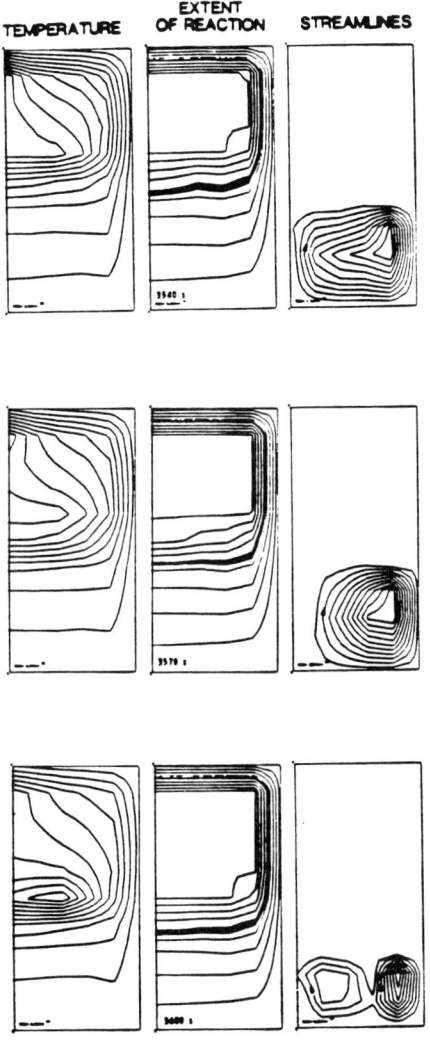

Figure 13. (Continued)

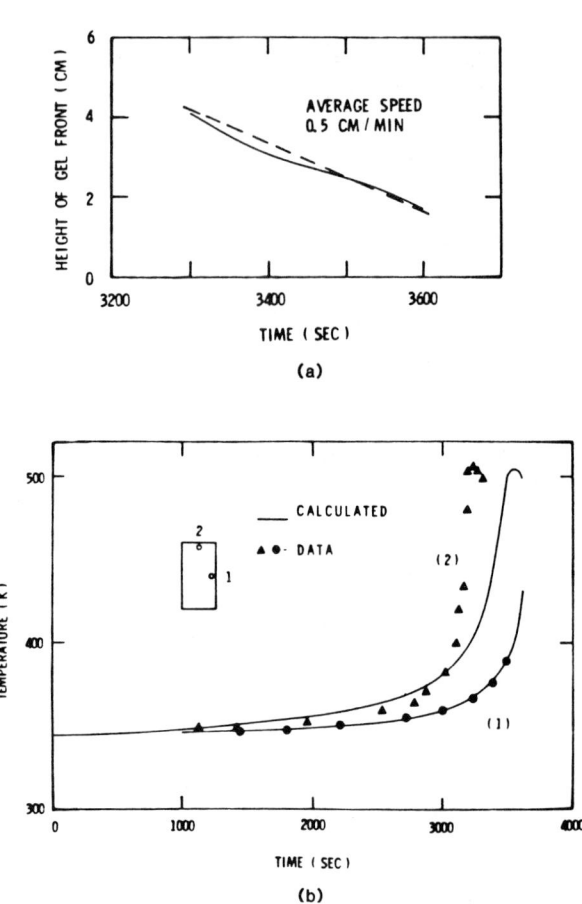

Figure 14. Comparison of numerical results with experimental data on the velocity of the solidification front (a) and temperature histories at two points in the contianer (b).

4 Closure

The examples presented here serve to demonstrate that in solving nonlinear partial differential equations numerically, it is quite easy to obtain multiple solutions and oscillatory phenomena. Furthermore, we mentioned an example involving a viscoelastic fluid [5] in which there was a lack of convergence of the numerical algorithm. Such computational difficulties are not atypical and can arise either due to the numerical method used or from the complexities of the physical model. Determining the source of the difficulties is extremely important and requires considerably more detailed mathematical analysis of the governing partial differential equations at the outset. Such analysis can assist the numerical analyst in choosing the proper numerical algorithm, identifying the appropriate physical solution, or at least enabling him to recognize that the problem posed does not lead to physically meaningful results.

Acknowledgement

We are grateful to the Institute of Mathematics and Its Applications (IMA), University of Minnesota, for the opportunity to participate in this workshop and for its partial support of this work. The comments of R. C. Givler and J. W. Swegle were extremely helpful in the preparation of the final manuscript. This research was also supported in part by the U. S. Department of Energy under contract number DE-AC04-76DP00789.

References

1. P. A. Persson and G. Bjarnholt, "A Photographic Technique for Mapping Failure Waves and Other Instability Phenomena in Liquid Explosives Detonation," *Proc. Fifth Symposium (International) on Detonation*, Office of Naval Research, 115 (1970).

2. M. E. Kipp and J. W. Nunziato, "Numerical Simulations of Detonation Failure in Nitromethane," *Proc. Seventh Symposium (International) on Detonation*, Office of Naval Research, 394 (1981).

3. P. Hartman, *Ordinary Differential Equations*, Wiley (1964).

4. D. D. Joseph, M. Renardy, J.-C. Saut, "Hyperbolicity and Change of Type in the Flow of Viscoelastic Fluids," *Arch. Rational Mech. Anal.* **87**, 213 (1985).

5. D. K. Gartling, "One-Dimensional Finite Element Solutions for a Maxwell Fluid," *J. Non-Newtonian Fl. Mech.*, **17**, 203 (1985).

6. J. von Neumann, "Theory of Detonation Waves," Progress Report PB31090 (1942).

7. R. Courant and K. O. Friedrichs, *Supersonic Flow and Shock Waves*, Interscience (1948).

8. M. L. Wilkins, "Calculation of Elastic-Plastic Flow," *Methods in Computational Physics*, Vol. 3, Academic Press, 211 (1964).

9. J. W. Swegle, "TOODY IV — A Computer Program for Two-Dimensional Wave Propagation," Sandia National Laboratories Report SAND-78-0552 (1978).

10. L. F. Shampine and M. K. Gordon, *Computer Solution of Ordinary Differential Equations: The Initial Value Problem*, Freeman (1974).

11. R. D. Richtmyer and K. W. Morton, *Difference Methods for Initial Value Problems*, Interscience, 2nd Edition (1967).

12. D. J. Hicks, *Math. Comp.* **32**, 1123 (1978).

13. M. E. Malin, LANSL, unpublished data cited by A. W. Campbell and R. Engelke, "The Diameter Effect in High-Density Heterogeneous Explosives," *Proc. Sixth Symposium (International) on Detonation*, Office of Naval Research, 642 (1976).

14. A. W. Campbell, M. E. Malin, and T. E. Holland, *Proc. Second Symposium on Detonation*, Office of Naval Research, 336 (1955).

15. K. R. Hirschbuehler, "An Interferometric Study of Epoxy Resin Gelation," *Proc. Crit. Rev.; Techniques for the Characterization of Composite Materials*, A.M.M.R.C., MS82-3 (1982).

16. D. K. Gartling, C. E. Hickox, and J. W. Nunziato, "Finite Element Analysis of an Epoxy Curing Process," *Proc. Fifth International Symposium on Finite Elements and Flow Problems*, Univ. of Texas at Austin, 113 (1984).

17. R. L. Sani, P. M. Gresho, R. L. Lee, and D. F. Griffiths, "The Cause and Cure (?) of the Spurious Pressures Generated by Certain FEM Solutions of the Incompressible Navier-Stokes Equations: Part 1," *Int. J. Num. Meth. Fluids* **1**, 17 (1981); "Part 2," *ibid.*, 171 (1981).

18. D. K. Gartling, "Finite Element Analysis of Convective Heat Transfer Problems with Change of Phase," *Computer Methods in Fluids*, Pentech Press, 257 (1980).

19. D. K. Gartling, "A Finite Element Analysis of Volumetrically Heated Fluids in an Axisymmetric Enclosure," *Finite Elements in Fluids*, Vol. 4, Wiley, 233 (1982).

20. M. R. Kamel, S. Sourour, and M. Ryan, "Integrated Thermo-Rheological Analysis of the Cure of Thermosets," *31st SPE ANTEC*, **19**, 187 (1973).

Very High Order Accurate TVD Schemes

Stanley Osher[*]
Mathematics Department
University of California
Los Angeles, CA 90024

and

Sukumar Chakravarthy[**]
Rockwell Science Center
1049 Camino Dos Rios
Thousand Oaks, CA 91360

Abstract

A systematic procedure for constructing semi-discrete families of $2m - 1$ order accurate, $2m$ order dissipative, variation diminishing, $2m + 1$ point band width, conservation form approximations to scalar conservation laws is presented. Here m is an integer between 2 and 8. Simple first order forward time discretization, used together with any of these approximations to the space derivatives, also results in a fully discrete, variation diminishing algorithm. These schemes all use simple flux limiters, without which each of these fully discrete algorithms is even linearly unstable. Extensions to systems, using a nonlinear field-by-field decomposition are presented, and shown to have many of the same properties as in the scalar case. For linear systems, these nonlinear approximations are variation diminishing, and hence convergent. A new and general criterion for approximations to be variation diminishing is also given. Finally, numerical experiments using some of these algorithms are presented.

Introduction

Recently there has been an enormous amount of activity related to the construction and analysis of "high resolution" schemes approximating hyperbolic systems of conservation laws. Some samples of the successful consequences of this activity can be found in the proceedings of the sixth AIAA Computational Fluid Dynamics Conference [3], [17], [31]. Extensive bibliographies can also be found in these papers.

Our aim here is to extend the use of these methods by making them even more accurate. We shall give a systematic procedure for constructing semi-discrete approximations to scalar conservation laws. Except for isolated critical parts, these schemes will have 2m-1 order accuracy, 2m order dissipation, and a bandwidth using 2m+1 points, for m any integer between two and eight. They are in conservation form and TVD - the variation of the discrete solution is non-increasing in time. Hence, no spurious oscillations are possible.

The high resolution schemes constructed earlier [13], [21], [28] use five points ($m = 2$), and have second order accuracy. Some of these were proven to satisfy a single entropy inequality and hence to be convergent when

---

[*]Research suppported by NSF Grant No. MCS 82-00788, ARO Grant No. DAAG 29-82- 0090, NASA Grant No. NAG -1-270, and NASA Consortium Agreement No. NCA 2- 1R390-403
[**]Research supported by NASA Grant No. NAG-1-269

$f(q)$ in (1.1) below is convex [20], [21]. It is possible that the piecewise parabolic method of Woodward and Colella [7], is third order accurate, and shares some of the properties discussed here when $m = 2$.

In a parallel work [4], we shall extend the construction below for $m = 2$, in order to approximate systems of conservation laws in multi-dimensions, using triangle-based algorithms. That work stresses the computational aspects of the algorithms, especially as they relate to the Euler equations of compressible gas dynamics.

Conventional schemes such as Lax-Wendroff even with an entropy fix [16] seem to lack a variation bound, although the convergence of this method for scalar convex $f(q)$ can now be proved (DiPerna, private communication). From a practical point of view, this lack of a variation bound seems to lead to a lack of robustness when computing complex flows with strong shock waves and steep gradients.

Another drawback of most finite-difference schemes is that discontinuities are approximated by discrete transitions, that when narrow, usually overshoot or undershoot, or when monotone, usually spread the discontinuity over many grid points.

Upwind schemes have been designed and used over the years, largely because of their success in treating this difficulty. Those based on solving the Riemann problem either exactly Godunov's method [9] or approximately, e.g. (Osher's [18], or Roe's [23] with an entropy fix [24], [4]), have been extremely successful, especially when put in a second-order accurate, high resolution framework, e.g. [3],[17], [31].

We should particularly mention the early investigations of van Leer [28], [29]. There he introduced the concepts of flux limiters, and higher order Riemann solvers. Recently Harten [13] obtained sufficient conditions which he showed to be compatible with second order accuracy, and which guarantee that a scalar one-dimensional approximation is TVD - total variation diminishing. He constructed a scheme having that property and formally extended it to systems, using a field-by-field limiter, and Roe's decomposition.

We would also like to mention the work of Boris and Book [34], and Zalesak [32], concerning FCT schemes. They used flux limiters to suppress oscillations in their schemes.

Harten's construction in [13] was done first for a fully discrete, explicit in time approximation. P. Sweby [26] has investigated the properties of various limiters in this context. We shall not use Sweby's ideas here since we seek higher order accuracy, and his symmetry restriction would make our approximations only second order accurate in the semi-discrete context.

We shall use the now-introduced term "high resolution scheme" to mean a formal extension to systems via a field-by-field decomposition, of a scalar, higher than first order accurate, variation diminishing scheme. These schemes do not, in general, satisfy the entropy condition - e.g. expansion shocks exist as stable solutions of high resolution schemes based on Roe's (unmodified) scheme. In [21] we used Osher's decomposition and certain limiters to prove that limit solutions of a class of second order accurate high resolution schemes do satisfy the entropy condition for hyperbolic systems of conservation laws. We also proved convergence of another class of high resolution approximations to scalar convex conservation laws in [20] as well as in [21]. We believe that the ideas concerning the entropy condition in these two papers can be extended to the high order accurate schemes constructed in the following sections, but we do not attempt this here. The interested reader might also consider the remarks on entropy fixes in [4], [19] and [16].

The high-order accurate TVD schemes are first obtained here for semi-discrete (continuous in time) approximations, and can thus serve as a guideline for a wide variety of time discretizations, both implicit and explicit. See [2] for efficient implicit calculations approximating Euler's equations in transonic and supersonic aeronautics. TVD schemes also have a certain diagonal dominance that is very useful in implicit methods [2], [12].

An interesting and useful fact concerning time discretization is the following (mentioned in Theorems (3.1) and (3.2) below). All of the semi-discrete approximations constructed below are unconditionally (even linearly) unstable when (a) they are used together with simple first order accurate forward Euler time discretizations, and (b) the flux limiters are removed. However, they are all conditionally stable when the limiters, which enforce the variation bound, are kept. Thus, although the limiters might not act at all on a resulting steady state solution, they act non-linearly during transient calculations to enforce the variation bound. This elementary time differencing is sometimes useful, e.g. when steady state calculations on coarse grids are to be obtained simply.

Goodman and Leveque have recently shown [10] that two space dimensional scalar approximations cannot be TVD and still be more than first order accurate, given that the associated flux functions are reasonably smooth. Nevertheless two dimensional schemes based on dimension by dimension TVD differencing have worked quite well, even for complex configurations with very strong shocks. See e.g. [3], [5], [7]. In particular, it seems that our remark in the previous paragraph about conditional stability of Euler forward time discretization is also experimentally valid here. Perhaps a more sophisticated, scheme dependent, notion of variation is needed for the theory in several space dimensions.

The format of this paper is as follows. In Section I, we review the relevant theory of weak solutions of conservation laws and their approximations. In Section II, we exemplify our general theory by constructing families of second and third order accurate TVD schemes using five points. In Section III, we perform the general construction for scalar conservation laws and state Theorems (3.1) and (3.2) which contain the main results of this paper. In Section IV we prove the theorems. In Section V we obtain an apparently new and general criterion for an approximation to be TVD, which we hope will be useful. In Section VI we extend our construction to high resolution schemes approximating systems. Section VII contains some numerical evidence demonstrating the utility of these methods. Many more experimental results are given in [5].

From the point of view of utility, it should be noted that all our TVD schemes degenerate to formal first order accuracy at isolated critical points. It appears that this leads to a maximal accuracy for steady state calculations, of second order in the L^∞ norm and third order in L^1. Nevertheless, it should also be noted that all the formulas of Section 3 give schemes having the claimed accuracy on strips for which $|u_x| > 0$ - hence any compact, accurate, upwind biased, scheme will be reduced to one of these in its unlimited form.

I. Review of Theory of Weak Solutions and Their Approximations

We shall consider numerical approximations to the initial value problem for nonlinear hyperbolic systems of conservation laws.

$$\frac{\partial q}{\partial t} + \frac{\partial}{\partial x} f(q) = 0, t > 0, -1 \le x < 1 \tag{1.1}$$

with periodic boundary conditions:

$$q(x+1,t) = q(x,t),$$

given initial conditions $q(x,0)$.

Here $q(x,t)$ is an d-vector of unknowns, and the flux function $f(q)$ is vector-valued, having d components. The system is hyperbolic when the Jacobian matrix has real eigenvalues.

It is well-known that solutions of (1.1) may develop discontinuities in finite time, even when the initial data are smooth. Because of this, we seek a weak solution of (1.1).

These weak solutions are not necessarily unique. For physical reasons, the limit of the viscous equation, as viscosity tends to zero is sought. This leads to an infinite family of inequalities in the scalar case which when satisfied by so-called "entropy" solutions to (1.1) yield well-posedness in L^1 of the evolution problem. This result is due to Kruzkov [15].

For systems of equations, Lax has defined an entropy inequality using an entropy function [35]. The entropy inequality satisfied by "entropy" solutions to systems has an important geometric consequence concerning admissible discontinuities.

This theory is quite well developed and often reviewed - see e.g. [19], section II. One new result is the following; in the scalar convex case, a single entropy inequality is equivalent to the required infinite number, if the solution is of bounded variation. (See [8].) This fact was crucial to the convergence results in [20] and [21].

Next we consider a semi-discrete, method of lines, approximation to (1.1). We break the interval $(-1,1)$ into subintervals:

$$I_j = \{x / (j-\tfrac{1}{2})\Delta x \leq x \leq (j+\tfrac{1}{2})\Delta x\}$$

$j = 0, \pm 1, \ldots, \pm N$, with $(2N+1)\Delta x = 2$

Let $x_j = j\Delta x$, be the center of each interval I_j, with end points $x_{j-\frac{1}{2}}, x_{j+\frac{1}{2}}$.

Define the step function for each $t > 0$, as

$$Q_{\Delta x}(x,t) = q_j(t),$$

for $x \in I_j$.

The initial data is discretized via the simple method:

$$q(x_j, 0) = q_j(0), \text{ for each } j$$

For any step function, we define the difference operators

$$\Delta_\pm q_j = \pm(q_{j\pm 1} - q_j)$$

$$D_\pm q_j = \frac{1}{\Delta x} \Delta_\pm q_j$$

A method of lines, conservation form, discretization of (1.1), is a system of differential equations

$$\frac{\partial}{\partial t} q_j + D_+ \hat{f}_{j-\frac{1}{2}} = 0, \, j = 0, \pm 1, \ldots, \pm N. \tag{1.2}$$

$$Q_\Delta(x,0) = T_{\Delta x} q(x,0) \text{ for } x \in I_j$$

Here, the numerical flux defined by:

$$\hat{f}_{j-\frac{1}{2}} = \hat{f}(q_{j+k-1}, \ldots, q_{j-k}), \tag{1.3}$$

for $k \geq 1$, is a Lipschitz continuous function of its arguments, satisfying the consistency condition:

$$\hat{f}(q, q, \ldots, q) = f(q)$$

It is well known that bounded a.e. limits as $\Delta x \to 0$, of approximate solutions converge to weak solutions of (1.1). This does not necessarily imply that limit solutions will satisfy any of the above-mentioned entropy conditions. Some restrictions on the numerical flux are required.

The most general class of scalar flux functions known to yield convergent approximations whose limit solutions will satisfy all entropy conditions, for general scalar $f(q)$ is the class of "E" fluxes, introduced in [18].

A consistent scheme whose numerical flux $h_{j-\frac{1}{2}}$ satisfies

$$\text{sgn } (q_j - q_{j-1})[h_{j-\frac{1}{2}} - f(q)] \leq 0 \tag{1.4}$$

for all q between q_{j-1} and q_j, is said to be an E flux.

Other equivalent definitions are given in [18] and [27].

Unfortunately these schemes are at most first order accurate [18]. We shall use three point E schemes as building blocks for our higher order accurate TVD schemes described in the next sections. We have already done this to get convergent, second order TVD schemes, approximating convex scalar conservation laws in [20], and [21].

Examples of three-point E schemes include three-point monotone schemes, e.g. Engquist-Osher's [33], Godunov's [9] (which is canonical - see [18]), or entropy fixes of Roe's scheme [24]. These are defined again at the end of this section.

Together with an entropy inequality, a key estimate involved in most convergence proofs is a bound on the variation. For any fixed $t \geq 0$, the x variation of scalar $Q_{\Delta x}(x,t)$ is defined as

$$B(Q_{\Delta x}) = \sum_j |\Delta_+ q_j|$$

If we can write for every

$$\Delta_+ \hat{f}_{j-\frac{1}{2}} = -C_{j+\frac{1}{2}} \Delta_+ q_j + D_{j-\frac{1}{2}} \Delta_- q_j \tag{1.5a}$$

$$C_{j+\frac{1}{2}} \geq 0 \tag{1.5b}$$

$$D_{j-\frac{1}{2}} \geq 0 \tag{1.5c}$$

then it is easy to show, [21], using an argument of [25], that for $t_1 \geq t_2 \geq 0$.

$$B(Q_{\Delta x}(\cdot, t_1)) \leq B(Q_{\Delta x}(\cdot, t_2)) \tag{1.6}$$

Harten in [13], pointed out for explicit methods, and in [12], for implicit methods, that this decomposition could be obtained for schemes which are higher than first order accurate. See also earlier work by van Leer [28]. In section V we obtain a more general criterion than (1.5), guaranteeing that (1.6) is valid. We shall use criterion

(1.5) here to get very high order accurate, TVD schemes of the type

$$\frac{\partial}{\partial t} q_j = -\Delta_+ \hat{f}_{j-\frac{1}{2}} = C_{j+\frac{1}{2}} \Delta_+ q_j - D_{j-\frac{1}{2}} \Delta_- q_j \tag{1.7a}$$

with

$$C_{j+\frac{1}{2}} = C(q_{j+m}, \ldots, q_{j-m+1}) \geq 0 \tag{1.7b}$$

$$D_{j-\frac{1}{2}} = D(q_{j+m-1}, \ldots, q_{j-m}) \geq 0 \tag{1.7c}$$

which are $2m-1$ order accurate, except at isolated critical points, for $2 \leq m \leq 8$.

In addition to (1.6) we have a maximum principle for (1.7)

$$\min_k q_k(0) \leq q_j(t) \leq \max_k q_k(0), \tag{1.8}$$

for each j and all $t \geq 0$, [21].

Moreover, in [21], we also showed a limit on the possible accuracy of approximations of type (1.17), for $m = 2$. A glance at the proof of that Lemma (2.3) shows that the result is also valid for general m, namely:

Approximation (1.7) is at most first order accurate at nonsonic critical points of q, i.e. points $\bar{q}$ at which $f'(\bar{q}) \neq 0 = \bar{q}_x$.

In spite of this local degeneracy, higher order accuracy, combined with TVD does improve performance, even when discontinuities are present. This is shown numerically in ref [5] and elsewhere.

As promised, we now present several useful three-point E fluxes.

Engquist-Osher

$$h^{EO}(q_j, q_{j-1}) = \int_0^{q_j} \min(f'(s), 0) ds + \int_0^{q_{j-1}} \max(f'(s), 0) ds + f(0) \tag{1.9}$$

Godunov

$$h^G(q_j, q_{j-1}) = \min_{q_{j-1} \leq q \leq q_j} f(q), \text{ if } q_{j-1} \leq q_j = \max_{q_{j-1} \geq q \geq q_j}, f(q) \text{ if } q_{j-1} > q_j \tag{1.10}$$

Roe with entropy fix, approximating a convex $f(q)$ i.e., $f'' \geq 0$ with $f'(q) = 0$ at a single sonic point $\bar{q}$.
Define

$$h^{Rf}(q_j, q_{j-1}) = \frac{1}{2}[(f(q_j) + f(q_{j-1})) - \left|\frac{\Delta_- f(q_j)}{\Delta_- q_j}\right| \Delta_- q_j] \tag{1-11}$$

unless

$$q_{j-1} < \bar{q} < q_j,$$

then take any Lipschitz function so that:

$$h^{R_f}(q_j, q_{j-1}) \le f(\bar{q})$$

See e.g. [4], [24], for various fixes of this type.

II. Second and Third Order Accurate TVD Schemes Which Use a Five-Point Module

We begin by exemplifying our general theory using a very important and convenient class of schemes. We shall approximate the scalar conservation law by a family of five point, semi-discrete method of lines, and TVD approximations.

Let $h(q_{j+1}, q_j)$ be the numerical flux corresponding to a three- point E scheme. Next we define

$$df^-_{j+\frac{1}{2}} = h(q_{j+1}, q_j) - f(q_j) \tag{2.1a}$$

$$df^+_{j+\frac{1}{2}} = f(q_{j+1}) - h(q_{j+1}, q_j) \tag{2.1b}$$

We can then write

$$h(q_{j+1}, q_j) = \frac{1}{2}\left[f(q_{j+1}) + f(q_j)\right] - \frac{1}{2}\left[df^+_{j+\frac{1}{2}} - df^-_{j+\frac{1}{2}}\right]$$

These new quantities df^- and df^+ denote the difference in flux across the waves with negative and positive velocities respectively in the interval under consideration. The subscript $j+\frac{1}{2}$ denotes the interface between two cells whose centroids are denoted by grid points with subscripts j and $j+1$ respectively. Thus $df^+_{j+\frac{1}{2}}$ denotes the difference (taken from right to left) in flux across all the positive (forward) breaking waves at the cell interface $j + \frac{1}{2}$, etc.

A general semi-discrete conservation form approximation to (1.1) can be given as

$$q_t + \frac{\left(\hat{f}_{j+\frac{1}{2}} - \hat{f}_{j-\frac{1}{2}}\right)}{\Delta x} = 0 \tag{2.2}$$

Here the quantity $\hat{f}$ is the representative for numerical flux.

With this notation the numerical flux of one new family of TVD schemes can be represented by

$$\hat{f}_{j+\frac{1}{2}} = h(q_{j+1},q_j) - \alpha(df^-_{j+3/2})^{(1)} - (\frac{1}{2}-\alpha)(df^-_{j+\frac{1}{2}})^{(0)} \qquad (2.3)$$
$$+ (\frac{1}{2}-\alpha)(df^+_{j+\frac{1}{2}})^{(0)} + \alpha(df^+_{j-\frac{1}{2}})^{(-1)},$$

for $0 < \alpha \leq \frac{1}{2}$.

The superscripts shown over the df denote flux-limited values of df, and are computed as follows:

$$\left(df^-_{j+3/2}\right)^{(1)} = \min \mathrm{mod}\left[df^-_{j+3/2}, b\, df^-_{j+\frac{1}{2}}\right] \qquad (2.4\mathrm{a})$$

$$\left(df^-_{j+\frac{1}{2}}\right)^{(0)} = \min \mathrm{mod}\left[df^-_{j+\frac{1}{2}}, b\, df^-_{j+3/2}\right] \qquad (2.4\mathrm{b})$$

$$\left(df^+_{j+\frac{1}{2}}\right)^{(0)} = \min \mathrm{mod}\left[df^+_{j+\frac{1}{2}}, b\, df^+_{j-\frac{1}{2}}\right] \qquad (2.4\mathrm{c})$$

$$\left(df^+_{j-\frac{1}{2}}\right)^{(-1)} = \min \mathrm{mod}\left[df^+_{j-\frac{1}{2}}, b\, df^+_{j+\frac{1}{2}}\right] \qquad (2.4\mathrm{d})$$

In the above, the operator "min mod" is defined by:

$$\min \mathrm{mod}\,[x,y] = (\mathrm{sgn}\, x)\,\max(0, \min|x|, y\,\mathrm{sgn}\, x) \qquad (2.5)$$

(see e.g. [26]), and b is a "compression" parameter chosen in the range

$$1 < b \leq 1 + \frac{1}{2\alpha} = b_{max}. \qquad (2.6)$$

The case $\alpha = 0$ also yields a TVD scheme, but this one is not time dissipative, so steady state solutions are difficult to obtain: We recommend that α be positive in all applications. The dissipation in our general algorithm is an increasing function of α.

The non-TVD or unlimited forms of the schemes in the new family can be obtained by replacing the $(df)^{(v)}$ terms appearing in (2.3) with the corresponding unlimited df values. The truncation error of the unlimited form (up to second order) is given by:

$$TE = (\frac{1}{6}-\alpha)(\Delta x)^2\,\frac{\partial^3}{\partial x^3}f(u) \qquad (2.7)$$

It is interesting to note that *TE* is independent of the particular E-scheme used, i.e. independent of h.

Particular schemes in the new family may be chosen by picking various values for the parameter α. Some special cases are summarized in Table 2.1. The *TE* shown in the last column corresponds to the unlimited forms. The names given to the TVD schemes are based on the names used in the literature, e.g. [29], for the corresponding unlimited schemes.

Value of α	Name of TVD Scheme	b_{max}	2nd order TE
1/6	Third-Order	4	0
1/2	Fully Upwind	2	$-1/3(\Delta x)^2 \frac{\partial^3}{\partial x^3} f(u)$
1/4	Fromm's	3	$\frac{-1}{12}(\Delta x)^2 \frac{\partial^3}{\partial x^3} f(u)$
1/8	Low TE second-order	5	$\frac{1}{24}(\Delta x)^2 \frac{\partial^3}{\partial x^3} f(u)$
0	Central	∞	$\frac{1}{6}(\Delta x)^2 \frac{\partial^3}{\partial x^3} f(u)$
1/3	No Name	5/2	$\frac{-1}{6}(\Delta x)^2 \frac{\partial^3}{\partial x^3} f(u)$

TABLE 2.1 Particular Cases of New Family of TVD Schemes

Semi-discrete notions of TVD schemes only show that, when a suitable time discretization is chosen, the overall algorithm is TVD, hence has a convergent subsequence as $\Delta t \to 0$. See e.g. [21]. There is always a CFL restriction on explicit schemes. For simplicity, we consider the explicit scheme given by forward Euler time discretization:

$$\frac{q_j^{n+1} - q_j^n}{\Delta t} + \frac{\hat{f}^n_{j+\frac{1}{2}} - \hat{f}^n_{j-\frac{1}{2}}}{\Delta x} = 0 \qquad (2.8)$$

This is only first order accurate in time.

As part of a general result Theorem (3.2), it follows that the *unlimited* versions of (2.8) are all *unstable* for any CFL number $\lambda = \frac{\Delta t}{\Delta x}$. However, Table (2.2) gives *stable* time steps for the *flux-limited* versions for $b = b_{max}$.

The general condition for (2-8) to be TVD is:

$$\frac{\Delta t}{\Delta x} \left(\frac{df^+_{j+\frac{1}{2}} - df^-_{j+\frac{1}{2}}}{\Delta_+ q_j} \right) \leq \frac{4\alpha}{1+4\alpha} \quad \text{if } b = b_{max}. \qquad (2.9a)$$

and for general b it is

$$\frac{\Delta t}{\Delta x}\left(\frac{df^+_{j+\frac{1}{2}} - df^-_{j+\frac{1}{2}}}{\Delta_+ q_j}\right) \leq \frac{1}{1+\alpha+b(\frac{1}{2}-\alpha)} \qquad (2.9b)$$

Value of α	b_{max}	$\left(\frac{\Delta t}{\Delta x}\left(\frac{df^+_{j+\frac{1}{2}} - df^-_{j+\frac{1}{2}}}{\Delta_+ q_j}\right)\right)_{max}$
1/6	4	2/5
1/2	2	2/3
1/4	3	1/2
1/8	5	1/3
0	∞	0
1/3	5/2	4/7

TABLE 2.2. Stable Time Steps for This New Class of TVD Schemes.

In equations (2.4a) and (2.4b) the flux-limited values of df are defined. This value is computed in some interval by comparing the original unlimited value with its neighboring value, after that neighbor has been multiplied by the "compression" parameter b. Assuming that the two values being compared are of the same sign, the "min mod" operator chooses the one whose absolute value is the smallest. If $b > 1$, the flux-limited value returned most often will be the unlimited value itself. Thus, for most grid points (away from high second-gradient regions where the unlimited value of slope df can be much greater than the unlimited value of the neighboring slope), the TVD scheme is identical to the corresponding unlimited scheme. (Having a larger value of b enhances this property.) At critical points of the fluxes, the neighboring values of df can be of opposite sign. There, the "min mod" operator returns the value zero. Thus, away from maxima, minima, and points of discontinuity, the TVD scheme reduces to its corresponding unlimited scheme.

We next present a class of schemes having the same five-point band width and which are all third-order accurate in their unlimited versions. However, the flux limiting is a bit more far-reaching than in the α class defined above. This may cause a slight deterioration of accuracy when we use a coarse grid to approximate solutions having many critical points.

The numerical flux is defined by:

$$\begin{aligned}\hat{f}_{j+\frac{1}{2}} = & h(q_{j+1},q_j) - (\frac{1}{12}+\beta)(df^-_{j+\frac{3}{2}})^{(1)} - (\frac{1}{2}-2\beta)(df^-_{j+\frac{1}{2}})^{(0)} \\ & + (\frac{1}{12}-\beta)(df^-_{j-\frac{1}{2}})^{(-1)} - (\frac{1}{12}-\beta)(df^+_{j+\frac{3}{2}})^{(1)} \\ & + (\frac{1}{2}-2\beta)(df^+_{j+\frac{1}{2}})^{(0)} + (\frac{1}{12}+\beta)(df^+_{j-\frac{1}{2}})^{(-1)}\end{aligned} \qquad (2.10)$$

Here we take $0 < \beta \le \frac{1}{12}$. Again, $\beta = 0$ corresponds to a non-dissipative, central difference, but TVD scheme.

The flux-limited values of df are defined through:

$$(df^-_{j+\frac{3}{2}})^{(1)} = \min \operatorname{mod} \left[df^-_{j+\frac{3}{2}}, b df^-_{j+\frac{1}{2}} \right] \tag{2.11a}$$

$$(df^-_{j+\frac{1}{2}})^{(0)} = \min \operatorname{mod} \left[df^-_{j+\frac{1}{2}}, b df^-_{j+\frac{3}{2}} \right] \tag{2.11b}$$

$$(df^-_{j-\frac{1}{2}})^{(-1)} = \min \operatorname{mod} \left[df^-_{j-\frac{1}{2}}, b df^-_{j+\frac{1}{2}}, b df^-_{j+3/2} \right] \tag{2.11c}$$

$$(df^+_{j+3/2})^{(1)} = \min \operatorname{mod} \left[df^+_{j+\frac{3}{2}}, b df^+_{j+\frac{1}{2}}, b df^+_{j-\frac{1}{2}} \right] \tag{2.11d}$$

$$(df^+_{j+\frac{1}{2}})^{(0)} = \min \operatorname{mod} \left[df^+_{j+\frac{1}{2}}, b df^+_{j-\frac{1}{2}} \right] \tag{2.11e}$$

$$(df^+_{j-\frac{1}{2}})^{(-1)} = \min \operatorname{mod} \left[df^+_{j-\frac{1}{2}}, b df^+_{j+\frac{1}{2}} \right] \tag{2.11f}$$

In the above, the operator "min mod" of three quantities is defined through

$$\min \operatorname{mod} [x,y,z] = \min \operatorname{mod} [\min \operatorname{mod} [x,y], z], \tag{2.12}$$

This is easily seen to be independent of the order of x, y, z. Again, b is a "compression" parameter. Here it is chosen in the range.

$$1 < b \le 3 + 12\beta \tag{2.13}$$

The non-TVD or unlimited forms of these schemes are obtained by replacing each $(df)^{(v)}$ term by its corresponding unlimited df value. The third order truncation error of the unlimited form coincides with the dissipation and is proportional to β. See the proof of Theorem (3.1) below.

For $\beta = 1/12$, this scheme (2.10),(2.11) coincides with (2.3),(2.4) for $\alpha = 1/6$. The limiting simplifies a bit here, since the coefficients of (2.11c) and (2.11d) vanish. For this reason, we prefer this scheme to any of the other third order "β" schemes.

Finally, we compute the CFL number guaranteeing that (2.8),(2.10),(2.11) is TVD. The results are:

(2.14a) (for general β satisfying (2.13))

$$\frac{\Delta t}{\Delta x} \frac{df^+_{j+\frac{1}{2}} - df^-_{j+\frac{1}{2}}}{\Delta_+ q_j} \le \left[\frac{13}{12} + \beta + b(\frac{7}{12} - 3\beta)\right]^{-1}$$

(2.14b) (for $b = 3 + 12\beta = b_{max}$)

$$\frac{\Delta t}{\Delta x} \frac{df^+_{j+\frac{1}{2}} - df^-_{j+\frac{1}{2}}}{\Delta_+ q_j} \le \left[\frac{34}{12} - \beta - 36\beta^2\right]^{-1}$$

III. General $2m-1$ and $2m-2$ Order Accurate TVD Schemes. Which Use a $2m+1$ Point Module for $m \le 8$.

We use the notation of the previous section to approximate (1.1) via a family of schemes of the type (2.2) where:

$$\hat{f}_{j+\frac{1}{2}} = \hat{f}^{m,\beta}_{j+\frac{1}{2}} = h(q_{j+1}, q_j) + \sum_{k=-m+1}^{m-1} \left(\mu_k^m + (-1)^k \beta \binom{2m-2}{k+m-1}\right) \left(df^-_{j+k+\frac{1}{2}}\right)^{(k)} \quad (3.1)$$

$$+ \sum_{k=-m+1}^{m-1} \left(v_k^m - (-1)^k \beta \binom{2m-2}{k+m-1}\right) \left(df^+_{j+k+\frac{1}{2}}\right)^{(k)}$$

Here m is an integer, $m \ge 2$, and β satisfies $0 < \beta < \left(m\binom{2m}{m}\right)^{-1}$. (The upper bound on β could be relaxed considerably, at a cost of complicating our calculations. We shall not do this here.) The binomial coefficient is defined for A,B integers with $0 \le B \le A$, as usual:

$$\binom{A}{B} = \frac{A!}{B!(A-B)!}$$

The coefficients v_k^m, μ_k^m can be defined recursively by:

$$v_{m-1}^m = (-1)^{m-1} \left(m\binom{2m}{m}\right)^{-1}, \quad \text{for } m \ge 2. \quad (3.2a)$$

$$v_0^m = \frac{1}{2} \quad (3.2b)$$

$$v_k^m = -v_{-k}^m, \quad k = 1, \ldots, m-1 \quad (3.2c)$$

and

$$v_k^{m+1} = v_k^m + (-1)^k \left((m+1)\binom{2m+2}{m+1}\right)^{-1} \frac{k}{m}\binom{2m}{m-k} \quad \text{for } k = 1, 2, \ldots, m-1. \tag{3.2d}$$

An alternative direct formulation comes by defining:

$$v_k^m = \sum_{j=k}^{m-1} \lambda_j^m = -v_{-k}^m \quad \text{for } k = 1, \ldots, m-1, \tag{3.3a}$$

where

$$\lambda_k^m = \sum_{j=k+1}^{m} (-1)^{j-1} \left(j\binom{2m}{m}\right)^{-1} \binom{2m}{m+j}, \tag{3.3b}$$

and

$$v_0^m = \frac{1}{2} \tag{3.3c}$$

Also:

$$\mu_k^m = v_k^m, \quad \text{for } k = \pm 1, \pm 2, \ldots, \pm(m-1), \tag{3.4a}$$

$$\mu_0^m = -\frac{1}{2} \tag{3.4b}$$

We define the flux limited quantities as follows. For each j:

$$\left[df_{j+\frac{1}{2}}^-\right]^{(k)} = \min \mathrm{mod} \left[df_{j+\frac{1}{2}}^-, bdf_{j-k+\frac{1}{2}}^-, bdf_{j-k+3/2}^+\right] \quad \text{for all } k \text{ with } 0 \neq k \neq 1. \tag{3.5a}$$

$$\left[df_{j+\frac{1}{2}}^-\right]^{(0)} = \min \mathrm{mod} \left[df_{j+\frac{1}{2}}^-, bdf_{j+3/2}^-\right] \tag{3.5b}$$

$$\left[df_{j+\frac{1}{2}}^-\right]^{(1)} = \min \mathrm{mod} \left[df_{j+\frac{1}{2}}^-, bdf_{j-\frac{1}{2}}^-\right] \tag{3.5c}$$

$$\left[df_{j+\frac{1}{2}}^+\right]^{(k)} = \min \mathrm{mod} \left[df_{j+\frac{1}{2}}^+, bdf_{j-k+\frac{1}{2}}^+, bdf_{j-k-\frac{1}{2}}^+\right] \quad \text{for all } k \text{ with } 0 \neq k \neq -1. \tag{3.5d}$$

$$\left[df_{j+\frac{1}{2}}^+\right]^{(0)} = \min \mathrm{mod} \left[df_{j+\frac{1}{2}}^+, bdf_{j-\frac{1}{2}}^+\right] \tag{3.5e}$$

$$\left[df_{j+\frac{1}{2}}^+\right]^{(-1)} = \min \mathrm{mod} \left[df_{j+\frac{1}{2}}^+, bdf_{j+\frac{3}{2}}^+\right] \tag{3.5f}$$

The compression parameter b is allowed to vary between:

$$0 < b < \left(\sum_{j=2}^{m} \frac{1}{2j-1}\right)^{-1} \left(1 + 2\beta \binom{2m-2}{m-1}\right) \quad (3.6)$$

We can now state the following:

Theorem (3.1) ("β" schemes).

The scheme (2.2), (3.1)-(3.6) has the following properties:

(a) It is TVD and satisfies the maximum principle.

(b) For any $\beta, 0 < \beta \leq \left(m\binom{2m}{m}\right)^{-1}$, and $m \leq 7$, b can be taken to be greater than one. For $m = 8$, there exists β_o such that for $0 < \beta_o \leq \beta \leq \left(8\binom{16}{8}\right)^{-1}$, b can be again taken to be greater than one.

(c) The unlimited version is $(2m-1)-order$ accurate and $2m-order$ dissipative, with truncation error and dissipation both proportional to β. Thus, for $b > 1$, the TVD scheme will return $(2m-1)-order$ accuracy except at critical points, or points of discontinuity, where it is formally only first-order accurate.

(d) The simple Euler forward difference time discretized version (2.8) is TVD if the CFL restriction:

$$\frac{\Delta t}{\Delta x}\left(\frac{df^+_{j+\frac{1}{2}} - df^-_{j+\frac{1}{2}}}{\Delta_+ q_j}\right) \leq \left[1+b\left[\frac{1}{2}\sum_{j=2}^{m}\left(\frac{1}{2j-1}\right) + \frac{1}{2} - \sum_{j=2}^{m}\frac{1}{2j(2j-1)}\right.\right.$$

$$\left.\left. - \beta\binom{2m-2}{m-1} - \beta\binom{2m-2}{m-2}\right] + \sum_{j=2}^{m}\frac{1}{2j(2j-1)} + \beta\binom{2m-2}{m-2}\right]^{-1}$$

is valid.

(e) This same forward difference time-discretized scheme, without flux limiters, is linearly unstable for any CFL number.

This theorem will be proven in the next section.

These β-schemes give one more order of accuracy per $2m+1$ module than the α-schemes defined next, except for a special case $\alpha = 2\beta = 2\left(m\binom{2m}{m}\right)^{-1}$ when they coincide.

Again using the notation of the previous section, we approximate (1.1) via a family of schemes of the type

(2.2), where

$$\hat{f}^{m,\alpha}_{j+\frac{1}{2}} = h(q_{j+1}, q_j) + \sum_{k=-m+2}^{m-1} [\mu_k^{m-1} + (-1)^k \alpha \binom{2m-3}{k+m-2}] \left(df^-_{j+k+\frac{1}{2}}\right)^{(k)} \quad (3.7)$$

$$+ \sum_{k=-m+1}^{m-2} [v_k^{m-1} - (-1)^k \alpha \binom{2m-3}{k+m-1}] \left(df^+_{j+k+\frac{1}{2}}\right)^{(k)}$$

Here m is an integer with $m \geq 2$, and $0 < \alpha < \left((m-1)\binom{2m-2}{m-1}\right)^{(-1)}$. (Again, we impose the upper bound on α for simplicity only.) The coefficients v_k^{m-1}, μ_k^{m-1} were defined in (3.2),(3.3),(3.4), and we also define

$$v^{m-1}_{-m+1} = 0 = \mu^{m-1}_{m-1} \quad (3.8)$$

The flux limited quantities are defined precisely as in (3.5), and now the quantity b is allowed to vary between

$$0 < b \leq \frac{1 + 2\alpha \binom{2m-3}{m-1}}{2\alpha \binom{2m-3}{m-1} + \sum_{j=2}^{m-1} (2j-1)^{-1}} \quad (3.9)$$

(where the sum $\sum_{j=2}^{1} (2j-1)^{-1} = 0$, by definition.)

We can now state the following:

Theorem 3.2 ("α" schemes).

The scheme defined via (2.2), (3.7) has the following properties

(a) It is TVD

(b) If $m \leq 8$, then b can be taken to be greater than one.

(c) If $\alpha = 2/m \binom{2m}{m}^{-1}$, then this scheme is identical to the "β" scheme for the same m, with $\beta = 1/m \binom{2m}{m}^{-1}$. Hence its unlimited version is $2m-1-order$ accurate.

(d) For all other admissible values of α, the unlimited version of the scheme is $2m-2$ order accurate with $2m$ order dissipation, which is proportional to α. The truncation error is equal to

$$TE = (-1)^m \left[\alpha - 2/m \binom{2m}{m}^{-1}\right] (\Delta x)^{2m-2} \left(\frac{\partial}{\partial x}\right)^{2m-1} f(u),$$

and is thus independent of the choice of h, the E flux.

(e) The simple Euler forward difference time discretized version, (2.8), is TVD if the CFL restriction:

$$\frac{\Delta t}{\Delta x}\left(\frac{df^+_{j+\frac{1}{2}} - df^-_{j+\frac{1}{2}}}{\Delta_+ q_j}\right) \le \left[1 + b\left(\frac{1}{2} - \sum_{j=2}^{m-1}\frac{1}{2j(2j-1)}\right)\right.$$

$$\left. + \frac{1}{2}\sum_{j=1}^{m-2}\frac{1}{2j-1}\left[-\alpha\binom{2m-3}{m-2}\right] + \alpha\binom{2m-3}{m-2} + \sum_{j=2}^{m-1}\frac{1}{2j(2j-1)}\right]^{-1}$$

(f) This same forward difference time discretized scheme, without flux limiters, is linearly unstable for any CFL number.

This theorem will be proven in the next section.

IV. Proof of Main Theorems

Let the kth power of the shift operator be defined as

$$S^k q_j = q_{j+k}$$

Define the central difference operator for $k=1,2,\ldots$

$$D_o(k\Delta x) = \frac{1}{2k\Delta x}\left(S^k - S^{-k}\right) \tag{4.1}$$

We shall use the following well-known formula - see e.g. [14]. Let q be any smooth function with $q_j = q(j\Delta x) = q(x)$. Then

$$q_x(x) = -2\sum_{k=1}^{m}\frac{(-1)^k(m!)^2}{(m+k)!(m-k)!}D_o(k\Delta x)q(x) \tag{4.2}$$

$$+ (-1)^m\frac{2(m!)(m+1)!}{(2m+2)!}(\Delta x)^{2m}\left(\frac{\partial}{\partial x}\right)^{2m+1}q(x) + O((\Delta x))^{2m+1}$$

Let C^{2m} denote the operator from which uniquely defines $2m^{th}$ order accurate differencing based on central difference operators using a module $(-m,m)$. We define

$$C^{2m} = -2\sum_{k=1}^{m}(-1)^k\binom{2m}{m-k}\binom{2m}{m}^{-1}D_o(k\Delta x) = \sum_{k=-m}^{m-1}\lambda_k^m D_+ S^k \tag{3a}$$

where

$$\lambda_k^m = \sum_{j=\max(-k,k+1)}^{m}(-1)^{j+1}\binom{2m}{m-k}\left(j\binom{2m}{m}\right)^{-1} \tag{4.3b}$$

For our purposes, a better formulation is

$$C^{2m} = D_+ S^{-1} + \sum_{k=-m+1}^{m-1} v_k^m D_+(\Delta_+ S^{k-1}), \text{ (where } D_+ = \frac{1}{\Delta x}\Delta_+), \tag{4.4a}$$

with

$$v_k^m = \sum_{j=k}^{m-1} \lambda_j^m, \quad k=1,\ldots,m-1 \tag{4.4b}$$

$$v_o^m = \frac{1}{2} \tag{4.4c}$$

$$v_k^m = -v_{-k}^m, k = -1,\ldots,-m+1 \tag{4.4d}$$

It is fairly simple to verify (4.3) and (4.4).

Next we recognize that (4.2) implies that the operator

$$C^{2m-2} + (-1)^{m-1} 2\left(m\binom{2m}{m}\right)^{-1}(\Delta x)^{2m-2} D_+^{m-1} D_-^m, \tag{4.5}$$

is a $2m-1$ *order* approximation to $\frac{\partial}{\partial x}$, with stencil $(-(m-1),m)$. Moreover, the same is true for the operator

$$C^{2m} - \lambda_{m-1}^m (\Delta x)^{2m-1} D_+^m D_-^m = C^{2m} - (-1)^{m-1}\left(m\binom{2m}{m}\right)^{-1}(\Delta x)^{2m-1} D_+^m D_-^m \tag{4.6}$$

This operator is easily shown to be unique. Thus, we have the important result:

$$C^{2m} = C^{2m-2} + (-1)^{m-1}\left(m\binom{2m}{m}\right)^{-1}(\Delta x)^{2m-2} D_+^{m-1} D_-^m (2 + \Delta x D_+) \tag{4.7}$$

This translates to (using (4.4))

$$\sum_{k=-m+1}^{m-1} v_k^m \Delta_+\left(\Delta_+ S^{k-1}\right) = \sum_{k=-m+2}^{m-2} v_k^{m-1} \Delta_+\left(\Delta_+ S^{k-1}\right) \tag{4.8}$$

$$+ \sum_{k=1-m}^{m-2} 2\left(m\binom{2m}{m}\right)^{-1}(-1)^{k-1}\binom{2m-3}{m+k-1}\Delta_+\left(\Delta_+ S^{k-1}\right)$$

$$+ \sum_{k=1-m}^{m-1} \left(m\binom{2m}{m}\right)^{-1}(-1)^k \binom{2m-2}{m+k-1}\Delta_+\left(\Delta_+ S^{k-1}\right)$$

which implies

$$v_k^m = v_k^{m-1} + (-1)^k \left(m\binom{2m}{m}\right)^{-1} \binom{2m-2}{m+k-1}(m-1)^{-1}k \qquad (4.9)$$

for $k=1,\ldots,m-1$. Thus we have proven (3.2) (again defining $v_m^m = 0$.)

It is now easy to see that

$$(-1)^k v_k^m > 0, \quad \text{for } k = 1,\ldots,m-1. \qquad (4.10)$$

Next we apply the identity (4.8) to the grid function defined by:

$$q_j = \frac{1 - (-1)^j}{2} = -q_{-j} \text{ for } j \geq 0. \qquad (4.11)$$

This leads us to the useful result:

$$\sum_{k=1}^{m-1} (-1)^k v_k^m = \sum_{k=1}^{m-2} (-1)^k v_k^{m-1} + \frac{1}{4}(2m-1), \quad m = 2,3,\ldots \qquad (4.12)$$

thus

$$\sum_{k=1}^{m-1} (-1)^k v_k^m = \frac{1}{4} \sum_{j=2}^{m} \frac{1}{(2j-1)}, \quad m = 2,3\ldots \qquad (4.13)$$

(The fact that the series above diverges as $m \to \infty$ explains why $b_{\max} \not\to 0$ as $m \to \infty$.)

We may rewrite:

$$C^{2m} = D_+ + \sum_{k=-m+1}^{m-1} \mu_k^m D_+ \left(\Delta_+ S^{k-1}\right) \qquad (4.14a)$$

This, together with (4.4a) gives us the identities:

$$\mu_o^m = -\frac{1}{2} \qquad (4.14b)$$

$$\mu_k^m = v_k^m, \, k \neq 0. \qquad (4.14c)$$

Next, we claim that we can rewrite:

$$C^{2m} f(q_j) = \frac{1}{\Delta x} df_{j+\frac{1}{2}}^- + \sum_{k=-m+1}^{m-1} \mu_k^m D_+ \left(df_{j+k-\frac{1}{2}}^-\right) + \frac{1}{\Delta x} df_{j-\frac{1}{2}}^+ + \sum_{k=-m+1}^{m-1} v_k^m D_+ \left(df_{j+k-\frac{1}{2}}^+\right) \qquad (4.15)$$

We verify this by rewriting the right side above as:

$$D_+ h(q_j, q_{j-1}) + \sum_{\substack{k=-m+1 \\ k \neq 0}}^{m-1} v_k^m D_+ (\Delta_- f(q_j))$$

$$+ \frac{1}{2}D_+\left[-h(q_j,q_{j-1}) + f(q_{j-1}) + f(q_j) - h(q_j,q_{j-1})\right]$$

$$= D_0(\Delta x)f(q) + \sum_{\substack{k=-m+1 \\ k \neq 0}}^{m-l} v_k^m D_+(\Delta_- f(q_j))$$

$$= D_+ S^{-1}f(q) + \sum_{k=-m+1}^{m-1} v_k^m D_+(\Delta_- f(q_j)) = C^{2m}f(q_j)$$

We have thus rewritten the $2m$th order, nondissipative approximation, C^{2m}, in terms of an arbitrary E flux, in a form convenient for the purpose of making it TVD.

Next we note that the approximation to

$$q_t = 0$$

of the form:

$$\frac{\partial q_j}{\partial t} = (-1)^{m-1}\beta(\Delta x)^{2m-1}D_+^m D_-^m q_j = \beta \sum_{k=1-m}^{m-1} (-1)^k \binom{2m-2}{m+k-1} D_+(\Delta_+ S^{k-1})q_j \quad (4.16)$$

is dissipative of order $2m$, and accurate of order $2m - 1$. Its Fourier transform is easily seen to satisfy

$$\frac{\partial}{\partial t}\hat{q}(\zeta) = -\frac{\beta}{\Delta x}[2-2\cos(\zeta \Delta x)]^m \hat{q}(\zeta) \quad (4.17)$$

Thus, for an arbitrary E flux, we may write a $2m - 1$ order scheme, with $2m$-order dissipation, approximating (1.1) as:

$$\frac{\partial}{\partial t}q_j = -C^{2m}f(q_j) + (-1)^{m-1}\beta(\Delta x)^{2m-1}D_+^m D_-^{m-1}\left[df_{j-\frac{1}{2}}^+ - df_{j-\frac{1}{2}}^-\right] \quad (4.18)$$

$$= -\left[\frac{df_{j+\frac{1}{2}}^+}{\Delta x} + \sum_{k=-m+1}^{m-1}\left[\mu_k^m + (-1)^k\beta\binom{2m-2}{m+k-1}\right]D_+\left(df_{j+k-\frac{1}{2}}^-\right)\right]$$

$$+ \left[\frac{df_{j-\frac{1}{2}}^+}{\Delta x} + \sum_{k=-m+1}^{m-1}\left[v_k^m - (-1)^k\beta\binom{2m-2}{m+k-1}\right]D_+(df_{j+k-\frac{1}{2}}^+)\right]$$

Thus, we have constructed the unlimited version of the numerical flux, $\hat{f}_{j+\frac{1}{2}}^{m,\beta}$, of (3.1), having the relevant desired properties.

For convenience, we require

$$\beta \binom{2m-2}{m+k-1} \leq |\mu_k^m| = |\nu_k^m| \text{ for each } k = 0, \pm 1, \ldots, +-(m-1) \tag{4.19}$$

We claim that this is true for all k, if it is true for $k = m - 1$, or if

$$\beta \leq \left(m\binom{2m}{m}\right)^{-1} = |\nu_{m-1}^m|. \tag{4.20}$$

We shall prove this using induction and (4.9). The result is obviously always valid for $k = m - 1$. Suppose (4.19) is valid for all $|k| \leq m - 1$ and all numbers up to m. Then we have, from (4.9)

$$(-1)^k \nu_k^{m+1} - \left((m+1)\binom{2m+2}{m+1}\right)^{-1}\binom{2m}{m+k} = (-1)^k \nu_k^m - \left(m\binom{2m}{m}\right)^{-1}\binom{2m-2}{m+k-1} \tag{4.21}$$

$$+ \left[\frac{k}{m}\binom{2m+2}{m+1}^{-1}(m+1)^{-1}\binom{2m}{m-k} + \left(m\binom{2m}{m}\right)^{-1}\binom{2m-2}{m+k-1} - (m+1)^{-1}\binom{2m+2}{m+1}^{-1}\binom{2m}{m+k}\right]$$

We shall show that the last expression in brackets above is always positive.

For $k = 0$, we have:

$$\frac{1}{2(2m-1)} - \frac{1}{2(2m+1)} > 0$$

For $k = 1$, we have:

$$\frac{1}{2(m+1)(2m-1)} + \frac{m-1}{2m(2m-1)} + \frac{-m}{2(m+1)(2m+1)}$$

$$> (m-1)\left[\frac{1}{2m(2m-1)} - \frac{1}{2(m+1)(2m-1)}\right] > 0$$

For $k \geq 2$, we have:

$$\frac{1}{2m(2m-1)} \frac{(m-1)\ldots(m-k)}{(m+k-1)\cdots(m+1)} \left[\frac{1}{(m^2-k^2)} km \binom{2m-1}{2m+1} + 1 - \frac{m^2(2m-1)}{(m^2-k^2)(2m+1)}\right]$$

$$= \frac{1}{(2m)(2m-1)} \frac{(m-1)\ldots(m-k)}{(m+k-1)\ldots(m+1)} \left[\frac{2}{2m+1} + \binom{2m-1}{2m+1}\frac{k}{(m+k)}\right] > 0$$

The claim is now proven.

Next we apply the flux limiter to (4.18), arriving at the scheme (2.2), (3.1)- (3.6). To verify that it decreases variation, we rewrite it as Equation (4.22):

$$\frac{\partial q_j}{\partial f} = D_+q_j \left[-\left(\frac{df^-_{j+\frac{1}{2}}}{\Delta_+q_j}\right)\left[1 + \sum_{k=-m+1}^{m-1}\left(\mu_k^m + (-1)^k\beta\binom{2m-2}{m+k-1}\right)\left(\frac{\left(df^-_{j+k+\frac{1}{2}}\right)^{(k)} - \left(df^-_{j+k-\frac{1}{2}}\right)^{(k)}}{df^-_{j+\frac{1}{2}}}\right)\right]\right]$$

$$- D_-q_j\left[\left(\frac{df^+_{j-\frac{1}{2}}}{\Delta_-q_j}\right)\left[1 + \sum_{k=-m+1}^{m-1}\left(v_k^m - (-1)^k\beta\binom{2m-2}{m+k-1}\right)\left(\frac{\left(df^+_{j+k-\frac{1}{2}}\right)^{(k)} - \left(df^+_{j+k-\frac{1}{2}}\right)^{(k)}}{df^+_{j-\frac{1}{2}}}\right)\right]\right]$$

$$= C_{j+\frac{1}{2}}D_+q_j - D_{j-\frac{1}{2}}D_-q_j$$

(4.22)

This is TVD and satisfies a maximum principle if, for each j:

$$C_{j+\frac{1}{2}}, \ D_{j+\frac{1}{2}} \geq 0. \tag{4.23}$$

See e.g. [21].

Thus, we need:

$$1 \geq \sum_{k=-m+1}^{m-1}\left(\mu_k^m + (-1)^k\beta\binom{2m-2}{m+k-1}\right)\frac{\left[\left(df^-_{j+k-\frac{1}{2}}\right)^{(k)} - \left(df^-_{j+k+\frac{1}{2}}\right)^{(k)}\right]}{df^-_{j+\frac{1}{2}}} \tag{4.24a}$$

$$1 \geq \sum_{k=-m+1}^{m-1}\left(v_k^m - (-1)^k\beta\binom{2m-2}{m+k-1}\right)\frac{\left[\left(df^+_{j+k-\frac{1}{2}}\right)^{(k)} - \left(df^+_{j+k+\frac{1}{2}}\right)^{(k)}\right]}{df^+_{j-\frac{1}{2}}} \tag{4.24b}$$

In (4.24b), we estimate the right side, using definition (3.5a,b,c), and recalling that the sign of the kth coefficient is $(-1)^k$ if $k \geq 0$, $(-1)^{k+1}$ if $k < 0$. Thus we need:

$$1 \geq \frac{1}{2} - \beta\binom{2m-2}{m-1} + \sum_{k=1}^{m-1}\left((-1)^k v_k^m - \beta\binom{2m-2}{m+k-1}\right)b + \sum_{k=-m+1}^{-1}\left((-1)^{k+1}v_k^m + \beta\binom{2m-2}{m+k-1}\right)b \tag{4.25}$$

or:

$$\tfrac{1}{2} + \beta \binom{2m-2}{m-1} \ge 2\left[\sum_{k=1}^{m-1}(-1)^k v_k^m\right]b. \tag{4.26}$$

or (using (4.13)):

$$b \le \frac{1 + 2\beta\binom{2m-2}{m-1}}{\sum_{k=2}^{m}\frac{1}{2k-1}} \le \frac{1 + \frac{2}{m}\binom{2m}{m}^{-1}\binom{2m-2}{m-1}}{\sum_{k=2}^{m}\frac{1}{2k-1}}, \tag{4.27}$$

which implies $D_{j-\tfrac{1}{2}} \ge 0$.

A similar argument shows that $C_{j+\tfrac{1}{2}} \ge 0$ for the same values of b.

A simple exercise on a pocket calculator shows us that

$$\sum_{j=2}^{7}\frac{1}{2j-1} \approx .9551 \tag{4.28a}$$

$$\sum_{j=2}^{8}\frac{1}{2j-1} \approx 1.0218 \tag{4.28b}$$

$$\sum_{j=2}^{9}\frac{1}{2j-1} \approx 1.0809 \tag{4.28c}$$

It is possible to choose $\beta \le \left(8\binom{16}{8}\right)^{-1}$ so that $b \ge 1$ because $\tfrac{1}{15} > .0218$. If it were possible to do this for $m = 9$, we would have $\tfrac{1}{17} > .0809$, which is false.

Thus, within our constraints, 15th order accuracy (in 17 points) is the highest possible.

Next, we obtain the CFL restriction for the explicit forward-Euler time discretization, which we write as

$$q_j^{n+1} - q_j^n = \frac{\Delta t}{\Delta x}\left[C_{j+\tfrac{1}{2}}^n \Delta_+ q_j^n - D_{j-\tfrac{1}{2}}^n \Delta_- q_j^n\right] \tag{4.29}$$

The precise restriction for the scheme to be TVD, in addition to (4.23) is, for each j:

$$\frac{\Delta t}{\Delta x}\left[C_{j+\tfrac{1}{2}}^n + D_{j+\tfrac{1}{2}}^n\right] \le 1 \tag{4.30}$$

(see [13])

We thus wish to obtain upper bounds for

$$\left[1 + \sum_{k=-m+1}^{m-1} \left(\mu_k^m + (-1)^k \beta \binom{2m-2}{m+k-1}\right) \frac{\left[\left(df_{j+k+\frac{1}{2}}^-\right)^{(k)} - \left(df_{j+k-\frac{1}{2}}^-\right)^{(k)}\right]}{df_{j+\frac{1}{2}}^-}\right] \quad (4.31a)$$

and

$$\left[1 + \sum_{k=-m+1}^{m-1} \left(\nu_k^m - (-1)^k \beta \binom{2m-2}{m+k-1}\right) \frac{\left[\left(df_{j+k+\frac{3}{2}}^+\right)^{(k)} - \left(df_{j+k+\frac{1}{2}}^+\right)^{(k)}\right]}{df_{j+\frac{1}{2}}^+}\right] \quad (4.31b)$$

A routine calculation using the definitions (3.5), noting the signs of the coefficients, gives us the result (d) in Theorem (3.1), modulo proving that

$$v_1^m = -\sum_{j=2}^{m} \frac{1}{2j(2j-1)}, \quad m \geq 2. \quad (4.32)$$

For $m = 2$, (3.2a) gives:

$$v_1^2 = \left(2\binom{4}{2}\right)^{-1} = \frac{1}{12}.$$

Assume that (4.32) is valid up to m. Then, (4.9) for $k = 1$, gives us

$$v_1^{m+1} = v_1^m - \frac{1}{m+1}\binom{2m-2}{m+1}^{-1}\binom{2m}{m+1}\frac{1}{m} = -\sum_{j=2}^{m} \frac{1}{2j(2j-1)} - \frac{1}{2(m+1)(2(m+1)-1)}$$

Finally, we check the stability of these linearized "β" schemes, without flux limiters, using explicit forward Euler time discretization. We linearize about a constant state $\bar{q}$, at which $f'(\bar{q}) = a \neq 0$.

This is:

$$\frac{q_j^{n+1} - q_j^n}{\Delta t} = -\left[C^{2m} a q_j^n + (-1)^{m-1}\frac{\beta}{\Delta x}\left(h_o(\bar{q}, \bar{q}) - h_1(\bar{q}, \bar{q})\right)(\Delta_+ \Delta_-)^m q_j^n\right] \quad (4.33)$$

We note that h is an E flux. In [19] it was shown for such fluxes that:

$$h_o(\bar{q}, \bar{q}) \geq 0 \geq h_1(\bar{q}, \bar{q})$$

If equality holds for both above, then consistency implies

$$0 = h_o(\bar{q}, \bar{q}) + h_1(\bar{q}, \bar{q}) = f'(\bar{q}) = a \neq 0$$

which is a contradiction. Thus we may define the positive quantity

$$B = \beta(h_o(\bar{q}, \bar{q}) - h_1(\bar{q}, \bar{q})) > 0$$

The amplification matrix for (4.33) is

$$1 - \frac{a\Delta t}{\Delta x}i(\zeta + C(\zeta)\zeta^{2m+1}) - \frac{\Delta t}{\Delta x}B(2 - 2\cos\zeta)^m = A(\zeta) \qquad (4.34)$$

for $-\pi \leq \zeta < \pi$, $C(0) \neq 0$, and $C(\zeta)$ real analytic for real ζ. Then the relation:

$$|A(\zeta)|^2 = 1 + a^2\left(\frac{\Delta t}{\Delta x}\right)^2\zeta^2 - \frac{2\Delta t}{\Delta x}B\zeta^{2m} + O(\zeta^{2m+1}) \leq 1. \qquad (4.35)$$

which implies:

$$|a^2|\frac{\Delta t}{\Delta x} \leq 2B\zeta^{2m-2} + O(\zeta^{2m-1}) \text{ as } |\zeta| \downarrow 0.$$

This is a contradiction, since $m \geq 2$.

Theorem (3.1) is now proven.

To construct the "α" schemes of Theorem (3.2) we first construct a dissipative approximation to $q_t = 0$:

$$\frac{\partial q_j}{\partial t} = (-1)^m \alpha \Delta_+^{m-1}\Delta_-^{m-1}D_-q_j, \text{ for } \alpha > 0 \qquad (4.36)$$

The operator on the right above has module (-m,m-1), and its symbol is:

$$\alpha(-1)^m 2^{m-1}(\cos\zeta - 1)^{m-1}(1 - e^{-i\zeta}) = \alpha 2^{m-1}[-(1-\cos\zeta)^m - i(1-\cos\zeta)^{m-1}\sin\zeta] \qquad (4.37)$$

Thus, this operator is dissipative of order $2m$ and accurate of order $2m - 2$. It may be rewritten as:

$$\frac{\partial q_j}{\partial t} = \alpha \sum_{k=-m+1}^{m-2}(-1)^k\binom{2m-3}{k+m-1}D_+(\Delta_+S^{k-1}q_j) \qquad (4.38)$$

Similarly, the operator on the right side of

$$\frac{\partial q_j}{\partial t} = (-1)^{m-1}\alpha\Delta_+^{m-1}\Delta_+^{m-1}D_+q_j, \text{ for } \alpha > 0, \qquad (4.39)$$

has module $(-m + 1, m)$, and its symbol is:

$$\alpha 2^{m-1}[-(1-\cos\zeta)^m + i(1-\cos\zeta)^{m-1}\sin\zeta] \qquad (4.40)$$

It is again dissipative of order $2m$, is $2m - 2$ order accurate, and it may be rewritten as:

$$\frac{\partial q_j}{\partial t} = \alpha \sum_{k=-m+2}^{m-1} (-1)^k \binom{2m-3}{k+m-2} D_+(\Delta_+ S^{k-1} q_j). \tag{4.41}$$

Thus, we may use (4.15), (replacing m by $m - 1$), (4.38) and (4.41) to obtain the unlimited "α" scheme.

$$\frac{\partial}{\partial t} q_j = -C^{2m-2} f(q_j) + (-1)^m \alpha (\Delta x)^{2m-2} \left[D_+^{m-1} D_-^{m-1} D_+ h(q_j, q_{j-1}) \right] \tag{4.42}$$

$$= - \left[\frac{1}{\Delta x} df_{j+\frac{1}{2}}^- + \sum_{k=-m+2}^{m-1} \left[\mu_k^{m-1} + (-1)^k \alpha \binom{2m-3}{k+m-2} \right] D_+ \left(df_{j+k-\frac{1}{2}}^+ \right) \right]$$

$$+ \left[\frac{1}{\Delta x} df_{j-\frac{1}{2}}^+ + \sum_{k=-m+1}^{m-2} \left[v_k^{m-1} - (-1)^k \alpha \binom{2m-3}{k+m-1} \right] D_+ \left(df_{j+k-\frac{1}{2}}^+ \right) \right]$$

Using (4.2), we see that the leading term of the truncation error of the right side of (4.42) is:

$$TE = (-1)^m \left[\alpha - 2 \frac{(m-1)! \, m!}{(2m)!} \right] (\Delta x)^{2m-2} \partial_x^{2m-1} f(q), \tag{4.43}$$

which is independent of the choice of the E flux, $h(q_{j+1}, q_j)$.

From (4.07), it is clear that α and β schemes coincide if $\alpha = \frac{2}{m} \binom{2m}{m}^{-1}$, and $\beta = \frac{1}{m} \binom{2m}{m}^{-1}$.

For convenience, we want (for $m \geq 2$):

$$\alpha \binom{2m-3}{k+m-2} \leq |\mu_k^{m-1}| = |v_k^{m-1}|, \quad k = 0, 1, \ldots, m - 2. \tag{4.44}$$

We claim that this is valid for all these k if it is valid for $k = m - 2$, or if

$$\alpha \leq \left((m-1) \binom{2m-2}{m-1} \right)^{-1} \tag{4.45}$$

This is trivial for $m = 2$. For $m > 2$, this reduces to showing that:

$$\left(m \binom{2m}{m} \right)^{-1} \binom{2m-1}{k+m-1} \leq |v_k^m|, \quad k = 0, 1, \ldots, m-1. \tag{4.46}$$

We obtained a stronger inequality in (4.19), (4.20), so the validity of (4.44) from (4.45) is obvious.

Next we apply the flux limiters to (4.42), arriving at the scheme. (2.2), (3.7). To verify that the scheme decreases variation, we rewrite it as Equation (4.47):

$$\frac{\partial q_j}{\partial t} = D_+ q_j \left[\frac{-df_{j+\frac{1}{2}}^-}{\Delta_+ q_j} \left[1 + \sum_{k=-m+2}^{m-1} \left(\mu_k^{m-1} + (-1)^k \alpha \binom{2m-3}{m+k-1} \right) \left(\frac{\left(df_{j+k+\frac{1}{2}}^-\right)^{(k)} - \left(df_{j+k-\frac{1}{2}}^-\right)^{(k)}}{df_{j+\frac{1}{2}}^-} \right) \right] \right]$$

$$- D_- q_j \left[\frac{df_{j-\frac{1}{2}}^+}{\Delta_- q_j} \left[1 + \sum_{k=-m+1}^{m-2} \left(v_k^{m-1} - (-1)^k \alpha \binom{2m-3}{m+k-1} \right) \left(\frac{\left(df_{j+k+\frac{1}{2}}^+\right)^{(k)} - \left(df_{j+k-\frac{1}{2}}^+\right)^{(k)}}{df_{j-\frac{1}{2}}^+} \right) \right] \right]$$

$$= C_{j+\frac{1}{2}} D_+ q_j - D_{j-\frac{1}{2}} D_- q_j$$

We must show that (4.23) is valid for this scheme. Thus we need:

$$1 \geq \sum_{k=-m+2}^{m-1} \left(\mu_k^{m-1} + (-1)^k \alpha \binom{2m-3}{k+m-2} \right) \left(\frac{\left(df_{j+k-\frac{1}{2}}^-\right)^{(k)} - \left(df_{j+k+\frac{1}{2}}^-\right)^{(k)}}{df_{j+\frac{1}{2}}^+} \right) \tag{4.48a}$$

$$1 \geq \sum_{k=-m+1}^{m-2} \left(v_k^{m-1} - (-1)^k \alpha \binom{2m-3}{k+m-1} \right) \left(\frac{\left(df_{j+k-\frac{1}{2}}^+\right)^{(k)} - \left(df_{j+k+\frac{1}{2}}^+\right)^{(k)}}{df_{j+\frac{1}{2}}^+} \right) \tag{4.48b}$$

In (4.48b), we estimate the right side, using the definition of v_k^m from (4.4b), (4.4c), and (4.4d). We recall that the kth coefficient has sign $(-1)^k$ if $k \geq 0$, $(-1)^{k+1}$ if $k < 0$. Thus we need:

$$1 \geq \left[\frac{1}{2} - \alpha \binom{2m-3}{m-1} \right] + \sum_{k=1}^{m-2} \left((-1)^k v_k^{m-1} - \alpha \binom{2m-3}{m+k-1} \right) b \tag{4.49}$$

$$+ \sum_{k=-m+1}^{-1} \left((-1)^{k+1} v_k^{m-1} + \alpha \binom{2m-3}{m+k-2} \right) b$$

or,

$$\frac{1}{2} + \alpha \binom{2m-3}{m-1} \geq 2 \left[\sum_{k=1}^{m-2} (-1)^k v_k^{m-1} \right] b + \alpha \binom{2m-3}{m-2} b \tag{4.50}$$

Using (4.13) gives us:

$$b \leq \frac{1 + 2\alpha \binom{2m-3}{m-1}}{\sum_{k=2}^{m-1} \left(\frac{1}{2k-1} \right) + 2\alpha \binom{2m-3}{m-2}} \tag{4.51}$$

The same inequality establishes (4.48a).

Using (4.28), we see that we can take $b > 1$ for $m \leq 8$, but not for $m = 9$. Thus 14th order (or 15th for $\alpha = \frac{2}{m} \binom{2m}{m}^{-1}$) in 17 points, is the best possible, as predicted.

Next we obtain the CFL restriction for the explicit time discretization (4.29), which requires inequality (4.30). This time it involves obtaining upper bounds for:

$$1 + \sum_{k=-m+2}^{m-1} \left(\mu_k^{m-1} + (-1)^k \alpha \binom{2m-3}{m+k-2} \right) \left(\frac{\left(df^-_{j+k+\frac{1}{2}}\right)^{(k)} - \left(df^-_{j+k-\frac{1}{2}}\right)^{(k)}}{df^-_{j+\frac{1}{2}}} \right) \tag{4.52a}$$

and:

$$1 + \sum_{k=-m+1}^{m-2} \left(v_k^{m-1} - (-1)^k \alpha \binom{2m-3}{m+k-1} \right) \left(\frac{\left(df^+_{j+k+\frac{1}{2}}\right)^{(k)} - \left(df^+_{j+k-\frac{1}{2}}\right)^{(k)}}{df^+_{j-\frac{1}{2}}} \right) \tag{4.52b}$$

A routine calculation using the definitions in (3.5) and (4.32), and the signs of each coefficient, gives us result (e) in the statement of Theorem (3.2).

Finally, we check the stability of these linearized "α" schemes. We again linearize about a constant state $\bar{q}$, at which $f'(\bar{q}) = a \neq o$. The resulting scheme is as follows in Equation (4.53):

$$\frac{q_j^{n+1} - q_j^n}{\Delta t} = - \left[C^{2m-2} a \, q_j + (-1)^{m-1} \alpha (\Delta x)^{2m-2} D_+^{m-1} D_-^{m-1} D_+ [h_1(\bar{q},\bar{q}) q_j + h_o(\bar{q},\bar{q}) q_{j-1}] \right]$$

We also know that:

$$h_o(\bar{q},\bar{q}) \geq 0 \geq h_1(\bar{q},\bar{q})$$

with at least one of these inequalities being strict.

The amplification matrix for (4.53) is:

$$1 - \frac{\Delta t}{\Delta x} ai[\zeta + C(\zeta) \zeta^{2m-1}] \quad (4.54)$$

$$- \frac{\Delta t}{\Delta x} \alpha \, 2^{m-1}(1 - \cos \zeta)^{m-1}[h_o(\bar{q},\bar{q})[1 - \cos \zeta + i \sin \zeta] - h_1(\bar{q},\bar{q})[1 - \cos \zeta - i \sin \zeta]]$$

The rest of the proof goes as in (4.35).

Theorem (3.2) is now proven.

V. A More General Class of TVD Schemes

Given a conservation form approximation to the scalar version of (1.1) of the type:

$$q_j^{n+1} = q_j^n - \lambda \left(\hat{f}_{j+\frac{1}{2}}^n - \hat{f}_{j-\frac{1}{2}}^n \right) \quad (5.1)$$

where

$$\hat{f}_{j+\frac{1}{2}}^n = h(q_{j+k}^n, \ldots, q_{j-k+1}^n)$$

Suppose we can rewrite

$$\hat{f}_{j+\frac{1}{2}}^n - \hat{f}_{j-\frac{1}{2}}^n = - \sum_{\nu=-k}^{k-1} A_{j+\frac{1}{2}}^{(\nu)} \Delta_+ q_{j+\nu}^n \quad (5.2)$$

where the $A_{j+1/2}^{(r)}$ are functions of $(q_{j+k}^n, \cdots, q_{j-k+1}^n)$, subject to the following restrictions for each j:

$$A_{j+\frac{1}{2}}^{(k-1)} \geq 0 \geq A_{j+\frac{1}{2}}^{(-k)} \quad (5.3a)$$

$$A_{j+\frac{1}{2}}^{(\nu-1)} \geq A_{j-\frac{1}{2}}^{(\nu)}, \quad \text{for } -k + 1 \leq \nu \leq k - 1, \; \nu \neq 0 \quad (5.3b)$$

$$1 \geq \lambda \left(A_{j-\frac{1}{2}}^{(0)} - A_{j+\frac{1}{2}}^{(0)} \right), \quad \text{(the CFL restriction)} \quad (5.3c)$$

Then we have the following:

Theorem (5.1)

Given an approximation to (1.1), of the form (5.1), satisfying (5.2),(5.3), then the scheme is TVD, i.e

$$\sum_j |\Delta_+ q_j^{n+1}| \leq \sum_j |\Delta_+ q_j^n|$$

Proof:

Using a, by now, standard argument -e.g. [1], [13] and [28], we first compute:

$$\Delta_+ q_j^{n+1} = \lambda A_{j+\frac{3}{2}}^{(k-1)} \Delta_+ q_{j+k}^n - \lambda A_{j+\frac{1}{2}}^{(-k)} \Delta_+ q_{j-k}^n + \left(1 + \lambda \left[A_{j+\frac{3}{2}}^{(-1)} - A_{j+\frac{1}{2}}^{(0)}\right]\right) \Delta_+ q_j^n \qquad (5.4)$$

$$+ \lambda \sum_{\substack{\nu=-k+1 \\ \nu \neq 0}}^{k-1} \left(-A_{j+\frac{1}{2}}^{(\nu)} + A_{j+\frac{3}{2}}^{(\nu-1)}\right) \Delta_+ q_{j+\nu}^n$$

Inequalities (5.3a,b, and c) were chosen so that each coefficient of $\Delta_+ q_{j+\nu}^n$ in (5.4) is non-negative. Thus we may take the absolute value of both sides, obtaining the inequality:

$$|\Delta_+ q_j^{n+1}| \le |\Delta_+ q_j^n| + \lambda A_{j+\frac{3}{2}}^{(k-1)} |\Delta_+ q_{j+k}^n| - \lambda A_{j+\frac{1}{2}}^{(-k)} |\Delta_+ q_{j-k}^n| \qquad (5.5)$$

$$+ \lambda \sum_{\nu=-k+1}^{k-1} \left(-A_{j+\frac{1}{2}}^{(\nu)} + A_{j+\frac{3}{2}}^{(\nu-1)}\right) |\Delta_+ q_{j+\nu}^n| = |\Delta_+ q_j^n| + \lambda \Delta_+ \sum_{\nu=-k}^{k-1} A_{j+\frac{1}{2}}^{(\nu)} |\Delta_+ q_{j+\nu}^n|$$

We sum the inequality (5.5) over j, the result follows.

Next we approximate (1.1) via a semi-discrete method (2.2), where

$$\hat{f}_{j+\frac{1}{2}} - \hat{f}_{j-\frac{1}{2}} = -\sum_{\nu=-k}^{k-1} A_{j+\frac{1}{2}}^{(\nu)} \Delta_+ q_{j+\nu} \qquad (5.6)$$

Here the $A_{j+\frac{1}{2}}^{(\nu)}$ satisfy (5.3a and b). Then we have the following:

Theorem (5.2)

Given an approximation to (1.1) of the form (2.2), satisfying (5.6) and (5.3a and b), then the scheme is TVD, i.e.:

$$\frac{\partial}{\partial t} \sum_j |\Delta_+ q_j| \le 0$$

Proof

We follow an idea of Sanders [25], used by us in [21]. Let

$$\chi_{j+\frac{1}{2}}(t) = \operatorname{sgn} \Delta_+ q_j$$

Then

$$\frac{\partial}{\partial t}|\Delta_+ q_j| = \chi_{j+\frac{1}{2}} \frac{\partial}{\partial t}\Delta_+ q_j = A^{(k-1)}_{j+\frac{3}{2}} \chi_{j+\frac{1}{2}} \Delta_+ q_{j+k} - A^{(-k)}_{j+\frac{1}{2}} \chi_{j+\frac{1}{2}} \Delta_+ q_{j-k} \qquad (5.7)$$

$$+ \sum_{\substack{v=-k+1 \\ v \neq 0}}^{k-1} \left[-A^{(v)}_{j+\frac{1}{2}} + A^{(v-1)}_{j+\frac{3}{2}} \right] \chi_{j+\frac{1}{2}} \Delta_+ q_{j+v} + |\Delta_+ q_j| \left[-A^{(0)}_{j+\frac{1}{2}} + A^{(-1)}_{j+\frac{3}{2}} \right]$$

Because of (5.3a and b), all the coefficients of $\chi_{j+\frac{1}{2}} \Delta_+ q_{j+v}$ for $v \neq 0$, are non-negative. Thus, we have

$$\frac{\partial}{\partial t}|\Delta_+ q_j| \leq \sum_{v=-k+1}^{k-1} \left[-A^{(v)}_{j+\frac{1}{2}} + A^{(v-1)}_{j+\frac{3}{2}} \right] |\Delta_+ q_{j+v}| + A^{(k-1)}_{j+\frac{3}{2}} |\Delta_+ q_{j+k}| - A^{(-k)}_{j+\frac{1}{2}} |\Delta_+ q_{j-k}| \qquad (5.8)$$

$$= \Delta_+ \sum_{v=-k}^{k-1} A^v_{j+\frac{1}{2}} |\Delta_+ q_{j+v}|$$

Summing this inequality gives us Theorem (5.2).

We hope that this very general approach to the construction of TVD schemes will lead to an even wider class of useful high order accurate methods. We shall discuss this in a future paper.

VI. Extensions to Hyperbolic Systems of Conservation Laws

We shall approximate such a system (1.1), using the scalar TVD approximations developed in sections II and III. The key tool in this construction is the use of a nonlinear field-by-field decomposition, which effectively decouples the system. Several such decompositions exist - Godunov's [9], Osher's [22], and Roe's [23]. For simplicity of exposition, we shall only use the last here.

Although the formal version of the last method is well known to violate the entropy condition - i.e. to have stable expansion shocks, it is possible to remove this difficulty, e.g. [19], by changing the differencing slightly near sonic points - points where $\lambda_k(q) = 0$ for some eigenvalue k. What we shall do here can be viewed as an extension of some of the work in [13] and [21] to higher order, non-oscillatory methods. For simplicity of exposition only, we shall ignore the entropy difficulty.

Given two states q_j, q_{j+1}, Roe [23] defines a matrix $A_{j+\frac{1}{2}}$ satisfying the equality:

$$f(q_{j+1}) - f(q_j) = A_{j+\frac{1}{2}}(q_{j+1} - q_j) \qquad (6.1)$$

This matrix is supposed to depend continuously on q_j, q_{j+1}, to have only real eigenvalues, and to satisfy

$$\lim_{q_{j+1} \to q_j} A_{j+\frac{1}{2}} = \partial f(q_j)$$

Such a matrix exists if a convex entropy exists for the system [11]. See [23],[24] for some special properties of physical systems.

Let the eigenvalues of $A_{j+\frac{1}{2}}$ be denoted by $\lambda^{(p)}_{j+\frac{1}{2}}$, $p = 1,\ldots,d$. The corresponding left eigenvectors are $l^{(p)}_{j+\frac{1}{2}}$, and right eigenvectors are $r^{(p)}_{j+\frac{1}{2}}$, normalized so that

$$l^{(p)}_{j+\frac{1}{2}} \cdot r^{(q)}_{j+\frac{1}{2}} = \delta_{pq} = 1 \text{ if } p = q = 0 \text{ if } p \neq q$$

Then we may write

$$q_{j+1} - q_j = \sum_{p=1}^{d} \alpha^{(p)}_{j+\frac{1}{2}} r^{(p)}_{j+\frac{1}{2}} \tag{6.2a}$$

$$f(q_{j+1}) - f(q_j) = \sum_{p=1}^{d} \lambda^{(p)}_{j+\frac{1}{2}} \alpha^{(p)}_{j+\frac{1}{2}} r^{(p)}_{j+\frac{1}{2}} \tag{6.2b}$$

Let

$$x^+ = \max(x,0) \quad x^- = \min(x,0) \tag{6.3}$$

Next we define Roe's first order numerical flux:

$$h(q_{j+1}, q_j) = \frac{1}{2}(f(q_{j+1}) + f(q_j)) - \frac{1}{2}|A_{j+\frac{1}{2}}|(q_{j+1} - q_j) \tag{6.4}$$

so

$$df^-_{j+\frac{1}{2}} = \sum_{p=1}^{d} (\lambda^{(p)}_{j+\frac{1}{2}})^- \alpha^{(p)}_{j+\frac{1}{2}} r^{(p)}_{j+\frac{1}{2}} \tag{6.5a}$$

$$df^+_{j+\frac{1}{2}} = \sum_{p=1}^{d} (\lambda^{(p)}_{j+\frac{1}{2}})^+ \alpha^{(p)}_{j+\frac{1}{2}} r^{(p)}_{j+\frac{1}{2}} \tag{6.5b}$$

Now we use the notation of section III to construct the high order non-oscillatory scheme for systems of conservation laws. Let the quantities v^m_k, μ^m_k, b, β, and α, be defined as in those sections. The numerical fluxes used to construct semi-discrete approximations of the form (1.7a) are defined via:

(β schemes of $2m-1$ order accuracy):

$$\hat{f}_{j+\frac{1}{2}} = \hat{f}^{m,\beta}_{j+\frac{1}{2}} = h(q_{j+1}, q_j) + \sum_{k=-m+1}^{m-1} \left(\mu^m_k + (-1)^k \beta \binom{2m-2}{k+m-1}\right) \left(df^-_{j+k+\frac{1}{2}}\right)^{(k)} \tag{6.6}$$

$$+ \sum_{k=-m+1}^{m-1} \left(\nu_k^m - (-1)^k \beta \binom{2m-2}{k+m-1}\right) \left(df_{j+k+\frac{1}{2}}^+\right)^{(k)}$$

or

(α schemes of $2m-2$ or $2m-1$ order accuracy)

$$\hat{f}_{j+\frac{1}{2}} = \hat{f}_{j+\frac{1}{2}}^{m,\alpha} = h(q_{j+1}, q_j) + \sum_{k=-m+2}^{m-1} \left(\mu_k^{m-1} + (-1)^k \alpha \binom{2m-3}{k+m-2}\right) \left(df_{j+k+\frac{1}{2}}^-\right)^{(k)} \quad (6.7)$$

$$+ \sum_{k=-m+1}^{m-2} \left(\nu_k^{m-1} - (-1)^k \alpha \binom{2m-3}{k+m-1}\right) \left(df_{j+k+\frac{1}{2}}^+\right)^{(k)}$$

Now we define these vector valued flux limited quantities as follows:

$$[df_{j+\frac{1}{2}}^-]^{(k)} = \sum_{p=1}^{d} \min \bmod \, [(\lambda_{j+\frac{1}{2}}^{(p)})^- \alpha_{j+\frac{1}{2}}^{(p)}, b(\lambda_{j-k+\frac{1}{2}}^{(p)})^- \alpha_{j-k+\frac{1}{2}}^{(p)}, b(\lambda_{j-k+\frac{3}{2}}^{(p)})^- \alpha_{j-k+\frac{3}{2}}^{(p)}] r_{j+\frac{1}{2}}^{(p)}, \quad (6.8a)$$

for all k with $0 \neq k \neq 1$.

$$[df_{j+\frac{1}{2}}^-]^{(0)} = \sum_{p=1}^{d} \min \bmod \, [(\lambda_{j+\frac{1}{2}}^{(p)})^- \alpha_{j+\frac{1}{2}}^{(p)}, b(\lambda_{j+\frac{3}{2}}^{(p)})^- \alpha_{j+\frac{3}{2}}^{(p)}] r_{j+\frac{1}{2}}^{(p)} \quad (6.8b)$$

$$[df_{j+\frac{1}{2}}^-]^{(1)} = \sum_{p=1}^{d} \min \bmod \, [(\lambda_{j+\frac{1}{2}}^{(p)})^- \alpha_{j+\frac{1}{2}}^{(p)}, b(\lambda_{j-\frac{1}{2}}^{(p)})^- \alpha_{j-\frac{1}{2}}^{(p)}] r_{j+\frac{1}{2}}^{(p)} \quad (6.8c)$$

$$[df_{j+\frac{1}{2}}^+]^{(k)} = \sum_{p=1}^{d} \min \bmod \, [(\lambda_{j+\frac{1}{2}}^{(p)})^+ \alpha_{j+\frac{1}{2}}^{(p)}, b(\lambda_{j-k+\frac{1}{2}}^{(p)})^+ \alpha_{j-k+\frac{1}{2}}^{(p)}, b(\lambda_{j-k-\frac{1}{2}}^{(p)})^+ \alpha_{j-k-\frac{1}{2}}^{(p)}] r_{j+\frac{1}{2}}^{(p)} \quad (6.8d)$$

for all k with $0 \neq k \neq -1$

$$[df_{j+\frac{1}{2}}^+]^{(0)} = \sum_{p=1}^{d} \min \bmod \, [(\lambda_{j+\frac{1}{2}}^{(p)})^+ \alpha_{j+\frac{1}{2}}^{(p)}, b (\lambda_{j-\frac{1}{2}}^{(p)})^+ \alpha_{j-\frac{1}{2}}^{(p)}] r_{j+\frac{1}{2}}^{(p)} \quad (6.8e)$$

$$[df_{j+\frac{1}{2}}^+]^{(-1)} = \sum_{p=1}^{d} \min \bmod \, [(\lambda_{j+\frac{1}{2}}^{(p)})^+ \alpha_{j+\frac{1}{2}}^{(p)}, b(\lambda_{j+\frac{3}{2}}^{(p)})^+ \alpha_{j+\frac{3}{2}}^{(p)}] r_{j+\frac{1}{2}}^{(p)} \quad (6.8f)$$

It is easily seen that each of the unlimited versions of these semi-discrete algorithms does indeed have the desired accuracy, for general nonlinear systems of hyperbolic conservation laws.

In the special case of linear diagonalizable hyperbolic systems:

$$f(q) = A\,q = A_{j+\frac{1}{2}} q, \text{ for each } j+\frac{1}{2},$$

we have a great deal of theory. We may now use the $\alpha^{(p)}_{j+\frac{1}{2}}$ to help measure the variation.

Define

$$|\Delta_+ q_j| = \sum_{p=1}^{d} |\alpha^{(p)}_{j+\frac{1}{2}}| \qquad (6.9)$$

A scheme of this semi-discrete type is said to be TVD if

$$\frac{d}{dt} \sum_j |\Delta_+ q_j| \leq 0$$

Also the ratio needed below is defined by its value on the right:

$$\left| \frac{df^+_{j+\frac{1}{2}} - df^-_{j+\frac{1}{2}}}{\Delta_+ q_j} \right| = \sup_p |\lambda^{(p)}_{j+\frac{1}{2}}| \qquad (6.10)$$

We now have:

Theorem (6.1)

All the results of Theorems (3.1) and (3.2) go over word for word to the corresponding schemes for systems where the flux is defined by (6.4) to (6.8), and where $f(q) = Aq = A_{j+\frac{1}{2}} q$.

No theory of this type is known for nonlinear systems. The entropy condition was proven for bounded a.e. limits of a special second order TVD type approximation, using Osher's flux-decomposition in [21]. We find numerically that these schemes work quite well for compressible inviscid gas dynamical flows at widely varying Mach numbers. See [5] for the results of several numerical experiments.

VII. Results of Numerical Experiments

We now discuss some numerical results. Some members of the new family of schemes were programmed for a linear wave equation with a source term which drives the solution to a time-asymptotic steady state.

$$q_t + q_x - \pi \cos(\pi x) = 0. \qquad (7.1)$$

The semi-discrete TVD spatial differencing was combined with a family of multi-stage time differencing (which includes the simple one-stage scheme shown in Eq. (2.8)). The steady state exact solution of Eq. 7.1 is given by

$$q(x) = \sin(\pi x) \qquad (7.2)$$

The l_1 norm of the difference between the numerical and analytic steady-state solutions was computed and is presented below in Table (7.1) for a first-order accurate TVD scheme, and for the TVD and unlimited forms of some members of the new family of schemes.

We now discuss the results shown in Table (7.1). The last column entitled "global accuracy" is the order of accuracy measured from the numerical results. The order of accuracy is first measured based on the 20 interval and 30 interval solutions. Then, it is measured based on the 30 interval and the 40 interval solutions. Lastly, it is measured based on the 20 interval and the 40 interval solutions. The average of these three values have been entered in the last column of Table (7.1). The individual orders of accuracy (for every pair of intervals) is computed as follows: let the l_1 norm of the error for J intervals be denoted by E_J; then, the order of global (corresponding to the overall solution) error, O, is given by

$$O_{J1,J2} = \frac{\ln(E_{J2}) - \ln(E_{J1})}{\ln(J2) - \ln(J1)} \tag{7.3}$$

where $J1$ and $J2$ denote the number of intervals in the pair of solutions being considered.

Many facts stand out clearly in an analysis of the results of the 5-point schemes. The TE of the unlimited schemes is close to the theoretically derived values. The global accuracy of the TVD schemes

Scheme	\multicolumn{3}{c}{l_1 norm of Error}	Global Accuracy
	20 intervals	30 intervals	40 intervals	
\multicolumn{5}{c}{First-Order Accurate Monotone Upwind Scheme;}
First-Order	0.1496	0.1013	0.07662	$(\Delta x)^{0.97}$;
\multicolumn{5}{c}{Third-Order Scheme, $\alpha = 1/6, b = 4$;}
Unlimited	0.0024856	0.000744	0.0003162	$(\Delta x)^{2.97}$;
TVD Limited	0.004212	0.001348	0.00076338	$(\Delta x)^{2.42}$;
\multicolumn{5}{c}{Fully Upwind Scheme, $\alpha = \frac{1}{2}, b = 2$;}
Unlimited	0.019717	0.0089566	0.005091	$(\Delta x)^{1.95}$;
TVD Limited	0.017874	0.00784	0.004756	$(\Delta x)^{1.89}$;
\multicolumn{5}{c}{Fromm's Scheme, $\alpha = \frac{1}{4}, b = 3$;}
Unlimited	0.005685	0.00239	0.0013221	$(\Delta x)^{2.10}$;
TVD Limited	0.00862	0.002773	0.001554	$(\Delta x)^{2.43}$;
\multicolumn{5}{c}{Low TE Second-Order Scheme, $\alpha = \frac{1}{8}, b = 5$;}
Unlimited	0.003125	0.001256	0.000681	$(\Delta x)^{2.19}$;
TVD Limited	0.006767	0.0014528	0.001058	$(\Delta x)^{2.52}$;
\multicolumn{5}{c}{TVD Central Difference Scheme, $\alpha = 0$ $b \gg 1$;}
Smoothed	0.0100006	0.0045107	0.002556	$(\Delta x)^{1.97}$;
TVD Limited	0.02655	0.00559	0.0080886	$(\Delta x)^{1.275}$;
\multicolumn{5}{c}{Unnamed Scheme, $\alpha = \frac{1}{3}, b = 5/2$}
Unlimited	0.01028809	0.004565	0.0025733	$(\Delta x)^{2.00}$;
TVD Limited	0.0111646	0.0045198	0.00268	$(\Delta x)^{2.09}$;

TABLE (7.1) Error Computations for some of the New 2nd and 3rd Order TVD Schemes

shows some variation. In fact, the global accuracy of the TVD schemes based on Fromm's discretization and the Low TE Second-Order discretization compare quite favorably with the global accuracy of the Third-Order TVD scheme. In the case of the first two, the TVD scheme has better accuracy than the corresponding unlimited scheme. In the case of the Third-Order scheme, the accuracy suffers by going to the TVD form. When we consider the magnitude of error as opposed to the order of accuracy, the Third-Order scheme comes out ahead of all the others. The global order of accuracy of a TVD scheme depends on a number of factors, such as the number of maxima and minima, the ratio of this number to the overall number of intervals, implementation of boundary conditions, etc. Thus, the global accuracy of the TVD and the unlimited forms can be different. On the other hand, the fact that the Third-Order scheme is indeed third-order accurate in its unlimited form and that it consistently has a lower magnitude of error seems to imply that the Third-Order scheme may be the most preferable of the lot. The other second-order schemes having a low truncation error also suggest themselves as schemes which must be given serious consideration. We do not recommend the use of the unlimited forms of the TVD schemes whether the order of accuracy of these is higher or lower than the corresponding TVD formulation. The errors of the unlimited forms are shown here only for comparison. The TVD Central scheme is also highly unreliable as shown by the fact that its error for 30 intervals was actually better than the error of 40 intervals. This is due to the lack of dissipation. It has already been mentioned that the orders of global accuracy given in the table are the average of three values. It is quite instructive to actually look at the individual values that are averaged. Some schemes show a wider variation than others. The last remark here is that the Fully Upwind scheme, that many researchers (including the present authors) have been using in the recent past, is just about the worst of the lot (excluding the highly unreliable TVD Central scheme). In fact, to obtain the same level of accuracy as the 20 interval solution using the Third-Order scheme, the Fully Upwind scheme would need to use 40 intervals. The purely centrally differenced scheme shown in Table (7.1) as the Smoothed Central scheme (non-TVD central differencing along with a very small amount of third-order fourth-difference smoothing) does not fare much better when compared with the other third-order and second-order accurate schemes. The fifth order accurate, seven-point scheme leads to the following results:

l_1 norm of Error

$$\text{Fifth Order, } \beta = \frac{1}{60} = \beta_{max}, b = \frac{9}{4}$$

	20 Intervals	30 Intervals	40 Intervals	80 Intervals;
unlimited	.0000489	.00000662	.000001181;	
TVD limited	.0148	.00142	.0014	.000168;
(TVD limited)*	.0104	.00046	.00056;	

TABLE (7.2) Error Computation for 5th Order TVD Scheme

Here the (TVD limited)* line denotes calculating the l_1 norm of the error computed only at points where limiting does not occur - i.e., at which the scheme is of minimal order of accuracy. This measures the effect of pollution into high-accuracy regions.

Next in Figures (7.1a-e) we test the compressive properties of various approximations. We solve $q_t = -q_x$ with an initial Heaviside function. The third-order accurate, Fromm's, and the low error second-order scheme appear to be extremely accurate - as accurate as the scheme favored by Sweby in [26]. The fully upwind scheme has more smearing, while the first order accurate upwind method smears the profile excessively.

Acknowledgment

The authors would like to thank Eitan Tadmor for his careful reading of a first draft.

Bibliography

[1] L. Abrahamsson and S. Osher, Monotone difference schemes for singular pertubation problems, SIAM J. Num. Anal., V. 19 (1982), pp. 979-991.

[2] S. R. Chakravarthy, Relaxation Methods for Unfactored Implicit Upwind Schemes, AIAA-84-0165, Reno, NA (1984)

[3] S. R. Chakravarthy and S. Osher, High resolution applications of the Osher upwind scheme for the Euler equations, Proc. AIAA Comp. Fluid Dynamics Conf., Danvers, Mass (1983), pp. 363-372.

[4] S. R. Chakravarthy and S. Osher, Computing with High Resolution Upwind Schemes for Hyperbolic equations, to appear in Proceedings of AMS-SIAM, 1983 Summer Seminar, La Jolla, CA.

[5] S. R. Chakravarthy and S. Osher, A new class of High Accuracy Total Variation Diminishing Schemes for Hyperbolic Conservation Laws, In Preparation.

[6] S. R. Chakravarthy, K. Y. Szema, S. Osher, J. Gorski, A new class off High Accuracy Total Variation Diminishing Schemes for the Navier-Stokes Equations, In Preparation.

[7] P. Colella and P. R. Woodward, The piecewise-parabolic method (PPM) for gas-dynamical simulations, LBL report #14661, (July 1982).

[8] R. J. DiPerna, Convergence of approximate solutions to conservation laws, Arch., Rat. Mech. and Analysis, 82 (1983), pp. 27-70.

[9] S. K. Godunov, A finite difference method for the numerical computation of discontinuous solutions of the equations of fluid dynamics, Mat. Sb., 47 (1959). pp. 271-290.

[10] J. B. Goodman and R. J. LeVeque, On the accuracy of stable schemes for two dimensional conservation laws, Math. Comp., (to appear).

[11] A. Harten, On the Symmetric Form of Systems of Conservation Laws with Entropy, ICASE Rep. No. 81-34, (1981), NASA Langley Research Center, Va.

[12] A. Harten, On a class of High Resolution Total-Variation-Stable Finite- Difference Schemes, SINUM, v.21, pp. 1-23 (1984).

[13] A. Harten, High resolution schemes for hyperbolic conservation laws, J. Comp. Phys., 49(1983), pp. 357-393.

[14] H. O. Kreiss and J. Oliger, Methods for the Approximate Solution of Time Dependent Problems, GARP Publication series No. 10, (1973).

[15] S. N. Kruzkov, First order quasi-linear equations in several independent variables, Math. USSR Sb., 10 (1970), pp. 217-243.

[16] A. Majda and S. Osher, Numerical viscosity and the entropy condition, Comm. Pure Appl. Math., V. 32 (1979), pp. 797-838.

[17] W. A. Mulder and B. Van Leer, Implicit upwind computations for the Euler equations, AIAA Comp. Fluid Dynamics Conf., Danvers, Mass., (1983), pp. 303- 310.

[18] S. Osher, Numerical solution of singular perturbation problems and hyperbolic systems of conservation laws, North Holland Mathematical Studies #47, eds. S. Axelsson, L. S. Frank, and A. van der Sluis, pp. 179-205.

[19] S. Osher, Riemann solvers, the entropy condition, and difference approximations, SINUM, v. 21, (1984), pp. 217-235.

[20] S. Osher, Convergence of Generalized MUSCL Schemes, NASA Langley Contractor Report 172306, (1984), Submitted to SINUM.

[21] S. Osher and S. R. Chakravarthy, High resolution schemes and the entropy condition, SINUM, (to appear).

[22] S. Osher and F. Solomon, Upwind schemes for hyperbolic systems of conservation laws, Math. Comp., V. 38 (1982), pp. 339-377.

[23] P. L. Roe, Approximate Riemann solvers, parameter vectors, and difference schemes, J. Comp. Phys., V. 43 (1981), pp. 357-372.

[24] P. L. Roe, Some contributions to the modelling of discontinuous flows, to appear in Proceedings of AMS-SIAM 1983 Summer Seminar, La Jolla, CA.

[25] R. Sanders, On convergence of monotone finite difference schemes with variable spatial differencing, Math. Comp., v.40 (1983), pp. 91-106.

[26] P. K. Sweby, High resolution schemes using flux limiters for hyperbolic conservation laws, SINUM (to appear).

[27] E. Tadmor, Numerical viscosity and the entropy condition for conservative difference schemes, NASA Contractor Report 172141, (1983), NASA Langley, Math Comp. (to appear).

[28] B. Van Leer, Towards the ultimate conservative scheme, II. Monotonicity and conservation combined in a second order scheme, J. Comp. Phys. 14 (1974), pp. 361-376.

[29] B. Van Leer, Towards the Ultimate Conservative Finite Difference Scheme III. Upstream-Centered Finite-Difference Schemes for Ideal Compressible Flow, J. Comp. Phys., v. 23, (1977), pp. 263-275.

[30] B. Van Leer, Towards the ultimate conservative difference scheme. IV. A New approach to numerical convection, J. Comp. Phys., 23 (1977), pp. 276-298,

[31] H. C. Yee, R. F. Warming, and A. Harten, Implicit total variation diminishing (TVD) shemes for steady state calculations, Proc. AIAA Comp. Fluid Dynamics Conf., Danvers, Mass., (1983), pp. 110-127.

[32] S. T. Zalesak, Fully Multidimensional Flux-Corrected Transport, J. Comp. Phys., v. 31, (1979), pp. 335-362.

[33] B. Engquist and S. Osher, Stable and entropy condition satisfying approximations for transonic flow calculations, Math. Comp., 34 (1980), pp. 45-75.

[34] J. P. Boris and D. L. Book, Flux-Corrected Transport I - SHASTA, A fluid transport algorithm that works, J. Comp. Phys., v. 11, (1973), pp. 38- 69.

[35] P. D. Lax, Hyperbolic Systems of Conservation Laws and the Mathematical Theory of Shock Waves, SIAM Regional Conference Lectures in Applied Mathematics No. 11 (1972)

THIRD ORDER

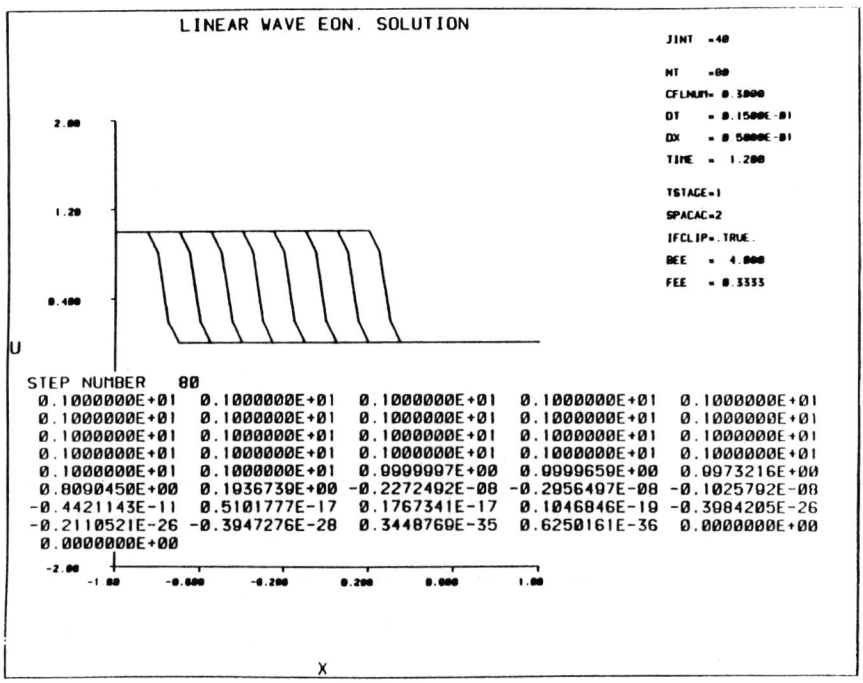

Figure (7.1a)

Third Order Accurate TVD Solution to $q_t = -q_x$

FROMM'S SCHEME

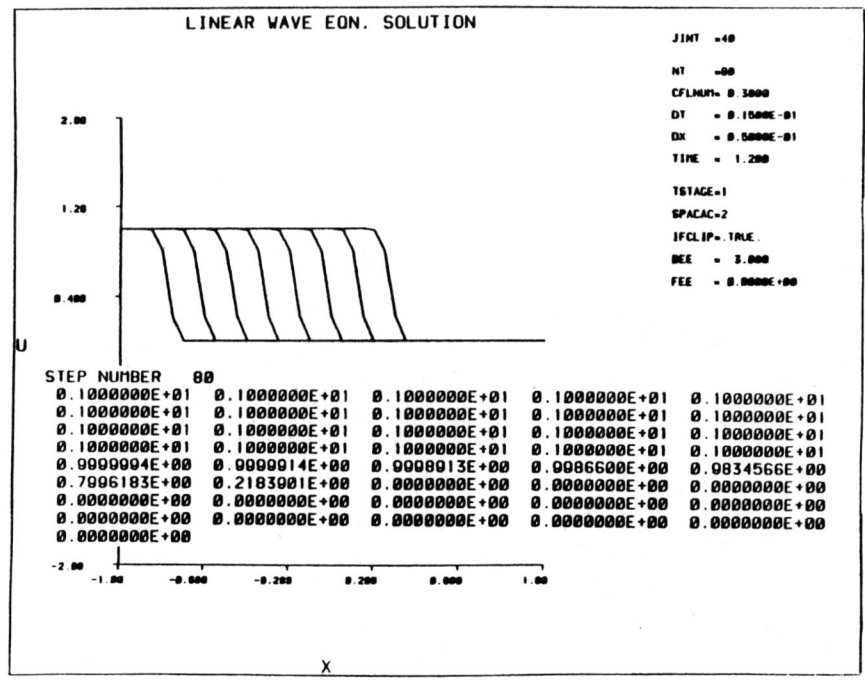

Figure (7.1b)

Fromm's TVD Solution to $q_t = -q_x$

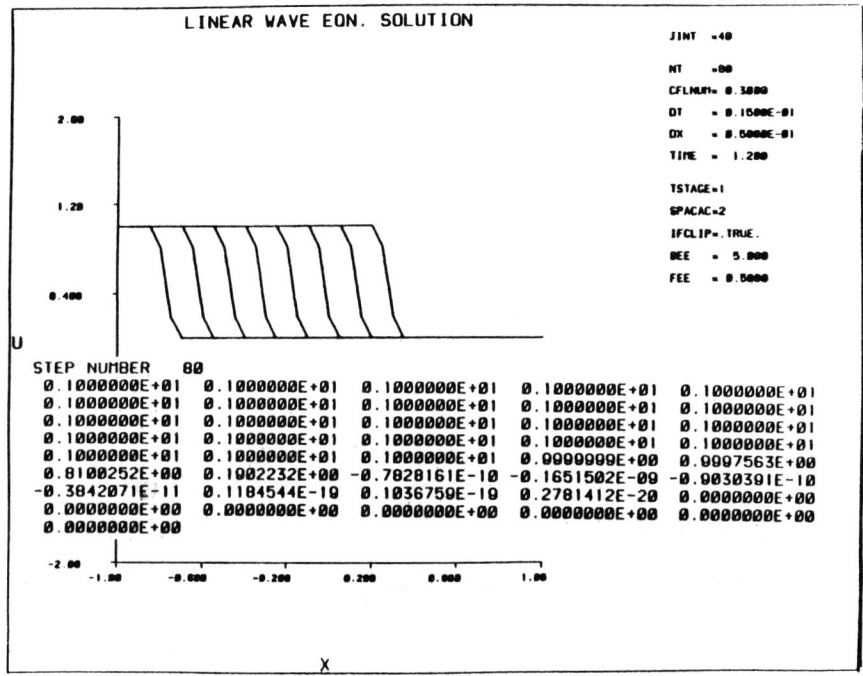

Figure (7.1c)

High Accuracy Second Order TVD Solution to $q_t = -q_x$

FULLY UPWIND WITH min mod (x,2y)

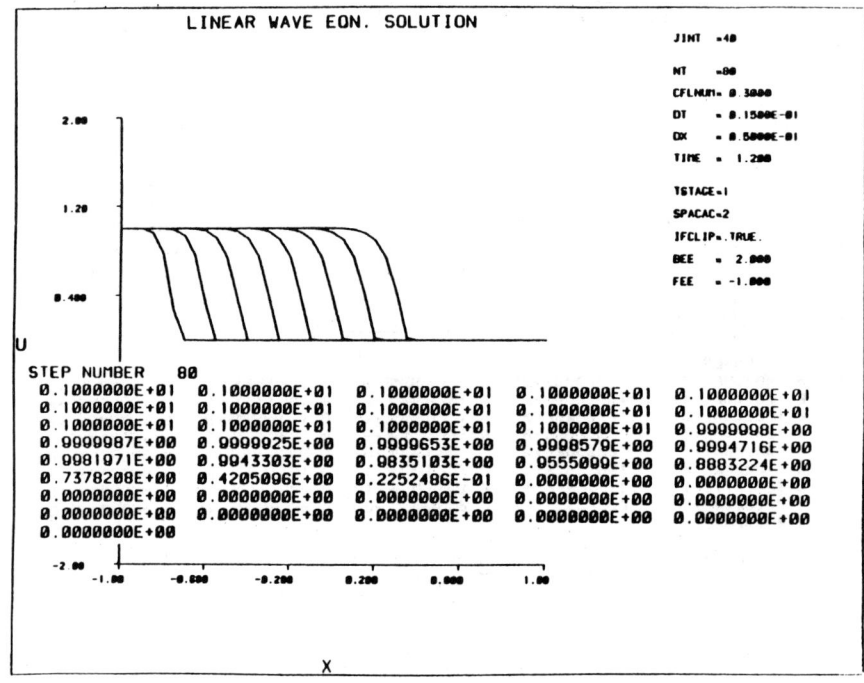

Figure (7.1d)
Fully upwind TVD Solution to $q_t = -q_x$

FULLY UPWIND, WITH SWEBY'S COMPRESSIVE LIMITER

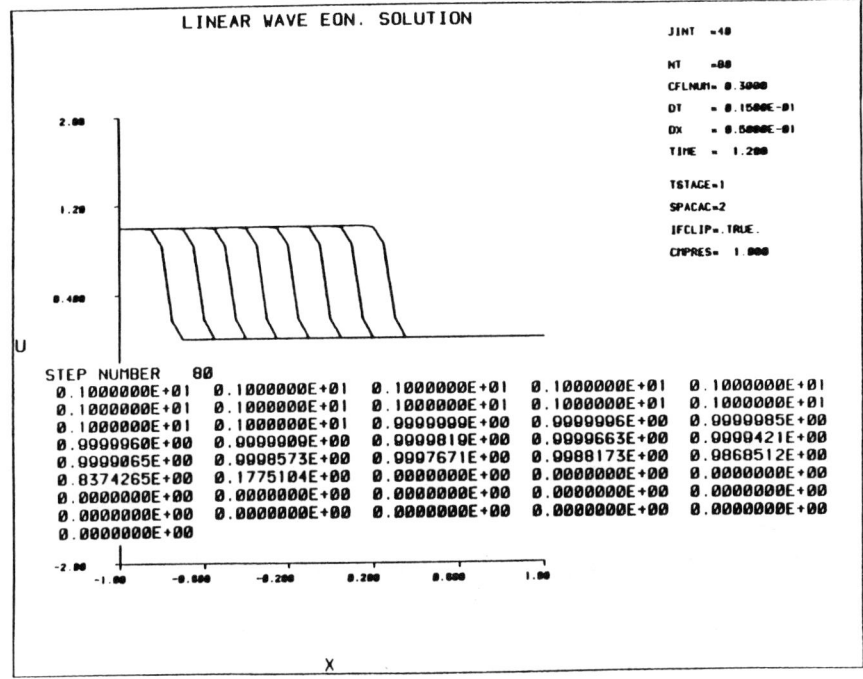

Figure (7.1e)

Fully upwind, with Sweby's Compressive Limiter, Solution to $q_t = -q_x$

FIRST ORDER UPWIND SCHEME

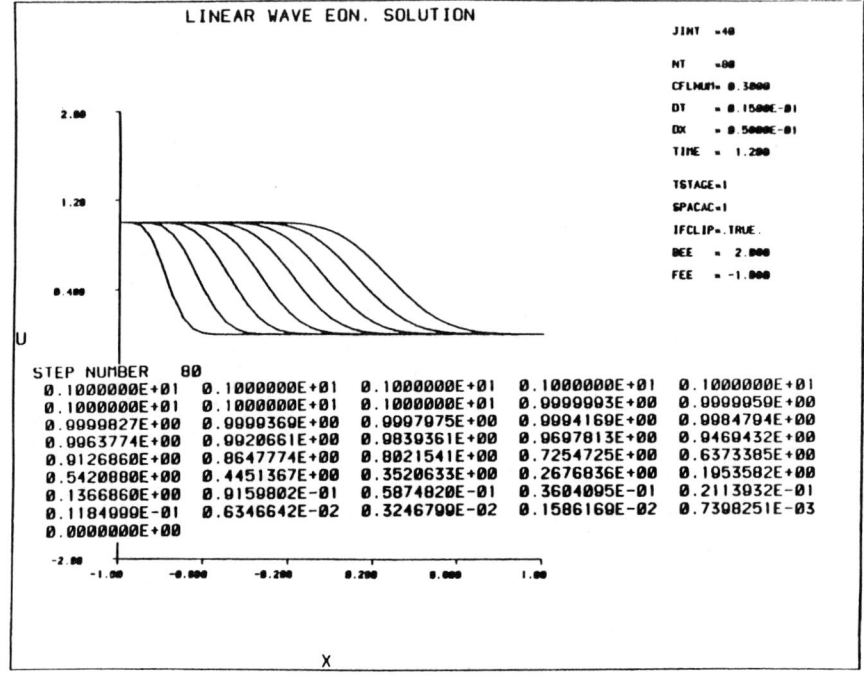

Figure (7.1f)
First order upwind solution to $q_t = -q_x$

CONVERGENCE OF APPROXIMATE SOLUTIONS TO SOME SYSTEMS OF CONSERVATIVE LAWS: A CONJECTURE ON THE PRODUCT OF THE RIEMANN INVARIANTS

Michel Rascle

Analyse Numérique
Université de St. Etienne
42023 St. Etienne Cedex (FRANCE)

and

Lefschetz Center for Dynamical Systems
Division of Applied Mathematics
Brown University
Providence, RI 02912

1. INTRODUCTION

The idea of applying the Compensated Compactness theory to hyperbolic systems of conservation laws was originated by L. Tartar [12]. He treated the scalar case (without any information on the derivatives) and proposed a strategy for the 2×2 case.

In [1] (and [2]), R.J. DiPerna succeeded in applying these ideas to a general 2×2 strictly hyperbolic genuinely nonlinear system and to the non-strictly hyperbolic case of the isentropic gas dynamics equations.

Our approach is based on ideas of "compensated compactness theory with varying directions". Before developing it, we have to recall the initial idea of L. Tartar.

Let us consider the Cauchy Problem

$$(P) \begin{cases} u_t + (f(u))_x = 0 & (1.1) \\ u(x,0) = u_0(x), \quad \forall x \in R & (1.2) \end{cases}$$

where (1.1) is a 2×2 hyperbolic system of conservation laws (f is a smooth function from R^2 to R^2).

Now, let us consider any "reasonable" approximation of this problem: (vanishing) viscosity, viscosity + dispersion (capillarity ..., with a suitable ratio of the vanishing coefficients), finite differences scheme etc...

Typically, let us consider the approximation by viscosity

$$(P_\varepsilon) \begin{cases} u_t^\varepsilon + (f(u^\varepsilon))_x = \varepsilon(Du_x^\varepsilon)_x, & \varepsilon \downarrow 0 \qquad (1.3) \\ u^\varepsilon(x,0) = u_0(x) & \qquad (1.4) \end{cases}$$

where D is a positive-(semi) definite diffusion matrix.

Suppose, for simplicity, that the sequence (u^ε) remains bounded in L^∞, uniformly with respect to ε (actually, that is often a realistic assumption, but it is almost always very difficult to prove). Then, by extracting a sub-sequence, still denoted by (u^ε), we have

$$u^\varepsilon \to u \qquad f(u^\varepsilon) \to f^* \qquad (1.5)$$

where $\to$ denotes the weak-star convergence in L^∞, when $\varepsilon \downarrow 0$. The problem is to prove that

$$f^* = f(u) \quad \text{a.e.} \qquad (1.6)$$

This would be obvious if we could prove that

$$u^\varepsilon \to u \quad \text{strongly in } L^p, p < +\infty. \qquad (1.7)$$

But, unfortunately, in general, there is no available estimate on the first order derivatives of the approximate solutions u^ε to guarantee (1.7). The only celebrated exception is the convergence of the Glimm scheme for "small" initial data, see [3].

However, the equation (1.3) provides - with a natural energy estimate - information on some linear combinations of the first order derivatives of u^ε. In particular, <u>any</u> pair (ϕ,ψ) of entropy and flux in the sense of Lax [4] will satisfy, by Murat's Lemma,

$$\frac{\partial}{\partial t} \phi(u^\varepsilon) + \frac{\partial}{\partial x} \psi(u^\varepsilon) \varepsilon K = K(\phi) \qquad (1.8)$$

where $K(\phi)$ is a strongly compact set of the Sobolev space H_{loc}^{-1}, and does not depend on ε.

Now, if $(\overline{\phi},\overline{\psi})$ is another entropy-flux pair, i.e. another additional conservation law of (1.1), it will also satisfy (1.8). Then we can apply the classical div-curl Lemma in R^2, to obtain

$$\overline{\phi^\varepsilon \psi^\varepsilon} - \overline{\psi^\varepsilon \phi^\varepsilon} \to \overline{\phi^* \psi^*} - \overline{\psi^* \phi^*} \quad \text{where} \quad \phi^\varepsilon = \phi(u^\varepsilon) \to \phi^* \text{etc...} \tag{1.9}$$

We can express (1.9) in terms of Young measures [13] [12]. For the sequel, it is more convenient to express all the functions with respect to the Riemann Invariants w, z of system (1.1) (see Section 2 for these basic definitions). So we assume that there is a global one-to-one map: $u = (u_1, u_2) \to (w, z)$.

Then, if $w^\varepsilon = w(u^\varepsilon)$ and $z^\varepsilon = z(u^\varepsilon)$ the associated Young measure family is a family of probability measures $\{\nu_{x,t}\}$ on the phase plane (w,z), such that, for any continuous function g

$$g(w^\varepsilon, z^\varepsilon) \to g^*$$

with

$$g^*(x,t) = \langle \nu_{x,t}, g(\cdot,\cdot) \rangle = \iint g(w,z) d\nu_{x,t}(w,z). \tag{1.10}$$

It is well-known that the sequences $(w^\varepsilon, z^\varepsilon)$ are strongly convergent if and only if $\nu_{x,t}$ is a delta-function (for almost all (x,t)).

Now, in terms of Young measures, we can rewrite (1.9) under the form

$$\langle \nu, \overline{\phi\psi} - \overline{\psi\phi} \rangle = \langle \nu, \phi \rangle \langle \nu, \overline{\psi} \rangle - \langle \nu, \psi \rangle \langle \nu, \overline{\phi} \rangle \tag{1.11}$$

where $\nu = \nu_{x,t}$.

The interesting fact in the case of a 2×2 system is that there is a lot of entropy-flux pairs: so we have an infinity of equations (1.11) to identify the probability measure $\nu = \nu_{x,t}$ at any fixed (x,t).

Actually R.J. DiPerna [1] used the family of Lax entropies [4] to prove that ν is a delta-function if the system is genuinely non-linear (see Section 2), except possibly on a strictly increasing curve $z = z(w)$ in the phase plane.

Now, we can explain our approach. Let us consider the (important) particular cases of the non-linear 1-D elasticity system

$$\begin{cases} u_t - (\sigma(v))_x = 0, \quad \sigma'(v) > 0 \\ v_t - u_x = 0 \end{cases} \tag{1.12}$$

and of the isentropic gas dynamics equations

$$\begin{cases} \rho_t + (\rho u)_x = 0 \\ (\rho u)_t + (\rho u^2 + p(\rho))_x = 0 \end{cases} \quad (1.13)$$

Let us consider the Riemann invariants $w^\varepsilon, z^\varepsilon$ of the solution of the viscous approximation of either of these systems. Then we have a nice information on the derivatives of w^ε or z^ε in the associated characeristic directions: roughly speaking,

$$\begin{aligned} \frac{\partial w^\varepsilon}{\partial t} + \lambda_2(w^\varepsilon, z^\varepsilon) \frac{\partial w^\varepsilon}{\partial x} &\in K \\ \frac{\partial z^\varepsilon}{\partial t} + \lambda_1(w^\varepsilon, z^\varepsilon) \frac{\partial z^\varepsilon}{\partial x} &\in K' \end{aligned} \quad (1.14)$$

where λ_1, λ_2 are the eigenvalues of the system and K, K' are strongly compact subsets of H_{loc}^{-1}, still independent of ε.

So it is reasonable to hope that (1.14) will provide a nice information on the product $w^\varepsilon z^\varepsilon$ (this would be obvious in the strictly hyperbolic case where λ_1 and λ_2 would be two different constants).

The right <u>conjecture</u> (see M. Rascle [6] for the particular case (1.12) and D. Serre [9], [10], [11] for the general case) is the following:

(C) $\begin{cases} \text{For a general } 2\times 2 \text{ system (1.1) if (1.14) is satisfied then, at} \\ \text{least for a sub-sequence:} \\ \\ (w - \lim p^\varepsilon w^\varepsilon z^\varepsilon)(w \lim p^\varepsilon) = (w - \lim p^\varepsilon w^\varepsilon)(w - \lim p^\varepsilon z^\varepsilon) \quad (1.15) \\ \\ \text{where the weight-function } p = p(w,z) \text{ is given by} \\ \\ p = \exp(-H) \\ \frac{\partial^2 H}{\partial w \partial z} = (\lambda_2 - \lambda_1)^{-2} \frac{\partial(\lambda_1, \lambda_2)}{\partial(w, z)} \quad (1.16) \\ \\ \text{thus } p \text{ is defined up to a tensor-product, in terms of } \lambda_1, \lambda_2 \text{ and their} \\ \text{jacobian with respect to } w, z. \end{cases}$

(Actually, the notations of D. Serre are slightly different.) Now, as any function of w (or z) is still a Riemann invariant, it is almost equivalent (in fact, a little stronger) to write the Conjecture (C) under the form

(C')
$$\begin{cases} \text{For a general 2×2 system (1.1), if (1.14) is satisfied, then} \\ \text{for almost all } (x,t), \, \nu = \nu_{x,t} \text{ satisfies} \\ \nu = p^{-1} \, \mu_1 \times \mu_2 \\ \text{where } p \text{ is given by (1.16).} \end{cases} \quad (1.17)$$

Now, for the elasticity system (1.12), the eigenvalues satisfy $\lambda_1 \equiv -\lambda_2$, so that $p \equiv 1$, and (C) becomes in this case (see [6], [7]):

$$w^\varepsilon z^\varepsilon \to w^* z^* = (w - \lim w^\varepsilon)(w - \lim z^\varepsilon) \, . \quad (1.18)$$

For the isentropic gas dynamics system (1.13), the weight-function $p(w,z)$ - not to be confused with the pressure $p(!)$ - turns out to be the density ρ, so that (C) becomes (see [8]).

$$(w - \lim \rho^\varepsilon w^\varepsilon z^\varepsilon)(w - \lim \rho^\varepsilon) = (w - \lim \rho^\varepsilon w^\varepsilon)(w - \lim \rho^\varepsilon z^\varepsilon) \, . \quad (1.19)$$

In both cases, if this conjecture is true, then we can exhibit a sequence of <u>negative</u> functions X^ε whose weak-limit is <u>nonnegative</u>. Therefore, the weak limit is identically zero, and the convergence is <u>strong</u> - moreover, the functions X^ε are negative due to a Cauchy-Schwarz inequality, whose case of equality simply corresponds to the <u>linearly degenerate case</u>, which allows oscillating solutions. So, if (C) is true, the convergence of the method becomes quite easy for systems (1.12) and (1.13).

In Section 2, we recall some basic definitions, in Sections 3 and 4 we simply show the above functions X^ε for (1.12) and (1.13), which is of course the crucial point refer to [6], [8] for the details). Finally, in Section 5, we mention some current progresses towards the proof of the (difficult) Conjecture (C). For the proof of this Conjecture when (at least) one of the eigenvalues is linearly degenerate, we refer to [10].

2. PRELIMINARIES

We first recall some known definitions. The 2×2 system of conservation laws

$$u_t + f(u)_x = 0$$

is called strictly hyperbolic if the Jacobian matrix $f'(u)$ has two real distinct eigenvalues $\lambda_1(u) < \lambda_2(u)$. The Riemann invariants $w(u)$ and $z(u)$ satisfy - because (1.1) is a 2×2 system -

$$\frac{\partial}{\partial t} w(u) + \lambda_2(w(u),z(u)) \frac{\partial w(u)}{\partial x} = 0$$

$$\frac{\partial}{\partial t} z(u) + \lambda_1(w(u),z(u)) \frac{\partial z(u)}{\partial x} = 0$$
(2.1)

for <u>smooth</u> solutions of (1.1) (w is the 2-Riemann invariant in the sense of Lax). We assume (this is realistic) that there exists a global one-to-one map: $u = (u_1, u_2) \to (w,z)$, so that all functions of u can be expressed in terms of (w,z). Then the eigenvalue λ_1 (resp. λ_2) is genuinely nonlinear if

$$\forall (w,z), \quad \frac{\partial \lambda_1}{\partial z} \neq 0 \quad (\text{resp.} \quad \frac{\partial \lambda_2}{\partial w} \neq 0)$$
(2.2)

on the contrary, λ_1 (resp. λ_2) is linearly degenerate if

$$\forall (w,z) \; \frac{\partial \lambda_1}{\partial z} \equiv 0, \; \text{i.e.} \; \lambda_1 \equiv \lambda_1(w) \quad (\text{resp.} \; \frac{\partial \lambda_2}{\partial w} \equiv 0, \; \text{i.e.} \; \lambda_2 \equiv \lambda_2(z))$$
(2.3)

An entropy-flux pair is a pair of functions (ϕ, ψ) such that, for any smooth solution of (1.1),

$$\frac{\partial}{\partial t} \phi(u) + \frac{\partial}{\partial x} \psi(u) = 0$$
(2.4)

which is possible if and only if (ϕ, ψ) satisfies the following hyperbolic linear system (in the phase plane).

$$\frac{\partial \psi}{\partial w} = \lambda_2(w,z) \frac{\partial \phi}{\partial w}$$

$$\frac{\partial \psi}{\partial z} = \lambda_1(w,z) \frac{\partial \phi}{\partial z} \qquad (2.5)$$

Now, we approximate (1.1) (1.2) by the "viscous" Cauchy Problem (1.3) (1.4) and for simplicity, we assume that the diffusion matrix D is the identity matrix. Then, using Murat's Lemma, we obtain

$$\frac{\partial}{\partial t} \phi(u^\epsilon) + \frac{\partial}{\partial x} \psi(u^\epsilon) = \epsilon(\phi(u^\epsilon))_{xx} - \epsilon \phi''(u^\epsilon)(u_x^\epsilon, u_x^\epsilon) \in K \qquad (2.6)$$

for any entropy-flux pair, and similarly

$$\frac{\partial w^\epsilon}{\partial t} + \lambda_2(w^\epsilon, z^\epsilon) \frac{\partial w^\epsilon}{\partial x} = \epsilon w''(u^\epsilon)(u_x^\epsilon, u_x^\epsilon) - \epsilon w''(u^\epsilon)(u_x^\epsilon, u_x^\epsilon) \in K'$$

$$\frac{\partial z^\epsilon}{\partial t} + \lambda_1(w^\epsilon, z^\epsilon) \frac{\partial z^\epsilon}{\partial x} = \epsilon z''(u^\epsilon)(u_x^\epsilon, u_x^\epsilon) - \epsilon z''(u^\epsilon)(u_x^\epsilon, u_x^\epsilon) \in K'' \qquad (2.7)$$

for the Riemann invariants $w^\epsilon = w(u^\epsilon)$, $z^\epsilon = z(u^\epsilon)$, where K, K', K" are strongly compact sets in H_{loc}^{-1}.

By the way, (2.6) is the formal justification of the <u>Lax Entropy condition</u> [4]

$$\frac{\partial}{\partial t} \phi(u) + \frac{\partial}{\partial x} \psi(u) \leq 0 \qquad (2.8)$$

for any <u>convex</u> entropy ϕ.

Of course, the huge difference between (2.6) and (2.7) is that (2.6) is in a <u>conservative</u> form. So, it is very easy to apply the compensated compactness theory - essentially, the div-curl Lemma in R^2 - to the family of relations (2.6), while our Conjecture (C) is very difficult.

However, we are going to show how powerful is this Conjecture for either the elasticity system (1.12) or the gas dynamics system (1.13).

3. THE ELASTICITY SYSTEM:

The equations are

$$\begin{cases} u_t - (\sigma(v))_x = 0 \\ v_t - u_x = 0. \end{cases} \quad (1.12)$$

The system is strictly hyperbolic for $\sigma'(v) > 0$, which we assume. Here, $u = y_t$ is the velocity, $y(x,t)$ is the position at time t of the point whose initial position was x, $v = y_x$ and $\sigma(v)$ is the stress.

The natural perturbation of this system is the <u>visco-elastic</u> system

$$\begin{cases} u_t^\varepsilon - (\sigma(v^\varepsilon))_x = \varepsilon u_{xx}^\varepsilon = \varepsilon v_{xt}^\varepsilon \\ v_t^\varepsilon - u_x^\varepsilon = 0. \end{cases} \quad (3.1)$$

The natural entropy-flux pair (ϕ, ψ) corresponds to the total energy

$$\phi = \frac{u^2}{2} + \Sigma(v), \quad \Sigma(v) = \int_0^v \sigma(s)\,ds. \quad (3.2)$$

We assume, as in the Introduction, that

$$\|u^\varepsilon\|_{L^\infty} \leq c \quad (3.3)$$

where c does not depend on ε. This is actually easy to prove in the (non-genuinely-non-linear) case where

$$\forall v \neq 0, \quad v \cdot \sigma''(v) > 0 \quad (3.4)$$

and is not proved in the general case.

The energy estimate given by (3.2) implies (2.7), but, without (3.3), it is not clear whether it implies (2.6) for any other entropy-flux pair.

The eigenvalues and the associated Riemann Invariants are given by

$$\lambda_1 = -\lambda_2 = -(\sigma'(v))^{1/2}$$
$$w = u - g(v), \quad z = u + g(v), \quad g(v) = \int_0^v (\sigma'(s))^{1/2}\,ds \quad (3.5)$$

Theorem 3.1 [6]

Suppose that Conjecture (C) is true for the system (1.12). Then the viscosity method is convergent: there exists a subsequence $(u^\varepsilon, v^\varepsilon)$ which converges weakly to an admissible weak solution (u,v) of (1.12). Moreover, the convergence is strong in L^p (for all finite p) if there is no non-trivial interval on which the function $v \to \sigma(v)$ is affine, i.e. on which the system is linearly degenerate.

Sketch of proof:

The Conjecture (C) implies (see (1.18))

$$\iint w^\varepsilon z^\varepsilon dx\, dt = \iint [(u^\varepsilon)^2 - (g(v^\varepsilon))^2] dxdt \to \iint w^* z^* dxdt = \iint [u^2 - (g^*)^2] dxdt \quad (3.6)$$

on the other hand, the div-curl Lemma, applied to (1.12) gives

$$\iint ((u^\varepsilon)^2 - v^\varepsilon \sigma(v^\varepsilon)) dxdt \to \iint (u^2 - v \cdot \sigma^*) dxdt \quad (3.7)$$

when $\varepsilon \downarrow 0$. We have of course extracted subsequences such that

$$u^\varepsilon \to u,\ v^\varepsilon \to v,\ g(v^\varepsilon) \to g^*,\ \sigma(v^\varepsilon) \to \sigma^*,\ w^\varepsilon \to w^*,\ z^\varepsilon \to z^*.$$

Then we subtract (3.6) from (3.7), to obtain

$$\iint ((g(v^\varepsilon))^2 - v^\varepsilon \sigma(v^\varepsilon)) dxdt \to \iint ((g^*)^2 - v \cdot \sigma^*) dxdt. \quad (3.8)$$

Now, let us define

$$X^\varepsilon = (g(v^\varepsilon) - g(v))^2 - (v^\varepsilon - v)(\sigma(v^\varepsilon) - \sigma(v)). \quad (3.9)$$

We develop the right-hand side, integrate, and pass to the limit. Due to (3.8), we obtain

$$\iint X^\varepsilon dxdt \underset{(\varepsilon \downarrow 0)}{\to} \iint ((g^*)^2 - 2g^* g(v) + (g(v))^2 - v\sigma^* + v\sigma^* - v\sigma(v)) + v\sigma(v)) dxdt =$$

$$= \iint (g^* - g(v))^2 dxdt \geq 0.$$

On the other hand, the functions X^ε are non-positive, thanks to the Cauchy-Schwarz inequality. Indeed:

$$(g(v^\varepsilon) - g(v))^2 = (\int_v^{v^\varepsilon} \sqrt{\sigma'}(s)ds)^2 \leq (v^\varepsilon - v)(\sigma(v^\varepsilon) - \sigma(v)). \tag{3.10}$$

Thus, we have exhibited a sequence of <u>non-positive functions</u> whose integral has a <u>non-negative limit</u>: therefore the limit is zero, so that

$$g^* = g(v), \quad w^* = w(u,v), \quad z^* = z(u,v)$$

and the <u>convergence is strong</u> in L^1. Hence a new sub-sequence $X \to 0$ a.e..

But X^ε vanishes if and only if $v^\varepsilon(x,t) = v(x,t)$ or σ' is constant on the interval $[v(x,t), v^\varepsilon(x,t)]$ (case of equality in the C.S. inequality).

Therefore, if there is no non-trivial interval on which σ is affine, we can prove that

$$v^\varepsilon \underset{\varepsilon \downarrow 0}{\to} v \text{ in } L^p, \quad \forall p < +\infty$$

and it is easy to conclude that the same property holds for (u^ε).

On the other hand, let us denote by $I(v(x,t))$ the largest interval, which contains $v(x,t)$, on which σ is affine, and suppose that this interval is non-trivial for some (x,t). Now, let η denote the (one-dimensional) Young's measure associated to the sequence (v^ε). Then it is possible to see [12] that

$$\text{supp}(\eta_{x,t}) \subset I(v(x,t)). \tag{3.11}$$

Hence

$$\sigma(v^\varepsilon) \to \sigma^* = \sigma(v) \tag{3.12}$$

and

$$\sigma'(v^\varepsilon) \to (\sigma')^* = \sigma'(v). \tag{3.13}$$

4. THE 2×2 GAS DYNAMICS SYSTEM, IN EULERIAN COORDINATES

Of course, (1.12) is also the system of (isentropic) gas dynamics equations in mass Lagrangian coordinates. Now, in Eulerian coordinates, we classically obtain the system (1.13), where ρ is the density, u the velocity, and $p = p(\rho)$

the pressure.

For this sytem, we have

$$\lambda_1 = u - c(\rho), \quad \lambda_2 = u + c(\rho), \quad c(\rho) = (p'(\rho))^{1/2}$$
$$w = u + h(\rho), \quad z = u - h(\rho),$$
$$h(\rho) = \int_0^\rho \frac{c(s)}{s} ds$$

<u>Theorem 4.1</u> [8]

If (C) is true for system (1.13), then the viscosity method is convergent for this system. In particular

$$w - \lim_{\varepsilon \to 0} (\rho^\varepsilon (u^\varepsilon)^2 + p(\rho^\varepsilon)) = \frac{(w - \lim_{\varepsilon \to 0}(\rho^\varepsilon u^\varepsilon))^2}{(w - \lim_{\varepsilon \to 0} \rho^\varepsilon)} + p(w - \lim_{\varepsilon \to 0} \rho) \qquad (4.1)$$

and the convergence is strong if there is no non-trivial interval on which the function $\rho \to \rho p(\rho)$ is affine, i.e. on which the system is linearly degenerate.

<u>Sketch of proof</u>:

Again, we use the Conjecture (C), under the form (1.19), to obtain

$$<\nu, \rho u^2 - \rho h^2(\rho)><\nu, \rho> = (<\nu, \rho u>)^2 - (<\nu, \rho h(\rho)>)^2 \qquad (4.2)$$

(it is more convenient here to use the notations of Young measures). Then we apply the div-curl Lemma to (1.13)

$$<\nu, \rho p(\rho)> = <\nu, \rho><\nu, \rho u^2 + p(\rho)> - (<\nu, \rho u>)^2 \qquad (4.3)$$

Again, we subtract (4.2) from (4.3), to obtain

$$<\nu, \rho><\nu, \rho h^2(\rho)> - <\nu, \rho p(\rho)> = (<\nu, \rho h(\rho)>)^2 - <\nu, \rho><\nu, p(\rho)> . \qquad (4.4)$$

Then, we define

$$X^\varepsilon = \rho^\varepsilon \rho (h(\rho^\varepsilon) - h(\rho)^*)^2 - (\rho^\varepsilon - \overset{*}{\rho})(p(\rho^\varepsilon) - \overset{*}{p(\rho)}) .$$

Again, we develop the right hand-side, we multiply by any function ϕ (e.g.

$\phi \equiv 1$) we integrate, and we take the limit when $\varepsilon \downarrow 0$. Thanks to (4.4), we obtain

$$\iint X^\varepsilon dxdt \to \iint [(\rho h(\rho))^* - \overset{*}{\rho}(h(\rho))^*]^2 dxdt \geq 0$$

with obvious notations.

Again, due to the Cauchy-Schwarz inequality, the functions X^ε are non-positive, and vanish if and only if $\rho^\varepsilon(x,t) = \rho(x,t)$ or $\rho \to \rho p(\rho)$ is affine on the interval $[\rho(x,t), \overset{*}{\rho^\varepsilon}(x,t)]$. Therefore, the sequence (X^ε) is again strongly convergent to the zero function, and we can prove (4.1) even in the linearly degenerate case.

Remark: The (forthcoming?) proof of Conjecture (C) could be more difficult in this case, because the system in non strictly hyperbolic for $\rho = 0$ (see R.J. DiPerna [2]). For a study of the mixed type transonic flow equations, see C. Morawetz [5].

5. REMARKS ON CONJECTURE (C)

So, we have seen in Sections 3 and 4 how powerful is this Conjecture (C). We refer to D. Serre [10] for the proof of (C) if at least one eignevalue is linearly degenerate.

Now, to prove (C) in all cases is much more difficult. An approach to do that is to construct two families of singular (entropy, flux) pairs, of the form

$$\begin{aligned} \phi &= \delta(w - \overline{w})V_0 + H(w - \overline{w})A \\ \psi &= \delta(w - \overline{w})\lambda_2 V_0 + H(w - \overline{w})B \end{aligned} \quad (5.1)$$

and

$$\begin{aligned} \hat{\phi} &= \delta(z - \overline{z})\hat{V}_0 + H(z - \overline{z})C \\ \hat{\psi} &= \delta(z - \overline{z})\lambda_1 \hat{V}_0 + H(z - \overline{z})D \end{aligned} \quad (5.2)$$

where $\delta(\cdot)$ is the delta-function and $H(\cdot)$ the Heaviside function. The functions V_0, $\hat{V}_0$ are the same ones as in the families of Lax entropies [4]

$\phi_k = e^{kw}(V_0 + \frac{V_1}{k} + \ldots)$; $\Phi_\ell = e^{\ell z}(\hat{V}_0 + \frac{\hat{V}_1}{k} + \ldots)$, and w, z are arbitrary.

Now, let us suppose we can define A, B, C, D such that we can neglect the action of the Young measure on all the terms involving these functions. Then apply the div-curl Lemma to the pairs (5.1) (5.2). We remain (formally) with

$$\langle \nu, \phi\hat{\psi} - \psi\hat{\phi} \rangle = \langle \nu, \delta(w - \overline{w})\delta(z - \overline{z}) V_0 \hat{V}_0 (\lambda_1 - \lambda_2) \rangle = $$
$$= \langle \nu, \phi \rangle \langle \nu, \hat{\Phi} \rangle - \langle \nu, \psi \rangle \langle \nu, \hat{\phi} \rangle . \quad (5.3)$$

Actually, ν can be (hopefully) singular so that it can be difficult to define the left-hand side of (5.3). However, we can approximate (ϕ, ψ) by smooth functions (ϕ_h, ψ_h), $h \to 0$, and normalize this pair to give a (formal) sense to (5.3).

Now, using a result of D. Serre, the right-hand side of (5.3) is equal to $(\hat{c} - c)\langle \nu, \phi \rangle \langle \nu, \hat{\Phi} \rangle$. We have assumed that $\langle \nu, A \rangle = \langle \nu, C \rangle = 0$. Therefore

$$\langle \nu, \phi \rangle \langle \nu, \hat{\Phi} \rangle = \langle \nu, \delta(w - \overline{w})V_0 \rangle \langle \nu, \delta(z - \overline{z})\hat{V}_0 \rangle = f(\overline{w}) g(\overline{z}).$$

Finally,

$$\langle \nu, \delta(w - \overline{w})\delta(z - \overline{z})V_0 \hat{V}_0 (\lambda_2 - \lambda_1) \rangle = (c - \hat{c})f(\overline{w})g(\overline{z}) \quad (5.4)$$

But it is easy to see that the weight-function p into (C) satisfies

$$p = (\lambda_2 - \lambda_1)V_0 \hat{V}_0 \quad (5.5)$$

up to a tensor-product.

Now, suppose to simplify that ν has a continuous density μ with respect to the Lebesgue measure in R^2, i.e. that

$$d\nu = \mu(w, z) dw dz.$$

We have thus expressed in (5.4) that

$$\mu(\overline{w}, \overline{z}) = \frac{(c - \hat{c})}{p(\overline{w}, \overline{z})} f(\overline{w}) g(\overline{z}) \quad (5.6)$$

which is precisely the Conjecture (C). So the proof of this Conjecture is reduced to the (non-trivial) justification of the present calculations, i.e. to a problem of non-linear geometric optics.

Note added to the proofs:

Since this workshop, we have proved the conjecture (C) for any 2×2 strictly hyperbolic system of conservation laws whose approximate solutions remain uniformly bounded in L^∞ : see M. Rascle, Un résultat de compacité par compensation a coefficients variables. Application a ℓ'elasticité non linéaire, to appear in C.R. Acad. Sc., Paris.

REFERENCES

[1] R.J. DiPerna, Convergences of Approximate Solutions to Conservation Laws, Arch. Rat. Mech. Anal., 82, (1983), 27-70.

[2] R.J. DiPerna, Convergence of the Viscosity Method for Isentropic Gas Dynamics, Comm. Math Physics, 91 (1983),1-30

[3] J. Glimm, Solutions in the large for non-linear hyperbolic systems of equations, Comm. Pure Appl. Math., 15(1965), 697-715.

[4] P.D. Lax, Shock Waves and Entropy, in Contributions to Non-Linear Analysis, E. Zarantonello editor, Academic Press, New York (1971).

[5] C. Morawetz, Weak Solution of Transonic Flow by Compensated Compactness, IMA Volumes in Mathematics and its Applications: Dynamical Problems in Continuum Physics, to appear.

[6] M. Rascle, Perturbations par viscosité de certains systémes hyperboliques non-linéaires, Thése, Université Lyon 1 (1983).

[7] M. Rascle, On the convergence of the viscosity method for the system of non-linear 1-D elasticity, A.M.S.-SIAM Summer Seminar on Non-Linear Systems of P.D.E. in Applications, Santa Fé (1984). To appear.

[8] M. Rascle, D. Serre, Compacité par compensation et systémes hyperboliques de lois de conservation, C.R. Acad. Sc., Paris, t. 299, Série I,(1984), 673-676.

[9] D. Serre, Compacité par compensation et systémes hyperboliques de lois de conservation, C.R. Acad. Sc., Paris, t. 299, Série I, (1984), 555-558.

[10] D. Serre, to appear.

[11] D. Serre, this workshop.

[12] L. Tartar, Compensated compactness and Applications to P.D.E., Heriot-Watt Symposium, vol. IV, R. Knops editor, Pitman (1979).

[13] L.C. Young, Lectures on the Calculus of Variations and Optimal Control Theory, W.B. Saunders, Philadelphia (1969).

APPLICATIONS OF THE THEORY OF COMPENSATED COMPACTNESS

M.E. Schonbek
Department of Mathematics
Duke University
Durham, NC 27706

§1 **Singular limits for dissipative and dispersive equations**

We shall discuss two applications of the method of compensated compactness. The first application is concerned with the zero dissipative limit and the zero dispersion limit for scalar conservation laws, the former corresponding to the Burgers equation

(1.1) $$u_t + (u^2/2)_x = \varepsilon u_{xx}$$

and the latter to the Korteveg de Vries equation

(1.2) $$u_t + (u^2/2)_x = \delta u_{xxx} .$$

It is well known that solutions of (1.1) converge strongly to a solution of the hyperbolic law

(1.3) $$u_t + (u^2/2)_x = 0,$$

as the parameter ε vanishes. On the other hand, if the initial data of the Cauchy problem for (1.2) are smooth, then the solutions converge strongly to a solution of (1.3) before shocks develop. After shocks have formed, solutions of (1.2) converge weakly to a function u which is governed by modulation equations [4]. We are interested in the relation between the two limiting processes. To this end, we study the singular limit of the K d V-Burgers equation

(1.4) $$u_t + (u^2/2)_x = \varepsilon u_{xx} - \delta u_{xxx} .$$

We show that if $\delta = O(\varepsilon^2)$ then there exists a subsequence of solutions of (1.4) which converges almost everywhere to a solution (1.3). We recall that the standard approach to problems of this type is to obtain uniform bounds for the amplitude and the derivatives and then pass to the limit using standard compactness results, i.e., extract a subsequence which converges strongly to the solution of the limiting equation. We note that in nonlinear problems it is necessary in general to establish strong convergence since nonlinear maps are typically not continuous with respect to the weak topology, i.e., if u^ε converges weakly to $\bar{u}$, $f(u^\varepsilon)$ need not converge to $f(\bar{u})$.

Using the theory of compensated compactness we are able to pass to the limit without control on the derivatives. Specifically we have established the following result

Theorem 1.1 Let $\Omega = R \times [0,T]$ for some $T>0$. Let $u^0_{\varepsilon,\delta}$ be a sequence of smooth compactly supported functions which lie in a bounded set of $H^2(\Omega)$ $L^4(\Omega)$. Let $u^\varepsilon_\delta: \Omega \to R$ denote the corresponding solutions of (1.4) with data $u^0_{\varepsilon,\delta}$. If $\delta = O(\varepsilon^2)$ then there exists a subsequence u^{ε_k} such that in the strong topology of L^q, $q < 4$

$$\lim_{\varepsilon_k \to 0} u^{\varepsilon_k}(x,t) = \bar{u}(x,t)$$

and $\bar{u} \in L^4(\Omega)$ satisfies the inviscid Burgers equation (1.3). As mentioned above, the proof is based on the theory of compensated compactness and also on results in measure theory. The results in measure theory describe the weak limit of continuous functions as the expected value of a family of probability measure. Specifically

Theorem 1.2 Let $u^\varepsilon: R^n \to R^m$ where $|u^\varepsilon|_{L^\infty} < M$. Then there exists a subsequence u^{ε_k} and an associated family of probability measures $\{v_x(\lambda): x \in R^n, \lambda \in R^m\}$ such that for any $f \in C(R^n, R)$

$$\lim f(u^{\varepsilon_k})(x) = \langle v_x, f(\lambda) \rangle = \int f(\lambda) dv_x(\lambda)$$

for almost all x in R^n. Here the limit is taken L^∞ weak $*$.

From this theorem follows that strong convergence is equivalent to having point masses as associated measures, i.e.:

Corollary 1.1: Under the hypothesis of theorem 1.2 u^{ε_k} converges strongly to u if and only if $\{v_x\} = \delta_{\overline{u}(x)}$.

For the proofs of Theorem 1.2 and Corollary 1.1 we refer the reader to [2,5]. The proof of Theorem 1.1 follows by establishing first some apriori estimates in L^4 and H^2 for the amplitude. There bounds together with an extension to L^4 of Theorem 1.2 and the theory of compensated compactness are sufficient to show that the associated measures are point masses. By Corollary 1.1 the convergence is strong. For a detailed proof we refer the reader to [7].

We next mention two theorems and a conjecture corresponding to large dispersion parameter δ in relation to the dissipation parameter ε.

Theorem 1.3: Let u^ε be a sequence of smooth solutions of (1.4) vanishing at infinity satisfying

$$|u^\varepsilon_\delta|_{L^\infty(\Omega)} \leq M.$$

If $\varepsilon = O(\delta)$ then no subsequence of u^ε_δ converges strongly to an entropy solution.

Theorem 1.4: Let u^ε be a subsequence of smooth solutions of (1.3) vanishing at infinity. If $\varepsilon = o(\delta^3)$ no subsequence will converge strongly to an entropy solution.

These theorems are proved by energy methods. Finally we state the following conjecture

Conjecture: Let u^ε_δ be a sequence of smooth solutions of (1.3) vanishing at infinity. If $\varepsilon^2 < \delta < \varepsilon^{1/3}$ then there exists no subsequence which converges strongly to an entropy solution.

The proof of this conjecture would imply that the quantity δ/ε^2 describes the boundary between the regions of weak and strong convergence to solutions of (1.3).

The reasons for the conjecture are heuristic. We expect that the parameter

ε^2/δ marks the transition between weak and strong convergence. Geometrically it corresponds to the case where the equation admits a 1 parameter family of scaled travelling waves

$$u(x,t) = \phi\left(\frac{x-\sigma t}{\varepsilon}\right)$$

The results obtained by G. Forest and D. McLaughlin [3,6] also point to $\delta = O(\varepsilon^2)$ as the possible cutoff between weak and strong convergence.

2§ Existence of solutions to singular conservation laws

The second problem we consider is the existence of singular conservation laws of the form

(2.1) $$u_t + f(u)_x + \frac{\phi(u)}{x} = 0, \quad x > 0,$$

where $f: R^n \to R^n$, $\phi: R^n \to R^n$. Algebraic singularities of this type arise, for example, in the equations of fluid dynamics with spherical or cylindrical symmetry. We recall that bounded initial data will not give rise to bounded solution, due to the focusing of waves at the origin. This is the main difficulty in estimating solutions. Here we will only discuss the scalar case with f and g smooth maps from R to R. The essential feature of waves focusing at the origin is retained. We establish existence of solutions by regularizing (2.1) and passing to the limit. We use two forms of regularization according to the sign of ϕ at infinity. More precisely, if $u\phi(u) > 0$ for large values of u then the regularization removes the singularity at $x = 0$ and adds a dissipative term:

(2.2) $$u_t + f(u)_x + \frac{\phi(u)}{x+\delta} = \varepsilon u_{xx}, \quad \varepsilon, \delta > 0,$$

If $u\phi(u) < 0$ for u large, we only remove the singularity at $x = 0$:

(2.3) $$u_t + f(u)_x + \frac{\phi(u)}{x+\delta} = 0.$$

We have established the existence of global weak solutions to (2.1) for the scalar Cauchy problem under the following hypothesis.

Theorem 2.1 Let $u_0(x)$ be a smooth function vanishing at zero and infinity and let f and ϕ be smooth. Suppose u_δ^ϵ be a sequence of solutions of (2.2) with initial data $u_\delta^\epsilon(x,0) = u_0(x)$ and boundary data $u_\delta^\epsilon(0,t) = 0$. If for $|u| > M$, $u\phi(u) > 0$, then there exists a subsequence u_k which converges in L^∞ weak * to a weak solution of

(2.4) $$u_t + f(u)_x + \frac{\phi(u)}{x} = 0,$$

$$u(x,0) = u_0(x),$$

which satisfies the Lax entropy condition. Moreover if $f'' \neq 0$ then $u_R \to \bar{u}$ in $L^P([0,T]\times(\alpha,\beta))$ for all $0 < \alpha < \beta < \infty$, $p < \infty$.

Theorem 2.2: Let f, ϕ, u_0 be smooth functions with the following properties:

1. u_0 has compact support contained in $(0,\infty)$

2. $f' \neq 0$ and $f'' \neq 0$

3. There exist constants α and M such that if $|u| \to M$ then

$$u\phi(u) < 0 \quad \text{and} \quad \frac{|f'|}{|\phi|} > \alpha > 0.$$

If u_δ is a sequence of solutions of (2.3), then there exists a subsequence u_k which converges in L^∞ weak * to an entropy solution of (2.4).

Theorem 2.3: Let u_δ be a sequence of weak solutions of (2.3) with smooth initial data $u_0(x)$ which has compact support in $(0,\infty)$. If there exist constants M, C_0, C_1 and r such that for $|u| > M$

$$|\phi(u)| < C_0|u| \quad \text{and} \quad f'(u) < -C_1|u|^r,$$

then there exists a subsequence u_k which converges in L^∞ weak * to an entropy solution of (2.4).

The proof of these theorems can be found in [8]. They are based on the theory of compensated compactness and the measure theory results mentioned in section 1. The hypothesis on f and ϕ together with an analysis on generalized backward characteristics allow us to establish uniform apriori bounds on the amplitude of the soution. These bounds together with the theory of compensated compactness will show that the associated measures are point masses. And Corollary (1.1) implies that the convergence is strong.

Finally we would like to mention that the problem of existence for systems is open. In the case of $n = 2$ an important example of systems of the type (2.1) arise in the description of symmetrical flow

$$\rho_t + (u\rho)_r + j\frac{u}{r}\rho = 0$$

$$(u\rho)_t + (u^2\rho+p)_r + j\frac{u^2}{r}\rho = 0,$$

where r is the distance from the center and $j = 1,2$ describe systems with cylindrical and spherical symmetry respectively.

References

[1] J. Bona and M. Schonbek. Travelling wave solutions of the Korteweg-de Vries Burgers equation. To appear in Proceedings of the Royal Society of Edinborough.

[2] B. Dacorogna. Weak continuity and Weak lower Semicontinuity of Nonlinear functionals. Lecture notes in Mathematics, no. 922. Springer-Verlag.

[3] G. Forest and D. McLaughlin. Modulation of perturbed K.d.V. trains. Preprint.

[4] P. Lax and C.D. Levermore. The zero dispersion limit for the Korteweg - de Vries equation. Proc. Nat. Acad. Scie. U.S.A. (76) 1979 No. 8 3602-3606.

[5] F. Murat. Compacité par Compensation. Ann. Scuola Norm. Sys. Pisa. Sci. Fis. Math. 5 (1978), 489-507.

[6] D. McLaughlin. On the construction of a modulation multiphase wave train for a perturbed K.d.V. equation.

[7] M. Schonbek. Convergence of solutions to nonlinear dispersive equations. Comm in P.D.E. 7(8) 959-1000, 1982.

[8] M. Schonbek. Existence of solutions to singular conservation laws. To appear in Journal of Nonlinear Analysis.

[9] L. Tartar. Compensated compactness and applications to partial differential equations. Nonlinear Analysis and Mechanics: Herriot Watt Symposium, Volume IV. Research Notes in Mathematics 39 R.J. Knops, Ed., Pitman Publishing Inc., 1979.

[10] G.B. Whitham. Linear and Nonlinear Waves, Wiley Interscience 1974.

A GENERAL STUDY OF A COMMUTATION RELATION GIVEN BY L. TARTAR

Denis Serre

U.E.R de Sciences
23, rue du Dr Paul Michelon
42023 St.-Etienne, Cedex FRANCE

I. Introduction

Let (S) be a strictly hyperbolic system of two conservation laws with two unknown functions:

$$(S) \quad \begin{cases} u_t + f(u,v)_x = 0 \\ \\ v_t + g(u,v)_x = 0 \end{cases} \quad x \in \mathbb{R}, \ t > 0$$

L. Tartar [1] studied the parabolic approximation of this system, intending to prove the existence of a weak entropy to the Cauchy problem:

$$(S_\varepsilon, \ \varepsilon > 0) \quad \begin{cases} u_t^\varepsilon + f(u^\varepsilon, v^\varepsilon)_x = \varepsilon u_{xx}^\varepsilon, \\ \\ v_t^\varepsilon + g(u^\varepsilon, v^\varepsilon)_x = \varepsilon v_{xx}^\varepsilon. \end{cases}$$

It is well known that we cannot expect more than uniform L^∞ estimates, which are not precise enough to pass to the limit in the nonlinear terms when ε goes to zero. Consequently, L. Tartar used the Compensated - Compactness theory (due to Murat and him [2]) in order to get more information. The most that we can get in this way is the following:

For all entropies ϕ_1 and ϕ_2, with fluxes ψ_1 and ψ_2 respectively, one has

$$\lim (\phi_1^\varepsilon \psi_2^\varepsilon - \phi_2^\varepsilon \psi_1^\varepsilon) = \lim \phi_1^\varepsilon \ \lim \psi_2^\varepsilon - \lim \phi_2^\varepsilon \ \lim \psi_1^\varepsilon$$

where the limits are taken with respect to the weak-star topology of $L^\infty_{loc}(R \times R^+)$.
This relation is a direct consequence of the div-curl Lemma provided that

$$\frac{\partial}{\partial t}\phi_i^\varepsilon + \frac{\partial}{\partial x}\psi_i^\varepsilon$$

lie in a compact set of $W^{-1,2}_{loc}$, which is true in many cases.

Let $\nu_{x,t}$ denote the Young measures of the sequence $(u^\varepsilon, v^\varepsilon)$ (one can assume that a suitable subsequence has been extracted):

$$\lim F(u^\varepsilon, v^\varepsilon) = a.e. = \langle \nu_{x,t}, F \rangle.$$

The probability measures $\nu_{x,t}$ are defined on the (u,v)-plane and have bounded support because of the L^∞ estimates. Then the previous relation can be rewritten as a commutation relation:

(E) For all entropies ϕ_1 and ϕ_2 with fluxes ψ_1 and ψ_2, one has a-e in (x,t)

$$\langle \nu_{x,t}, \phi_1\psi_2 - \phi_2\psi_1 \rangle = \langle \nu_{x,t}, \phi_1 \rangle \langle \nu_{x,t}, \psi_2 \rangle - \langle \nu_{x,t}, \phi_2 \rangle \langle \nu_{x,t}, \psi_1 \rangle.$$

Such an equation contains a great deal of information, since there are many pairs (ϕ, ψ) of entropy and flux: they span an infinite dimensional vector space. One can expect under reasonable assumptions that (E) implies that ν is a Dirac measure. In this case, the convergence would be strong in $L^1(R \times R^+)$ and $u = \lim u^\varepsilon$, $v = \lim v^\varepsilon$ would be a weak solution of (S). Such a result has been proved by R. Di Perna for nonlinear elasticity [3] and isentropic gas dynamics [4] by assuming a suitable behavior of the state function.

In this paper, we study systematically the equation (E). We begin by showing that ν need not be a Dirac measure in linearly degenerate cases. Next we study special families of entropies called East - (or W,S,N) - type entropies. By applying (E) to these entropies, one gets non trivial properties which generalize those of R. Di Perna, and one obtains information about the interaction of different kinds of waves. After solving (E) in several cases, we make a conjecture about the structure of ν in the general case. We then apply this conjecture to nonlinear elasticity and isentropic gas dynamics. In both these examples, one shows that it is valid to pass to the limit in the nonlinear terms, which is the result that we want. This last part is a joint work with Michel Rascle.

Abstracts of these results have been published [5],[6]. Proofs of them will be omitted here in many cases, but the details will appear elsewhere.

II. Facts about hyperbolic systems with two conservation laws

In this section, we recall some well-known properties and definitions about hyperbolic systems with two conservation laws. Consider the system (5). Strict hyperbolicity means that the eigenvalues $\lambda_1(u,v)$ and $\lambda_2(u,v)$ of the Jacobian matrix

$$A(u,v) = \begin{pmatrix} \dfrac{\partial f}{\partial u} & \dfrac{\partial f}{\partial v} \\ \dfrac{\partial g}{\partial u} & \dfrac{\partial g}{\partial v} \end{pmatrix}$$

are real and distinct.

The corresponding left eigenvectors can be chosen as gradients of functions $w_1(u,v)$ and $w_2(u,v)$. These two functions are called the Riemann invariants (Lax [7]). For classical solutions, system (5) is equivalent to its characteristic form:

$$\begin{cases} \left(\dfrac{\partial}{\partial t} + \lambda_1 \dfrac{\partial}{\partial x}\right) w_1 = 0, \\ \left(\dfrac{\partial}{\partial t} + \lambda_2 \dfrac{\partial}{\partial x}\right) w_2 = 0. \end{cases}$$

The characteristic speeds λ_1 and λ_2 can be viewed as functions of (w_1, w_2), and one says that the i-th characteristic field is genuinely nonlinear if $\dfrac{\partial \lambda_i}{\partial w_i}$ never vanishes, and linearly degenerate if it vanishes everywhere.

The entropies and fluxes are pairs $(\phi(u,v), \psi(u,v))$ such that, for all classical solutions of (S), one has

$$\frac{\partial \phi}{\partial t} + \frac{\partial \psi}{\partial x} = 0 .$$

These are the solutions of the linear hyperbolic system with variable coefficients

$$\frac{\partial \psi}{\partial w_i} = \lambda_i \frac{\partial \phi}{\partial w_i} , \quad i = 1,2. \tag{1}$$

We can eliminate the flux and keep only one equation whose solutions are the entropies:

$$\frac{\partial^2 \phi}{\partial w_1 \partial w_2} + a_1 \frac{\partial \phi}{\partial w_1} + a_2 \frac{\partial \phi}{\partial w_2} = 0 , \tag{2}$$

with

$$a_1 = \frac{1}{\lambda_1 - \lambda_2} \frac{\partial \lambda_1}{\partial w_2} , \quad a_2 = \frac{1}{\lambda_2 - \lambda_1} \frac{\partial \lambda_2}{\partial w_2} . \tag{3}$$

III. - Systems with a linearly degenerate field.

We consider systems for which one has $\frac{\partial \lambda_2}{\partial w_2} \equiv 0$, i.e. $\lambda_2 = \lambda_2(w_1)$. In that case, one computes explicitly all the entropies. They are

$$\phi(w) = \phi_0(w) \{ g(w_2) + \int_{w_1^*}^{w_1} a(\eta) K(\eta, w_2) \, d\eta \} .$$

Here, ϕ_0 and K are known functions of w, and g, a are arbitrary functions of one variable.

By applying (E) to these entropies and their associated fluxes, one easily gets that there exist two positive measures Y_i, $i = 1,2$, defined on intervals of $\mathbb{R}$, such that

$$\nu = \frac{1}{\phi_0} Y_1 \otimes Y_2 . \tag{4}$$

Moreover, the function

$$\eta \to \langle Y_2, K(\eta,\cdot)\rangle \qquad (5)$$

is a constant function.

Conversely, any probability measure satisfying (4) and (5) is a solution of (E). So, in this case, this equation is completely solved.

One sees in this example that the expected result (ν - Dirac measure) does not hold. It is interesting to know if such complicated measures can in fact be Young measures for sequences ($u^\varepsilon, v^\varepsilon$). The answer to this question is <u>yes</u>. One can construct entropy solutions (u,v) of any system with a linearly degenerate field such that:

i) (u,v) has two independent periods in the (x,t) - plane,

ii) the Young measure $\nu_{x,t}$ of the sequence

$$u^\varepsilon(x,t) = u(\tfrac{x}{\varepsilon}, \tfrac{t}{\varepsilon}), \quad v^\varepsilon(x,t) = v(\tfrac{x}{\varepsilon}, \tfrac{t}{\varepsilon}),$$

which does not depends on x,t, has the form $\nu = \dfrac{1}{\phi_0} Y_1 \times Y_2$,

iii) neither Y_1 nor Y_2 is a Dirac measure.

Then it seems that equation (E) contains all the constraints on the oscillations of the sequence ($u^\varepsilon, v^\varepsilon$). Let us point out that this is not true when (S) has three or more equations, since then there are only finitely many entropies.

IV. Special entropies

We consider the Goursat problem for the equation (2):

$$\begin{cases} \phi(w_1^*, w_2) = \theta_2(w_2), \\ \phi(w_1, w_2^*) = \theta_1(w_1), \end{cases}$$

where θ_1 and θ_2 are given functions of one variable, w^* is a given point, and

$\theta_1(w^*) = \theta_2(w^*)$.

It is well known that this problem has a unique solution, which is as regular as the data θ_1 and θ_2. A special case is provided by assuming $\theta_2 \equiv 0$ and $\theta_1(w_2) = 0$ for $w_1 < w_1^*$. Then it is easy to show that

$$\phi(w) = 0, \text{ for all } w \text{ s.t. } w_1 < w_1^*.$$

We call such an entropy an East-type entropy with limit w_1^*. Its associated flux, which is known up to an additive constant, can be chosen so that

$$\psi(w) = 0, \text{ for all } w \text{ s.t. } w_1 < w_1^*.$$

Then we say that (ϕ, ψ) is an East-type entropy pair of entropy-flux with limit w_1^*. Note that all $\overline{w}_1$ less than w_1^* is also a limit for (ϕ, ψ).

Similarly, there are West-type pairs with limit w_1^*, for which $\phi(w) = \psi(w) = 0$ if $w_1 > w_1^*$. There are also North-type and South-type pairs with limit w_2^*.

These families describe all solutions of (2). Actually, each entropy ϕ has a unique decomposition (w* being fixed) as

$$\phi = \phi(w^*) + \phi_E + \phi_W + \phi_S + \phi_N,$$

ϕ_E being East-type with limit w_1^*, and so on.

So it suffices to apply (E) to take special pairs in order to get all available information.

It will be useful to know the behavior of an East-type entropy ϕ_E. In the neighbourhood of its limit w_1^*, it behaves like

$$(w_1 - w_1^*)^+ \exp \int_{w_2}^{w_2^*} a_1(w_1, n) \, dn$$

for some convenient w_2^*, provided $\frac{\partial \phi}{\partial w_1}$ do not vanish identically on the line $w_1 = w_1^*$. Other formulas occur for other types.

V. Results related to the nonlinearity of characteristic fields.

In this section, we apply (E) to East and West-type pairs of entropy-flux only.

In the remainder of this article, we denote $R = [w_1^-, w_1^+] \times [w_2^-, w_2^+]$, the smallest characteristic rectangle containing the support of the measure ν.

Since (E) has been solved by Tartar [1] when $w_1^- = w_1^+$ or $w_2^- = w_2^+$, we suppose throughout this section that

$$w_1^- < w_1^+ \quad \text{and} \quad w_2^- < w_2^+ .$$

Choose $w_1^* \in (w_1^-, w_1^+)$ and (ϕ_1, ψ_1), (ϕ_2, ψ_2) respectively of East and West type with limit w_1^*. Then the crossed product $\phi_1 \psi_2 - \phi_2 \psi_1$ vanishes identically and an application of (E) yields

$$\langle \nu, \psi_2 \rangle \langle \nu, \phi_1 \rangle = \langle \nu, \psi_1 \rangle \langle \nu, \phi_2 \rangle . \tag{7}$$

An immediate consequence is that there is a real number $c_1(w_1^*)$ such that for all such pairs:

$$\langle \nu, \psi_i \rangle = c_1(w_1^*) \langle \nu, \phi_i \rangle .$$

But using the fact that the limit of a pair (ϕ, ψ) is not unique, one sees that $c_1(w_1^*)$ is a constant C_1 independent of w_1^*.

Theorem: There exist constants c_1, c_2 such that

i) for all East (or West) - type pairs (ϕ, ψ) with limit in $[w_1^-, w_1^+]$,

$$\langle \nu, \psi \rangle = c_1 \langle \nu, \phi \rangle ,$$

ii) for all North (or South) - type pairs (ϕ, ψ) with limit in $[w_2^-, w_2^+]$,

$$\langle \nu, \psi \rangle = c_2 \langle \nu, \phi \rangle$$

□

The next step is to apply (E) with one pair of East-type, and another of West-type, with limits such that the supports overlap. Using the previous theorem, we obtain

$$\langle \nu, \phi_1 \psi_2 - \phi_2 \psi_1 \rangle = 0$$

Now, we use the equivalent (6) for the entropies, and a related formula for the fluxes. After tedious calculations one obtains the following result, a generalization of DiPerna's.

<u>Theorem</u>: If Supp ν intersects the line $w_1 = \alpha$, then for each trace of ν on this line (in a convenient sense), namely Y, one has

$$\langle Y, \frac{\partial \lambda_1}{\partial w_1} \exp 2 \int_{w_2}^{w_2^*} a_1(\alpha, \eta) \, d\eta \rangle = 0 \ . \tag{8}$$

□

The function a_1 was the same one defined in (3). Equivalently:

<u>Proposition</u>: For all continuous function $F(w_1)$, one has

$$\langle \nu, F(w_1) \frac{\partial \lambda_1}{\partial w_1} \exp 2 \int_{w_2}^{w_2^*} a_1(w, \eta) \, d\eta \rangle = 0. \tag{8'}$$

□

The condition (5) can be viewed as this equality when $\lambda_2 = \lambda_2(w_1)$. As a consequence, if $\frac{\partial \lambda_1}{\partial w_1}$ is strictly positive on a vertical strip, then Supp ν cannot intersect this strip, unless $w_1^- = w_1^+$.

As usual, there is a symmetric result about the traces of ν on horizontal lines.

VI. Results and a conjecture related to the interaction of different kinds of waves

If we take an East-type and a North-type pair, one has, by virtue of the previous section

$$\langle \nu, \phi_1\psi_2 - \phi_2\psi_1 \rangle = (c_2 - c_1) \langle \nu, \phi_1 \rangle \langle \nu_1 \phi_2 \rangle.$$

Let their supports tend to the right and the upper edges of R respectively. Then one sees that

Proposition: $A^{++} = (w_2^+, w_2^+)$ lies in the supports of the traces of ν on the lines $\{w_1 = w^+\}$ and $\{w_2 = w_2^+\}$.

As a consequence A^{++} lies in the support of ν. Similarly results hold for the other corners $A^{\pm\pm}$ of R.

Now, using simultaneously one pair of each of the four types, and letting the supports tend to the four edges of R, one gets a formula which links the values of ν near the vertices of R. For example

Theorem: Suppose that ν has continuous density $N(w)$ on R. Then

$$\text{Log} \frac{N(w_1^+, w_2^+) \, N(w_1^-, w_2^-)}{N(w_1^+, w_2^-) \, N(w_1^-, w_2^+)} = - \int_{w_1^-}^{w_1^+} \int_{w_2^-}^{w_2^+} \frac{\partial(\lambda_1, \lambda_2)}{\partial(w_1, w_2)} \frac{dw_1 \, dw_2}{(\lambda_2 - \lambda_1)^2}. \qquad \square$$

Here $\frac{\partial(\lambda_1, \lambda_2)}{\partial(w_1, w_2)}$ denotes the Jacobian of the mapping $w \to (\lambda_1, \lambda_2)$. For convenience, we shall denote by $H(w)$ any solution of the following equation

$$\frac{\partial^2 H}{\partial w_1 \partial w_2} = - \frac{1}{(\lambda_2 - \lambda_1)^2} \frac{\partial(\lambda_1, \lambda_2)}{\partial(w_1, w_2)} \qquad (9)$$

Then the theorem expresses a property of $M = N \exp(-H)$:

$$M(A^{++}) \, M(A^{--}) = M(A^{+-}) M(A^{-+}). \qquad (10)$$

Such a formula and related results lead us to the following <u>conjecture</u>

(C) $\begin{cases} \text{There exist two positive measures } Y_i, \; i = 1,2, \text{ with supports in } [w^-, w^+] \\ \text{respectively, such that} \\ \qquad \nu = \exp H(w) \; Y_1 \otimes Y_2 \,. \end{cases}$

This conjecture expresses the fact that (10) is still true if $A^{\pm\pm}$ are the corners of any characteristic rectangle, not only R.

We have further motivations for conjecture (C):

i) It is consistent with the results of section III. In this case, it reduces to the formula (4).

ii) H(w) is defined up to the addition of $h_1(w_1) + h_2(w_2)$, but (C) does not depend on such a transformation.

iii) A linear change of the (x,t) - coordinates would act as a homothethy on the speeds λ_i, but it does not change the value of H, nor the form of ν.

iv) M. Rascle [8] claims that the conjecture (C) is true at least if ν has a finite support.

It must be pointed out that the weight exp H(w) plays a central role when oscillations occur in $u^\varepsilon, v^\varepsilon$. Take as an example the following system

$$w_t + zw_x = 0,$$
$$z_t + wz_x = 0,$$

whose characteristic fields are linearly degenerate. Then large oscillations with small frequencies in fact propagate with macroscopic speeds λ, μ which are not the averages of the microscopic ones z^ε and w^ε (ε denotes here the wave length of the oscillations). Actually, (see Serre [9]):

$$\lambda = \lim \frac{z^\varepsilon}{w^\varepsilon - z^\varepsilon} \Big/ \lim \frac{1}{w^\varepsilon - z^\varepsilon},$$

$$\mu = \lim \frac{w^\varepsilon}{w^\varepsilon - z^\varepsilon} \Big/ \lim \frac{1}{w^\varepsilon - z^\varepsilon},$$

and w-z is just the weight exp H in this case.

VII. Miscellaneous systems for which (E) has been solved

Besides the cases studied in section III, we can solve (E) in the following cases, by various ideas related to the section IV.

1) Suppose that $\frac{\partial \lambda_1}{\partial w_1}$ and $\frac{\partial \lambda_2}{\partial w_2}$ are both positive inside a circle $(w_1)^2 + (w_2)^2 = 1$, and are both negative outside, vanishing only on the circle. Then one proves that ν is a Dirac mass, which is the result that we want in general.

2) Suppose that the system (S) has the same quantities a_1 and a_2 (defined by (3)) than a system (S') whose one characteristic field is linearly degenerate, say $\lambda_2' = \lambda_2'(w_1)$.

Then one proves the conjecture (C), and that (S) equals (S') on

$$\text{Supp } Y_1 \times [w_1^-, w_1^+].$$

In these cases, the equation (E) is completely solved.

VIII. Applications of the conjecture

The most physically meaningful systems are, in fact, the following:

1-D Elasticity

(E1)
$$\begin{cases} u_t + \tau(v)_x = 0, \quad (\sigma' > 0) \\ v_t + u_x = 0. \end{cases}$$

1-D Isentropic Gas Dynamics

(IDG)
$$\begin{cases} \rho_t + q_x = 0, \\ q_t + (\frac{q^2}{\rho} + p(\rho))_x = 0, \quad (p' > 0). \end{cases}$$

R. Di Perna has solved them, using the compensated compactness theory, by assuming suitable conditions on σ or p. In the general case, the existence theory is still an open problem.

In the absence of further results, one can try to apply the conjecture (C) to these systems. M. Rascle and the author have successfully carried out this procedure (see also [10] for (E1)):

i) In (E1), the conjecture implies

$$\langle \nu, \sigma(\cdot) \rangle = \sigma(\langle \nu, u \rangle).$$

Actually, if ν is not a Dirac measure, then the system is linear for all the involved values of u.

ii) In (IGD), the conjecture implies

$$\langle \nu, \frac{q^2}{\rho} + p(\rho) \rangle = \frac{\langle \nu, q \rangle^2}{\langle \nu, \rho \rangle} + p(\langle \nu, \rho \rangle).$$

These results are the one that we want. They mean that the limits $u = \lim u^\varepsilon$ and $v = \lim v^\varepsilon$ are weak solutions of the system (S). Unfortunately, besides the fact that (C) has not been proved, we must note that for (IDG), the div-curl Lemma cannot be applied to every pair of entropy-flux, but only for half of them (R. Di Perna [4]).

Bibliography

[1] L. Tartar: Compensated Compactness and applications to PDE, in Research Notes in Maths, Nonlinear Analysis and Mechanics. Heriot-Watt Symposium, 4. 1979. Knops Ed. Pitman Press.

[2] F. Murat: Compacité par compensation, Ann. Scuola Norm. Sup. di Pisa, Sci. Fis. Mat. 5 (1978), 489-507.

[3] R. DiPerna: Convergence of Approximate solutions to Conservation Laws. Arch. Rat. Mech. Anal. (1983), 82, 27-70.

[4] R. DiPerna: Convergence of the viscosity method for isentropic gas dynamics, Comm. in Math. Phys. 91 (1983), 1-30.

[5] D. Serre: Compacité par compensation et systemes hyperboliques de lois de conservation. CRAS, 299 (1984), 555-558.

[6] M. Rascle - D. Serre: Compacité par compensation et systemes hyperboliques de lois de conservation. Applications. C.R.A..S. 299. Série I. 1984, n° 14, 673-676.

[7] P.D. Lax: Hyperbolic systems of conservation laws, II. Comm. in Pure and Appl. Math. 10 (1957), 537-566.

[8] M. Rascle: Private communication.

[9] D. Serre: Un systéme hyperbolique non linéaire avec des données oscillantes. C.R.A.S. 302 (1986) n°3, 115-118.

[10] M. Rascle: Thesis. 1983. S^t.Etienne. France.

Note added in proofs: The conjecture (C) has been proved recently by M. Rascle without any extra assumption. One needs only the existence of the rectangle R, which does not allow cases with cavitation in the gaz dynamics. The convergence of the viscosity method for nonlinear elasticity is thus proved for any stress-strain relation σ, up to L^∞ estimates. The proof is given in

[11] M. Rascle: Un resultat de "compracité par compensation" á coefficients variables. Application á ℓ'élasticité non linéarire. C.R.A.S. 302 91986) 311-314.

INTERRELATIONSHIPS AMONG MECHANICS NUMERICAL ANALYSIS, COMPENSATED COMPACTNESS, AND OSCILLATION THEORY.

M. Slemrod[#]

Department of Mathematical Sciences
Rensselaer Polytechnic Institute
Troy, N.Y. 12180

0. Introduction

This paper is the written version of my lecture delivered at the Institute for Mathematics and its Applications workshop on Oscillation Theory, Computation, and Methods of Compensated Compactness. As both the titles of the workshop and this paper suggest, I believe there is a continuum of ideas and methods relating these topics. Perhaps the unifying word is "regularization" for it is a goal of applied mathematics to understand the analytical and physical meanings of the various regularizations of the conservation laws of continuum mechanics. In particular the ability to pass to the limit as the regularization parameters vanish has been a long standing problem and it seems that major progress has been made on this question recently (indeed by several of the participants of this I.M.A. workshop).

These notes are intended only as a rough initiation to the subject matter of the workshop. For related ideas I recommend the recent review article of Lax [1].

The paper is divided into ten sections:

Section 1 discusses the balance laws of mass and momentum. We see how the usual equations for the motion of an elastic fluid and an elastic solid are altered by the inclusion of the effects of viscosity and capillarity.

[#] Research sponsored in part by the Air Force office of Scientific Research, Air Force Systems Command, USAF, Contract/Grant AFOSR-81-0172. The U.S. Government's right to retain a nonexclusive royalty-free license in and to the copyright covering this paper, for governmental purposes, is acknowledged.

Section 2 considers regularization via finite differences. We interpret the physical significance of the Lax-Friedrichs scheme in line with the ideas of Section 1.

Section 3 discusses traveling wave solutions for the mass-momentum conservation laws when viscosity and capillarity are present. We see that for a special Riemann problem we can obtain a weak solution of the original conservation law via passage to the zero limit in the viscosity and capillarity coefficients.

Section 4 continues the ideas of Section 3. We note that if viscosity is absent the limit procedure of Section 3 fails in general.

Section 5 considers the problem of passage to the limit for general initial data in the special case of Burgers' equation. The issues here are simplified by the ability to integrate the equation via the Hopf-Cole transformation.

Section 6 presents an elementary introduction to weak limits of oscillatory functions.

Section 7 expands on Section 4. We use formal matched asymptotic expansions (following in part the article of D. Mc Laughlin in this volume) to examine in detail the weak limit of the KdV equation. More details in the issue will be found on the articles of Forest, McLaughlin, Lax, and Venakides in this volume.

Section 8 summarizes the ideas of earlier sections and shows how weak continuity of nonlinear functions will be crucial in passing to the limit for regularized conservation laws.

Section 9 introduces the Young measure and shows how weak limits may be quantified with its use. This section amplifies on perhaps obvious details for the benefit of newcomers. V. Roytburd was of considerable assistance to me in its preparation.

Section 10 presents Luc Tartar's argument [2] for passing to the viscous limit for scalar conservation laws. For simplicity only Burgers' equation is considered.

More recently compensated compactness has been applied by R. DiPerna to 2x2 systems of conservation laws [3], [4], M. Schonbek to the KdV-Burgers equation [5].

The interested reader should consult these papers as well as their articles in this volume.

1. Mechanics

Let us begin by considering the balance of mass and linear momentum in Lagrangian coordinates

(0)
$$u_t = \tau_x ,$$
$$w_t = u_x ,$$

where

$$u(x,t) = \text{velocity},$$

$$w(x,t) = \begin{cases} \text{specific volume in a compressible fluid}, \\ \\ \text{deformation gradient in a solid}, \end{cases}$$

$$\tau(x,t) = \text{stress}.$$

Example 1: $\tau = -p(w)$ is a constitutive relation which defines either an <u>elastic solid</u> or an <u>elastic fluid</u>.

If we insert the relation $\tau = -p(w)$ into (0) we find (u,w) satisfy the system

(1)
$$u_t + p'(w)w_x = 0 ,$$
$$w_t - u_x = 0 .$$

If $p' < 0$, (1) is <u>strictly hyperbolic</u>.

If we impose initial values

$$u(x,0) = u_0(x), \quad w(x,0) = w_0(x)$$

then we have an initial value problem and we can ask the canonical questions: Do weak solutions exist for the initial value problem? If they do exist how should we compute them? We consider weak solutions since it is well known that shock formation will preclude existence of smooth solutions in general.

<u>Example 2</u>: $\tau = -p(w) + \mu u_x$, $\mu > 0$ (the viscosity), is a constitutive relation which defines either a type of visco-elastic solid or viscous fluid.

If we insert the relation $\tau = -p(w) + \mu u_x$ into (0) we find (u,w) satisfy the system

(2)
$$u_t + p'(w)w_x = \mu u_{xx},$$
$$w_t - u_x = 0.$$

At this point we ask our next question motivated by work dating to Rayleigh [6] though the question is usually associated with a paper of Gelfand [7].

Can we pass to the limit as $\mu \to 0+$ and obtain a weak solution of (1)?

<u>Example 3</u>. $\tau = -p(w) + \mu u_x - Aw_{xx}$ $\mu > 0$ (the viscosity), $A > 0$ (the capillarity coefficient) is a constitutive relation which generalizes Example 2 to include the effects of capillarity. Such ideas were introduced about eighty years ago in the papers of Korteweg [8] and van der Waals [9]. More recent discussion of Korteweg's theory may be found in [10], [11].

If we insert the relation $\tau = p(w) + \mu u_x - Aw_{xx}$ into (0) we find (u,w) satisfy the system

(3)
$$u_t + p'(w)w_x = \mu u_{xx} - Aw_{xxx},$$
$$w_t - u_x = 0.$$

For this system we can ask a similar question to that of Example 2.

Can we pass to the limit as $\mu \to 0+$, $A \to 0+$; $\mu = 0$, $A \to 0+$; $A = 0$, $\mu \to 0+$, to obtain a weak solution of (1)?

In (3) there are two competing mechanisms: viscosity which introduces diffusion and capillarity which introduces dispersion. To see the effect of this competition it is convenient to introduce the change of variables

$$v = u - D_2 w_x$$

where

$$\left.\begin{array}{c} D_1 \\ D_2 \end{array}\right\} \doteq \frac{u}{2} \mp \frac{1}{2}(\mu^2 - 4A)^{1/2} ,$$

$D_1 D_2 = A$, $\mu - D_1 = D_2$.

If we substitute $u = v + D_2 w_x$ into (3) we see

$$\begin{aligned} v_t &= u_t - D_2 w_{xt} \\ &= -p'(w)w_x + D_1 v_{xx} + ((D_1 + D_2)D_2 - D_1 D_2 - D_2^2) w_{xxx} . \end{aligned}$$

The third term on the right side vanishes and hence we see

(4)
$$v_t + p'(w)w_x = D_1 v_{xx} ,$$
$$w_t - u_x = D_2 w_{xx} .$$

Thus if

(5)
$$\mu^2 > 4A > 0$$

so that <u>viscosity dominates the capillarity</u> D_1 and D_2 are positive constants and (4) represents a complete parabolic regularization of (1). Contrast this to (2) which is an incomplete parabolic regularization of (1). In this context we see that if (5) holds the question as to the convergence of solutions (3) as

A → 0+, μ → 0+ to a solution of (1) becomes a question on the convergence of solutions of (4) to a solution of (1). In fact this was the program Gelfand originally proposed in [7]. Hence a physical interpretation of the Gelfand program is that weak solutions of the macroscopic system (1) should be limits of the more accurate microscopic system (3) (or (4)) which includes the effects of viscosity and capillarity, viscosity dominating capillarity as in (5).

2. Finite difference regularization

In the previous section we discussed regularizations of (1) by considering the effects of viscosity and capillarity. In this section we consider a second regularization via finite differences.

We write the Lax-Friedrichs finite difference scheme for (1):

(6)
$$\frac{1}{\Delta t} \{U(x,t+\Delta t) - \frac{U(x+\Delta x,t) + U(x-\Delta x,t)}{2}\} +$$

$$\frac{1}{2\Delta x} \{p(W(x+\Delta x,t)) - p(W(x-\Delta x,t))\} = 0,$$

$$\frac{1}{\Delta t} \{W(x,t+\Delta t) - \frac{W(x+\Delta x,t) + W(x-\Delta x,t)}{2}\}$$

$$-\frac{1}{2\Delta x} \{U(x+\Delta x,t) - U(x-\Delta x,t)\} = 0.$$

Set $\lambda = \frac{\Delta t}{\Delta x}$. Assume the difference equation has a smooth solution and expand u,w in a Taylor series about (x,t). We find u,w satisfy

$$U_t + p(W)_x = \frac{\Delta t}{2} (\phi(W,\lambda)U_x)_x + \ldots$$

$$W_t - U_x = \frac{\Delta t}{2} (\phi(W,\lambda)W_x)_x + \ldots$$

where ... denotes terms of $O((\Delta t)^2)$ and $\phi(W,\lambda) = p'(W) + \lambda^{-2}$. Thus we if $\phi(W,\lambda) > 0$ (which is the Courant-Friedrichs - Lewy condition) the Lax-Friedrichs regularization (6) is of course first order accurate to (1) but also second order accurate to

$$U_t + p(W)_x = \frac{\Delta t}{2} (\phi(W,\lambda)U_x)_x ,$$

(7)

$$W_t - U_x = \frac{\Delta t}{2} (\phi(W,\lambda)W_x)_x .$$

We notice the similarity between (7) and (4). Hence a physical interpretation of the Lax-Friedrichs regularization of (1) is that it artificially introduces the effects of both viscosity and capillarity. From practical considerations the implication of these ideas is as follows: If one desires to numerically simulate the physical effects of viscosity and capillarity the Lax-Friedrichs method may be a reasonable choice. This seems particular important in problem where p is non-monotone. (See [12], [13].)

3. Limits for the equation with viscosity and capillarity.

Consider initial values for the Riemann problem

$$w_0(x) = \begin{cases} w_- & x < 0, \\ w_+ & x > 0 \end{cases} \qquad u_0(x) = \begin{cases} u_- & x < 0 \\ u_+ & x > 0 \end{cases}$$

where w_-, w_+, u_-, u_+ are constants and satisfy the Rankine-Hugoniot jump conditions

$$-s(u_+ - u_-) + p(w_+) - p(w_-) = 0 ,$$

(8)

$$-s(w_+ - w_-) - (u_+ - u_-) = 0 ,$$

for s as determined from (8). In this case the known admissible weak solution (say according to the Lax entropy criteria [1]) is

$$w^* = \begin{cases} w_- & x < st , \\ w_+ & x > st \end{cases} \qquad u^* = \begin{cases} u_- & x < st \\ u_+ & x > st \end{cases}$$

as shown in Figure 1.

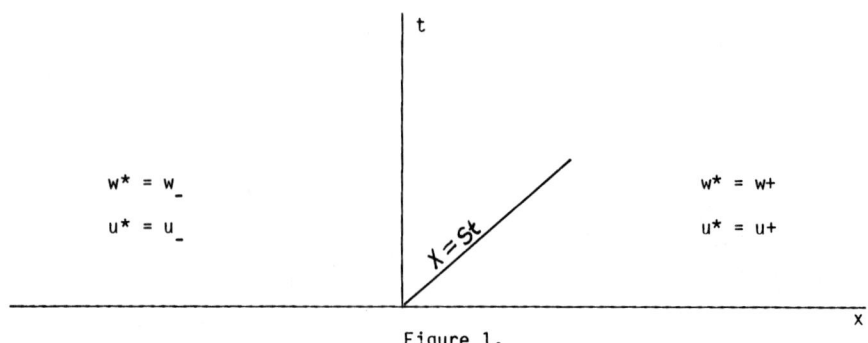

Figure 1.

We will try to see if this solution (w^*, u^*) of (1) is the limit of traveling wave solutions of (4)

$$w = \hat{w}\left(\frac{x-st}{\mu}\right), \quad u = \hat{u}\left(\frac{x-st}{\mu}\right).$$

In this case we see w must satisfy the ordinary differential equation

(9) $\quad \overline{A}\hat{w}'' + s\hat{w}' + p(\hat{w}) - p(w_-) + s^2(\hat{w}-w_-) = 0$

where $A \doteq \overline{A}\mu^2$. Equation (9) possesses equilibrium points $\hat{w}' = 0$, $\hat{w} = w_+$ and $\hat{w}' = 0$, $\hat{w} = w_-$. If we find a solution (9) connecting these equilibrium points so that

$$\hat{w}(+\infty) = w_+, \qquad \hat{w}(-\infty) = w_-,$$

$$\hat{w}'(+\infty) = 0, \qquad \hat{w}'(-\infty) = 0,$$

we see that $(w,u) \to (w^*,u^*)$ as $\mu \to 0+$. For simplicity we consider the case where p has the shape for an ideal fluid as shown in Figure 2.

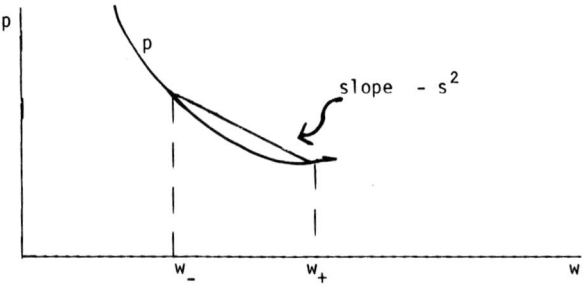

Figure 2.

The eigenvalues associated with (9) linearized about w_- and w_+ are respectively

$$\lambda_{w_-}^{\pm} = -\frac{s \pm (s^2 - 4(p'(w_-) + s^2)\bar{A})^{1/2}}{2\bar{A}},$$

so that $\lambda_{w_-}^+ > 0$, $\lambda_{w_-}^- < 0$ and $w = w_-$, $w' = 0$ is always a saddle, and

$$\lambda_{w_+}^{\pm} = -\frac{s \pm (s^2 - 4(p'(w_+) + s^2)\bar{A})^{1/2}}{2\bar{A}}.$$

If $\bar{A}$ is small, $\lambda_{w_+}^{\pm} < 0$ and $w = w_+$, $w' = 0$ is a stable node and if $\bar{A}$ is large, $\lambda_{w_+}^{\pm} = \alpha + i\beta$ where $\alpha < 0$, $\beta \neq 0$, and $w = w_+$, $w' = 0$ is a stable focus. Phase portraits in the two cases are shown below.

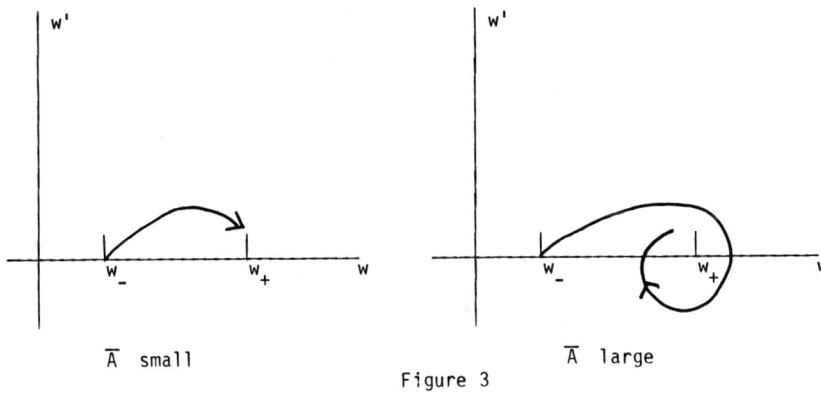

Figure 3

In either case we see

$$\hat{w}(\frac{x-st}{\mu}), \hat{u}(\frac{x-st}{\mu}) \to w^*, u^* \quad \text{as} \quad \mu \to 0+ .$$

In the case $\bar{A}$ small the limit is approached via a monotone wave, in the $\bar{A}$ large the limit is approached via an oscillatory wave. So we see in this simple case that we can recover the correct solution of (1) via a limit of solutions of (3) (or (4)).

4. Limit for the equation with capillarity alone

We consider in this section the question of reproducing our argument of Section 3 if viscosity is absent. In this case (3) becomes

(3')
$$u_t + p'(w)w_x = -Aw_{xxx},$$
$$w_t - u_x = 0.$$

We wish to look for a traveling wave solution of (3')

$$\hat{w} = w(\tfrac{x-st}{\varepsilon}), \quad u = \hat{u}(\tfrac{x-st}{\varepsilon}), \quad A = \varepsilon^2.$$

Substitution of w, u into (3') yields the ordinary differential equation

(10)
$$\hat{w}'' + p(\hat{w}) - p(w_-) + s^2(\hat{w} - w_-) = 0.$$

Equation (10) possesses equilibrium points $\hat{w} = w_+$, $\hat{w}' = 0$ and $\hat{w} = w_-$, $\hat{w}' = 0$. Again we would wish to find a solution which connects them. In this case passage to the limit as $\varepsilon \to 0$ would show $(w, u) \to (w^*, u^*)$ as $\varepsilon \to 0$. We once again see that $w = w_+$, $w' = 0$ is a saddle but now $w = w_-$, $w' = 0$ is a center. In fact (10) has a phase portrait of the form shown below,

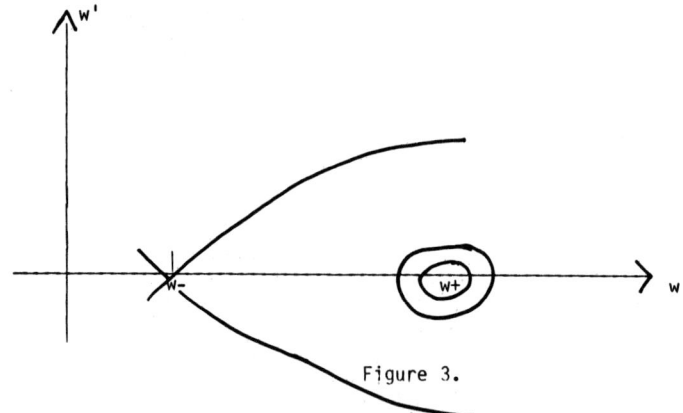

Figure 3.

Hence there is no way that we can connect $(w-, 0)$ to $(w+, 0)$ due to presence of complete orbits around $(w_+, 0)$. <u>Thus there is no hope of taking the capillarity limit $A \to 0+$ and recovering solution (w^*, u^*)</u>.

5. Passage to the limit for general data:
Burgers equation and the Hopf-Cole transformation

We have seen in Section 3 that if viscosity and capillarity were present we could pass to the limit of (3) to obtain the correct solution to Riemann problem for (1). There is a straightforward calculation that shows this approach works for the simple scalar conservation laws $\theta_t + \theta\theta_x = 0$.

Consider the heat equation

(11) $$u_t = \mu u_{xx}, \quad -\infty < x < \infty, \quad \mu > 0,$$

with initial data

(12) $$u(x,0) = u_0(x).$$

The solution of (11), (12) is given by

$$u(x,t) = (4\pi\mu t)^{-1/2} \int_{-\infty}^{\infty} \exp[-\frac{(x-y)^2}{4\mu t}] u_0(y)dy.$$

Now let $\theta(x,t) = -2\mu \frac{u_x}{u}$. Then we see

(13) $$\theta_t + \theta\theta_x = \mu\theta_{xx} \quad \text{(Burgers' equation)}$$

with initial data

(14) $$\theta_0(x) = -2\mu \frac{u_0'(x)}{u_0(x)}.$$

Hence $\theta(x,t)$ is given by the formula

$$\theta(x,t) = -2\mu \frac{u_x}{u} = \frac{\int_{-\infty}^{\infty} \exp[-\frac{(x-y)^2}{4\mu t}](\frac{x-y}{t}) \exp[-\frac{1}{2\mu}\int_{-\infty}^{y} \theta_0(\xi)d\xi]dy}{\int_{-\infty}^{\infty} \exp[-\frac{(x-y)^2}{4\mu t}]\exp[-\frac{1}{2\mu}\int_{-\infty}^{y} \theta_0(\xi)d\xi]dy}.$$

Careful analysis shows we can pass to the limit in this expression as $\mu \to 0+$ and obtain a limit $\theta^*(x,t)$ which is a weak solution of

$$\theta_t + \theta\theta_x = 0,$$

$$\theta(x,0) = \theta_0(x).$$

This idea was originally presented in [14]; see [1] and [15] for further discussion.

Hence if we view (13) as a simple scalar version of (4) we see that at least it is conceivable to expect that viscosity-capillarity limits will lead to correct weak solutions of our original conservation laws for an elastic solid or elastic fluid.

6. Weak limits

In Section 4 we saw that it is not generally possible to pass to the limit of (3) (or (4)) in the absence of viscosity. To better understand this issue in general, we note some elementary facts on weak limits.

Let $f(x)$ be a periodic function say in $L^2[0,2\pi]$ with Fourier representation

$$f(x) \sim \frac{a_0}{2} + \sum_{j=1}^{\infty} (a_j \cos jx + b_j \sin jx)$$

Define $f_n(x) = f(nx)$. Hence

$$f(nx) \sim \frac{a_0}{2} + \sum_{j=1}^{\infty} (a_j \cos jnx + b_j \sin jnx)$$

and for $\phi \in L^2[0,2\pi]$ we have

$$\int_0^{2\pi} \phi(x) f_n(x) dx = \frac{a_0}{2} \int_0^{2\pi} \phi(x) dx + \sum_{j=1}^{\infty} (a_j c_{jn} + b_j d_{jn})$$

where

$$c_{jn} = \int_0^{2\pi} \phi(x) \cos jnx \, dx,$$

$$d_{jn} = \int_0^{2\pi} \phi(x) \sin jnx \, dx.$$

From the Cauchy-Schwarz inequality we know

$$\left|\sum_{j=1}^{\infty} a_j c_{jn}\right| \leq \left[\sum_{j=1}^{\infty} a_j^2\right]^{1/2} \left[\sum_{j=1}^{\infty} c_{jn}^2\right]^{1/2}$$

where $jn = n, 2n, 3n \ldots$. Since

$$\sum_{j=1}^{\infty} c_{jn}^2 \leq \sum_{k=n}^{\infty} c_k^2 \to 0 \text{ as } n \to \infty$$

we see $\sum_{j=1}^{\infty} a_j c_{jn}$ (and similarly $\sum b_j d_{jn}$) $\to 0$

as $n \to \infty$. Hence

$$\lim_{n \to \infty} \int_0^{2\pi} \phi(x) f_n(x) \, dx = \int_0^{2\pi} \phi(x) \frac{a_0}{2} \, dx$$

or

$$f_n(\cdot) \rightharpoonup \bar{f}(\cdot) \text{ as } n \to \infty$$

where

$$\bar{f}(\cdot) = \frac{a_0}{2} = \frac{1}{2\pi} \int_0^{2\pi} f(x) dx$$

and "$\rightharpoonup$" denotes weak convergence in $L^2(0, 2\pi)$.

As simple examples consider:

1°. $f(x) = \sin x$, $f_n(x) = \sin nx$. Our result says

$$f_n(\cdot) \rightharpoonup \frac{1}{2\pi} \int_0^{2\pi} \sin x \text{ as } = 0 \text{ as } n \to \infty .$$

2°. $g(x) = \sin^2 x$, $g_n(x) = \sin^2 nx$. Our result says

$$g_n(\cdot) \rightharpoonup \frac{1}{2\pi} \int_0^{2\pi} \sin^2 x \, dx = \frac{1}{2} \text{ as } n \to \infty .$$

3°. Notice that while $f_n(\cdot) \to 0$ as $n \to \infty$

$$[f_n(\cdot)]^2 = g_n(\cdot) \rightharpoonup \frac{1}{2} \text{ as } n \to \infty ,$$

i.e.

$$\text{weak} \lim_{n \to \infty} [f_n(\cdot)]^2 \neq [\text{weak} \lim_{n \to \infty} f_n(\cdot)]^2.$$

7. Weak limit of KdV via formal two timing

In Section 4 we saw that in the absence of viscosity the capillarity limit alone of (3) will not generally be a weak solution of (1). In this section we examine this issue in more detail for the Korteweg-de Vries equation

$$\theta_t + \theta\theta_x + \varepsilon^2 \theta_{xxx} = 0 \tag{15}$$

which possesses only the dispersive (capillarity) term $\varepsilon^2 \theta_{xxx}$ and no diffusive (viscous) $\varepsilon \theta_{xx}$ term.

We shall follow an idea of Kruskal and Miura [16] and solve (15) by formal two timing. We look for a solution of (15) in the form

$$\theta(x,t) = \Theta\left(z = \frac{\psi(x,t)}{\varepsilon}, x, t\right)$$

$$= \theta_0(z,x,t) + \varepsilon\theta_1(z,x,t) + \varepsilon^2\theta_2(z,x,t) + \dots$$

We want $\Theta(z,x,t)$ periodic of period 2π in z. Set $\kappa(x,t) = \psi_x(x,t)$, $\lambda(x,t) = \psi_t(x,t)$,

and expand

$$\kappa = \kappa_0 + \varepsilon\kappa_1 + \dots$$
$$\lambda = \lambda_0 + \varepsilon\lambda_1 + \dots$$

We substitute these expansions into (15) and equate powers of ε to obtain

$$O(\varepsilon^{-1}): \quad \theta_{0_z}\lambda_0 + \theta_0\theta_{0z}\kappa_0 + \kappa_0^3 \theta_{0_{zzz}} = 0 \quad \text{or}$$

$$(\theta_0 - \theta_+)\lambda_0 + \left(\frac{\theta_0^2}{2} - \frac{\theta_+^2}{2}\right)\kappa_0 + \kappa_0^3 \theta_{0zz} = 0. \tag{16}$$

The phase portrait for (16) is shown below.

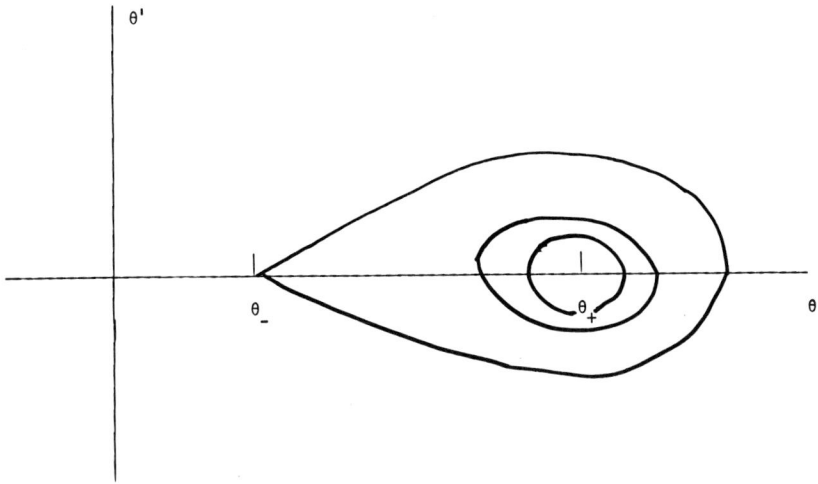

Figure 4.

Here $\theta_-(x,t)$, $\theta_+(x,t)$ are equilibrium points of (16), $\theta_-(x,t)$ is a saddle, $\theta_+(x,t)$ is a center.

Pick a periodic solution of (16) which is periodic of period 2π. This choice determines the dispersion relationship.

The $O(1)$ is relationship is

(17) $\theta_{0_t} + \theta_0 \theta_{0x} + \{\lambda_0 \theta_1 + \kappa_0(\theta_1 \theta_0) + \kappa_0^3 \theta_{1zz}$

$+ \kappa_0^2 \theta_{0_x} + 3\kappa_1 (-\theta_0 \frac{\lambda_0}{\kappa_0} - \frac{\theta_0^2}{2}) + (\kappa_0^2 \theta_0)_{xz} + \kappa_0^2 \theta_0\}_z = 0$

Since θ_0, θ_1 are to be 2π periodic in z integration of (17) with respect to z from 0 to 2π shows

(18) $\int_0^{2\pi} (\theta_{0t} + \theta_0 \theta_{0x}) \, dz = 0$

which is the average conservation law.

We can continue to solve for θ_1 2π periodic in z by use of the Fredholm alternative in (17). We can then follow these ideas through to the

$O(\varepsilon)$ equation. etc. So

$$\theta(x,t) = \Theta(z = \frac{\psi(x,t)}{\varepsilon}, x, t)$$

$$= \theta_0(z,x,t) + \varepsilon\theta_1(z,x,t) + \ldots$$

A calculation similar in spirit to the one we did in Section 6 (see the paper of D. McLaughlin in this volume) shows the weak limit $\varepsilon \to 0$ of $\theta_0(z,x,t)$ is once again the average of $\theta_0(z,x,t)$ over z, $0 \leq z \leq 2\pi$. Since as $\varepsilon \to 0$ the $O(\varepsilon^n)$, $n \geq 1$, terms approach zero we see

$$\theta(\cdot,\cdot) \to \frac{1}{2\pi} \int_0^{2\pi} \theta_0(z,x,t)dz = \langle\theta_0\rangle.$$

So our averaged conservation law (18) may be written as

(19) $$\langle\theta_0\rangle_t + \frac{1}{2}\langle\theta_0^2\rangle_x = 0$$

where now θ_0 is the weak limit as $\varepsilon \to 0$ as the KdV equation (15). Notice this is not the same as saying the weak limit $\langle\theta_0\rangle$ satisfies the original conservation law with $\varepsilon=0$ since we don't have $\langle\theta_0\rangle^2 = \langle\theta_0^2\rangle$ in general. Recall this is the same difficulty we had in 3° of Section 6.

Of course these results are formal. The rigorous theory has been given by Lax and Levermore [17] and Venakides [18]. Their articles in this volume plus the one of McLaughlin will undoubtedly be enlightening regarding the issues hinted at here.

8. Recapitulation of results and a general regularization philosophy for conservation laws

In earlier sections we have seen that in the extreme case when viscosity is present and capillarity is absent, say as represented by the Burgers equation

(20) $$\theta_t + \theta\theta_x = \mu\theta_{xx},$$

we can pass to the limit as $\mu \to 0+$ so that solutions θ^μ of (20) will provide the correct weak solution of the conservation law

(21) $$\theta_t + \theta\theta_x = 0 .$$

On the other hand we have seen that in the other extreme case when capillarity is present and viscosity is absent say as represented by the KdV equation

(22) $$\theta_t + \theta\theta_x = -\epsilon^2 \theta_{xxx}$$

we cannot generally expect the limit of solutions θ^ϵ of (22) to approach a weak solutions of (21) as $\epsilon \to 0$.

In the next sections we put forth two ideas which allows us to understand these extreme limiting cases and set us on the road to a general theory of limits of regularized conservation laws.
The two concepts are

(19) the Young measure (as developed by L.C. Young),

(20) the method of compensated compactness (as developed by L. Tartar and F. Murat).

First however let us see what we might desire in a regularization of a conservation law.

1° It should be consistent with our view of the physics of the problem.
2° It should aid us in obtaining a priori estimates which are independent of the regularization parameter.
3° Knowledge that the limits lead to physically meaningful solutions in cases which are comparatively easy to check, e.g. the existence of traveling wave solutions to Riemann problems as done in Section 3.
4° Knowledge that we don't have purely oscillatory traveling waves as in Section 4. Such waves of course preclude 3°.

As a simple example consider once again Burgers equation

$$\theta_t + \theta\theta_x = \mu\theta_{xx} .$$

This regularization $\mu\theta_{xx}$ shows a viscous diffusive domination and the maximum principle implies

(23) $\qquad \|\theta^\mu\|_\infty <$ const. (independent of μ).

Bound (23) shows $\{\theta^\mu\}$ has a subsequence also denoted by $\{\theta^\mu\}$ such that

$$\theta^\mu \to \bar\theta \quad \text{weak} \quad * \; L^\infty.$$

If we can show

(24) $\qquad (\theta^\mu)^2 \to (\bar\theta)^2 \quad \text{weak} \quad * \; L^\infty$

we see that by multiplying the Burgers equation (20) by a test function $\phi \in (C_0^\infty(\mathbb{R} \times \mathbb{R}^+)$

$$\int \phi \, \theta_t^\mu \, dxdt + \int \phi ((\theta^\mu)^2)_x \, dxdt$$

$$= \mu \int \phi \theta_{xx}^\mu \, dxdt = \mu \int \phi_{xx} \theta^\mu \, dxdt$$

where the integrals are taken over $\mathbb{R} \times \mathbb{R}^+$. If (24) holds then

$$\int \phi((\theta^\mu)^2)_x \, dxdt = -\int \phi_x (\theta^\mu)^2 \, dxdt$$

$$\to -\int \phi_x \bar\theta^2 \, dxdt \quad \text{as} \quad \mu \to 0+$$

and we see the limit $\bar\theta$ satisfies

$$\bar\theta_t + \bar\theta \, \bar\theta_x = 0$$

in the sense of distributions.

The ability to show (24) (which we know from Section 7 <u>cannot</u> be done in the case of oscillatory waves) is the crucial issue. It is to understand this limiting procedure that we turn to next.

9. Characterization of weak limits via the Young measure.

It this section we characterize weak limits in terms of the Young measure.

Let $\Omega \subset \mathbb{R}^m$, $K \subset \mathbb{R}^n$ be bounded open sets. Let $u_s : \Omega \to K$ be a measurable sequence of functions and denote by $C_0(\Omega \times \mathbb{R}^n)$ the continuous functions of compact support.

We introduce the Radon measure [19] on $C_0(\Omega \times \mathbb{R}^n)$:

$$\langle m_s, \phi(x,\lambda) \rangle \stackrel{\text{def.}}{=} \int_\Omega \phi(x, u_s(x)) dx$$

for $\phi \in C_0(\Omega \times \mathbb{R}^n)$.

Since $|u_s|$ are uniformly bounded in s, m_s has a weakly convergent subsequence, also denoted by m_s so that

$$\langle m_s, \phi \rangle \to \langle m, \phi \rangle$$

for all $\phi \in C_0(\Omega \times \mathbb{R}^n)$. This defines a measure m.

The measure m is thus seen to be a linear functional on $C_0(\Omega \times \mathbb{R}^n)$. We deduce a representation of m by the following argument.

Consider $\phi(x,\lambda) = f(\lambda)\psi(x)$, f and ψ continuous. For fixed f, $\langle m, f(\lambda)\psi(x) \rangle$ is a bounded linear functional on $\psi \in C_0(\Omega)$. Hence we can apply the Riesz representation theorem and write

(25) $$\langle m, f(\lambda)\psi(x) \rangle = \int_\Omega \psi(x)\, \omega(dx)$$

for a measure ω. This measure ω is absolutely continuous with respect to the Lebesgue measure on Ω. To see this note that if ψ_0 is a characteristic function of a set of Lebesgue measure zero then

$$\langle m, f(\lambda)\psi_0(x) \rangle = \int_\Omega \psi_0(x) f(u_s(x)) dx = 0$$

and therefore by the definition of m

$$\langle m, f(\lambda)\psi_0(x)\rangle = 0 = \int_\Omega \psi_0(x)\,\omega(dx).$$

which shows the absolute continuity of ω with respect to Lebesgue measure on Ω.

By the Radon-Nikodyn theorem, the absolute continuity of ω with respect to Lebesgue measure implies $\omega(dx)$ can be expressed as

(26) $$\omega(dx) = a_f(x)dx$$

where the subscript f denotes the dependence on f. In fact a_f depends linearly on f. This is seen from

$$\langle m, (c_1 f + c_2 g)(\lambda)\psi(x)\rangle =$$
$$= c_1 \langle m, f(\lambda)\psi(x)\rangle + c_2 \langle m, g(\lambda)\psi(x)\rangle$$
$$= c_1 \int_\Omega \psi(x) a_f(x)dx + c_2 \int_\Omega \psi(x) a_g(x)dx$$
$$= \int_\Omega \psi(x)(c_1 a_f(x) + c_2 a_g(x))dx$$

and also by (25) and (26)

$$\langle m, (c_1 f + c_2 g)(\lambda)\psi(x)\rangle = \int_\Omega \psi(x)\, a_{c_1 f + c_2 g}(x)\,dx,$$

so

$$\int_\Omega \psi(x)\,[c_1 a_f + c_2 a_g - a_{c_1 f + c_2 g}](x)dx = 0$$

for all $\psi \in C_0(\Omega)$. Hence we have

$$a_{c_1 f + c_2 g} = c_1 a_f + c_2 a_g \quad \text{a.e. in } \Omega.$$

Thus for almost all x in Ω the map $f \to a_f(x)$ is a linear functional easily seen to be bounded and hence by the Riesz representation theorem there is a representing measure $\nu_x(\lambda)$ so that

(27) $$a_f(x) = \int_{R^n} f(\lambda)d\nu_x(\lambda)$$

and by (25), (26), (27)

(28) $$\langle m, f(\lambda)\psi(x)\rangle = \int_\Omega \psi(x) \int f(\lambda) d\nu_x(\lambda) dx.$$

Finally if more generally

(29) $$\phi(x,\lambda) = \sum_{i=1}^{n} c_i\, f_i(\lambda)\, \psi_i(x)$$

we have by (28)

$$\langle m, \phi(x,\lambda)\rangle = \sum_{i=1}^{N} c_i \langle m, f_i(\lambda)\psi_i(x)\rangle$$

$$= \sum_{i=1}^{N} c_i \int_\Omega \psi_i(x) \int_{\mathbb{R}^n} d\nu_x(\lambda)\, f_i(\lambda)\, dx$$

$$= \int_\Omega \int_{\mathbb{R}^n} \left(\sum_{i=1}^{N} c_i \psi_i(x)\, f_i(\lambda) \right) d\nu_x(\lambda)\, dx$$

i.e.

(30) $$\langle m, \phi(x,\lambda)\rangle = \int_\Omega \int_{\mathbb{R}^n} \phi(x,\lambda)\, d\nu_x(\lambda)\, dx.$$

In fact since any continuous $\phi(x,\lambda)$ can be approximated as a sum (29) we see by continuity that (30) holds for all $\phi \in C_0(\Omega \times \mathbb{R}^n)$ and

$$m = \int_\Omega \nu_x(\lambda) dx.$$

Equation (30) is our fundamental representation formula for the measure m in terms of the Young measure ν_x.

We can now use the fundamental representation formula to categorize weak limits. Let f be a continuous $K \subset \mathbb{R}^n \to \mathbb{R}$ and $u_s : \Omega \to K$ be a measureable sequence of functions. Then $\|u_s\|_\infty <$ const. (independent of s) and we know there is a subsequence (also denoted by u_s) so that

$$u_s \to \bar{u} \text{ weak } * L^\infty.$$

Since $\| f(u_s(\cdot))\|_\infty <$ const. (independent of s) as well we can assume, by taking another subsequence of u_s if necessary, that

$$f(u_s(\cdot)) \to \overline{f} \text{ weak } * L^\infty.$$

Let $\psi \in C_0(\Omega)$ and consider

$$\langle m_s, \psi f \rangle = \int_\Omega \psi(x) f(u_s(x)) dx \quad \text{(by the definition of the Radon measure)}$$

$$\to \langle m, \psi f \rangle \quad \text{(by the definition of } m\text{)}$$

$$= \int_\Omega \int_{\mathbb{R}^n} \psi(x) f(\lambda) \, d\nu_x(\lambda) \, dx \quad \text{(by (30))}.$$

But since $f(u_s(\cdot)) \to \overline{f}$ weak $* L^\infty$ we have

$$\langle m_s, \psi f \rangle = \int_\Omega \psi(x) f(u_s(x)) dx \to \int_\Omega \psi(x) \overline{f}(x) \, dx.$$

Equating the two limits of $\langle m_s, \psi f \rangle$ we have

$$\int_\Omega \psi(x) \overline{f}(x) dx = \int_\Omega \int_{\mathbb{R}^n} \psi(x) f(\lambda) \, d\nu_x(\lambda) \, dx$$

and since ψ is any element of $C_0(\Omega)$ we see

(31) $$\overline{f}(x) = \int_{\mathbb{R}^n} f(\lambda) d\nu_x(\lambda) \quad \text{a.e. in } \Omega$$

$$= \langle f(\lambda), \nu_x \rangle$$

Equation (31) is our fundamental representation of weak limits in terms of the representing Young measure $\nu_x(\lambda)$. Notice that substitution of $f \equiv 1$ yields

$$1 = \int_{\mathbb{R}^n} d\nu_x(\lambda).$$

Since $\nu_x(\lambda)$ is a positive measure we see that in fact $\nu_x(\lambda)$ is a probability measure. In addition we see that if

$$\nu_x(\lambda) = \delta(\lambda - \overline{u}(x)),$$

δ the Dirac delta function centered at zero, then

$$\bar{f}(x) = f(\bar{u}(x))$$

and we have

$$\text{weak} * \lim f(u_s(\cdot)) = f(\lim \text{weak} * u_s(\cdot)).$$

Since this was the goal we posed at the end of Section 8 we see the ability to show the Young measure determined by a sequence u_s is the Dirac mass $\delta(\lambda - \bar{u}(x))$ will provide the key to the analysis of weak limits of regularized conservation laws.

10. L. Tartar's argument for Burgers' equation.

Consider Burgers' equation

(32)$_\mu$
$$\theta_t + \theta\theta_x = \mu\theta_{xx}.$$

We will see in this section how the ideas of the previous section were applied by L. Tartar in [2] to pass to the limit as $\mu \to 0+$ to obtain the correct weak solution of

(32)
$$\theta_t + \theta\theta_x = 0, \quad t>0, \quad -\infty < x < \infty,$$

$$\theta(x) = \theta_0(x).$$

1°. The first step in the argument is to consider $n(\theta)$ a C^2 convex function. Multiply (32) by $n'(\theta)$. We then see

(33)
$$n(\theta^\mu)_t + q(\theta^\mu)_x = \mu[n(\theta^\mu)_{xx} - n''(\theta^\mu)(\theta^\mu_x)^2]$$

where

(34)
$$q'(\theta) = \theta n'(\theta).$$

2°. Assume for any n convex and C^2 the right hand side of (33) is in a compact subset of H^{-1}_{loc}. This can be expressed in the form

$$\left.\begin{array}{l}\text{div}_{t,x}(n_1(\theta^\mu), q(\theta^\mu))\\ \text{curl}_{t,x}(q_2(\theta^\mu), -n_2(\theta^\mu))\end{array}\right\} \quad \text{lie in a compact subset of } H^{-1}_{loc} \quad (35)$$

3°. The sequence $\{\theta^\mu\}$ satisfies via the maximum principle $\|\theta^\mu\|_\infty <$ const. (independent of μ). So by considering appropriate subsequences we have

$$\theta^\mu \to \bar{\theta} \quad \text{weak } * L^\infty,$$

$$n_1(\theta^\mu) \to \bar{n}_1 \quad \text{weak } * L^\infty,$$

$$n_2(\theta^\mu) \to \bar{n}_2 \quad \text{weak } * L^\infty,$$

$$q_1(\theta^\mu) \to \bar{q}_1 \quad \text{weak } * L^\infty,$$

$$q_1(\theta^\mu) \to \bar{q}_2 \quad \text{weak } * L^\infty,$$

where by the fundamental representation formula for weak limits (31) we know

$$\begin{aligned}\bar{n}_1 &= \langle n_1(\lambda), \nu_{x,t}\rangle,\\ \bar{n}_2 &= \langle n_2(\lambda), \nu_{x,t}\rangle,\\ \bar{q}_1 &= \langle q_1(\lambda), \nu_{x,t}\rangle,\\ \bar{q}_2 &= \langle q_1(\lambda), \nu_{x,t}\rangle,\end{aligned} \quad (36)$$

4°. Apply the Div-Curl Lemma of Murat and Tartar [2] i.e. if (35) holds then

$$(n_1(\theta^\mu), q_1(\theta^\mu)) \cdot (q_2(\theta^\mu), -n_2(\theta^\mu))$$

is weakly continuous. That is we have

$$(n_1(\theta^\mu), q(\theta^\mu)) \cdot (q_2(\theta^\mu), -n_2(\theta^\mu)) \to$$

$$\bar{n}_1 \bar{q}_2 - \bar{q}_1 \bar{n}_2 \quad \text{as } \mu \to 0+ \quad (37)$$

in the sense of distributions.

Also apply the fundamental representation formula (31) to the left hand side of (37) so that

(38)
$$n_1(\theta^\mu) q_2(\theta^\mu) - q_1(\theta^\mu) n_2(\theta^\mu) \to$$
$$\langle n_1(\lambda) q_2(\lambda) - q_1(\lambda) n_2(\lambda), \nu_{x,t} \rangle \quad \text{weak} * L^\infty.$$

Equating the weak limits in (37) and (38) we find

(39)
$$\langle n_1(\lambda) q_2(\lambda) - q_1(\lambda) n_2(\lambda), \nu_{x,t} \rangle =$$
$$\langle n_1(\lambda), \nu_{x,t} \rangle \langle q_2(\lambda), \nu_{x,t} \rangle - \langle q_1(\lambda), \nu_{x,t} \rangle \langle y_2(\lambda), \nu_{x,t} \rangle.$$

Equation (39) holds for all n_1, n_2 C^2 convex. As noted by Lax in [1] since (39) is continuous in n_1, n_2 it also holds for all n_1, n_2 which are limits of n_1, n_2 convex in particular for

$$n_1(\lambda) = \lambda \quad , \quad n_2(\lambda) = |\lambda - \bar{\theta}|$$

$$q_1(\lambda) = \frac{\lambda^2}{2} \quad , \quad q_2(\lambda) = \frac{\bar{\theta}^2}{2} - \frac{\lambda^2}{2} , \quad \lambda < \bar{\theta} ,$$

$$\frac{\lambda^2}{2} - \frac{\bar{\theta}^2}{2} , \quad \bar{\theta} < \lambda.$$

In this case (39) can be rewritten as

$$(\bar{q}_1 - \frac{\bar{\theta}^2}{2}) \langle \nu_{x,t}, |\lambda - \bar{\theta}| \rangle = 0$$

where recall that

$$\bar{q}_1 = \text{weak} * L^\infty \text{ limit of } \frac{1}{2}(\theta^\mu)^2 .$$

So either $\bar{q}_1 = \frac{\bar{\theta}^2}{2}$ or

$\langle \nu_{x,t}, |\lambda - \bar{\theta}| \rangle = 0$, i.e.

$$\int_R |\lambda - \bar{\theta}(x,t)| \, d\nu_{x,t}(\lambda) = 0.$$

Since $\nu_{x,t}$ is a positive measure this can only be true if

$$\nu_{x,t}(\lambda) = \delta(\lambda - \bar{\theta}(x,t))$$

so that

$$\text{weak} * L^\infty \text{ limit of } (\theta^\mu)^2 = \overline{\theta^2}.$$

So either way the goal posed at the end of Section 8 is achieved.

5°. The only thing left to check is our assumption in 2° that the right hand side of (33) lies in a compact subset of H_{loc}^{-1} for n convex C^2. We do this as follows. Integration of (33) shows

$$\int_{-\infty}^{\infty} n(\theta^\mu(x,T))dx + \mu \int_0^T \int_{-\infty}^{\infty} n''(\theta^\mu(x,t)) \, \theta_x^\mu(x,t)^2 dx dt$$

$$= \int_{-\infty}^{\infty} n(\theta_0^\mu(x)) \, dx.$$

Since $n'' > 0$ we have

$$\{\mu n''(\theta^\mu(x,t)) \, \theta_x^\mu(x,t)\}^2$$

lies in a bounded (indep. of μ) subset of $L^1(\mathbb{R} \times [0,T])$. In fact we have

(40) $$\{\mu \, \theta_x^\mu(x,t)\}^2 \subset M,$$

M a bounded (indep. of μ) subset of $L^1(\mathbb{R} \times (0,T]$.

Next note

$$\int_0^T \int_{-\infty}^{\infty} (\mu n(\theta^\mu))_x^2 \, dx \, dt =$$

$$\int_0^T \int_{-\infty}^{\infty} \mu^2 (n'(\theta_x^\mu))^2 \, dx \, dt$$

$$\leq \mu \, (\text{const. indep. of } \mu) \int_0^T \int_{-\infty}^{\infty} \mu(\theta_x^\mu)^2 dx dt.$$

So by (40)

$$\mu n(\theta^\mu)_x \to 0 \text{ in } L^2(\mathbb{R} \times [0,T]) \text{ as } \mu \to 0+$$

so that the sequence $\{\mu n(\theta^\mu)_x\}$ is a compact subset of $L^2(\mathbb{R} \times (0,T]$ and hence

$\{\mu n(\theta^\mu)_{xx}\}$ is a compact subset of $H^{-1}(\mathbb{R} \times [0,T])$.

Thus the right hand side of (3.3) has the form of something in compact subset of $H^{-1}(\mathbb{R} \times [0,T])$ plus something in a bounded subset $L^1(R \times [0,T])$. Application of the following lemma of Murat delivers the desired result.

Lemma [2]. Let $\{\theta^\mu\}$ be such that

(i) $\{\theta^\mu\}$ belongs to a bounded set in $W^{-1,\infty}$,

(ii) each θ^μ can b decompsed into

$$\theta^\mu = A^\mu + B^\mu$$

where $\{A^\mu\}$ belongs to a bounded (indep. of μ) subset of L^1 and $\{B^\mu\}$ belongs to set in H^{-1}. Then $\{\theta^\mu\}$ lies in a compact subset of H^{-1}_{loc}.

In our example we have already shown (ii). To see (i) we simply use the left hand side (33) and the fact that $\{\theta^\mu\}$ is in a bounded set of $L^\infty(R \times [0,T])$ and the smoothness of η and q.

Acknowledgement

I would thank Victor Roytburd for his valuable assistance in the preparation of this paper.

References

1. P.D. Lax, Shock waves, increase of entropy and loss of information, in Seminar on Nonlinear Partial Differential Equations, Mathematical Sciences Research Institute Publications, ed. S.S. Chern, Springer-Verlag (1984), 129-173.

2. L. Tartar, Compensated compactness and applications to partial differential equations, in Research Notes in Mathematics 39, Nonlinear analysis and mechanics: Heriot-Watt Symposium, Vol. 4, ed. R.J. Knops, Pittman Press (1975), 136-211.

3. R. DiPerna, Convergence of approximate solutions to conservation laws, Archive for Rational Mechanics and Analysis, Vol. 82 (1983), 27-70.

4. R. DiPerna, Convergence of the viscosity method for isotropic gas dynamics, Comm. on Math. Physics Vol. 91 (1983), 1-30.

5. M. Schonbek, Convergence of solutions to nonlinear dispersive equations, Comm. in Partial Differential Equations Vol. 7 (1982), 959-1000.

6. Lord Rayleigh, Aerial plane waves of finite amplitude, Proc. Royal Soc. London, Series A Vol. 84 (1910), 247-284.

7. I.M. Gel'fand, Some problems in the theory of quasilinear equations, Uspehi. Mat. Nauk. Vol. 14 (1959), 87-158, English translation in Amer. Math. Soc. Translations, Series 2, Vol. 29 (1963), 295-381.

8. D.J. Korteweg, Sur la forme que prennent les equations de mouvement des fluides si l'on tient compte des forces capillaires par des variations de densite, Archives Neerlandaises des Sciences exactes er Naturelles, Series II Vol. 6, (1901), 1-24.

9. J.S. Rowlinson, Translation of J.D. van der Waals' "The thermodynamic theory of capillarity under the hypothesis of a continuous variation of density", J. Statistical Physics Vol. 20 (1979), 197-244.

10. C.A. Truesdell and W. Noll, The nonlinear field theories of mechanics, Vol. IV/3 of the Encyclopedia of Physics, S. Flugge, editor, Springer-Verlag (1965).

11. R. Hagan and J. Serrin, One-dimensional shock layers in Korteweg fluids, in Phase Transformations and Material Instabilities in Solids, ed. M. Gurtin, Academic Press (1984), 113-128.

12. M. Slemrod, Lax-Friedrichs and the viscosity-capillarity criteria, to appear Proc. of the Univ. of W. Va. Converence on Physical Partial Differential Equations, July, 1983, eds. J. Lightbourne and S. Rankin, Marcel-Dekker (1985).

13. M. Slemrod and J.E. Flaherty, Numerical integration of a Riemann problem for a van der Waals fluid, to appear Res. Mechanica.

14. E. Hopf, The partial differential equation $u_t + uu_x = \mu u_{xx}$, Comm. Pure and Appl. Math. Vol. 3 (1950), 201-230.

15. P.D. Lax, Hyperbolic systems of conservation Laws, II, Comm. Pure and Applied Math. Vol. 10 (1957), 537-566.

16. R. Miura & M. Kruskal, Application of a nonlinear WKB method to the Korteweg - de Vries equation, SIAM J. Appl. Math. Vol. 26 (1974), 376-395.

17. P.D. Lax and C.D. Levermore, The small dispersion limit for the KdV equation I, Comm. Pure Appl. Math Vol. 36 (1983), 253-290; II, C.P.A.M. Vol. 36 (1983), 571-594; III, C.P.A.M. Vol. 36 (1983), 809-829.

18. S. Venakides, The zero dispersion limit of the Korteweg - de Vries equation for initial potentials with non-trivial reflection coefficient, Comm. Pure and Appl. Math. Vol. 38 (1985), 125-156.

19. G. Choquet, Lectures on Analysis, Vol. 1, W.A. Benjamin, Inc. (1969).

THE SOLUTION OF COMPLETELY INTEGRABLE SYSTEMS
IN THE CONTINUUM LIMIT OF THE SPECTRAL DATA

Stephanos Venakides

Department of Mathematics
Stanford University
Stanford, California 94305

1. Introduction:

In this talk I wish to outline a procedure for studying the solution of completely integrable evolution equations in a distinguished limit. The procedure is applicable to equations which can be solved by the method of the inverse spectral transformation and have discrete spectral data. The distinguished limit corresponds to the above data tending to a continuum.

Such equations include the Korteweg de Vries, the cubic Schrödinger, and the sine-Gordon equations with periodic initial data. In these equations complete integrability manifests itself in the following form:

(i) The solution $u(x,t)$ where x is the space and t is the time variable is given by:

$$(1) \qquad u(x,t) = F(\nu_1, \nu_2, \ldots)$$

where $\nu_i(x,t)$ are relevant spectral data of an associated family of operators.

(ii) The functions $\nu_i(x,t)$ are determined from the solution of a system of algebraic equations

$$(2) \qquad G_i(\nu_1, \nu_2, \ldots; x, t) = 0 \qquad i = 1, 2 \ldots$$

Equations (2) should be interpreted as a complete set of first integrals of the completely integrable structure of the original equation. They contain integration constants which are computed from the initial data.

The central idea in our procedure is that as the ν_i's approach a continuum limit, equations (2) should tend to an integral equation for some corresponding density function. We call this equation the basic integral equation. In the same limit, equation (1) will yield some projection of $u(x,t)$ (e.g. a weak limit) as a functional of the solution to the basic integral equation.

We demonstrate the method by applying it to the Korteweg de Vries (kdV) equation with small dispersion:

(3)
$$u_t - 6uu_x + \varepsilon^2 u_{xxx} = 0$$
$$u(x,0) = -v(x)$$

where $v(x)$ is p-periodic and $\varepsilon > 0$ tends to zero. In this case the projection of the solution $u(x,t,\varepsilon)$ obtained from equation (2) is the weak limit

(4)
$$\bar{u}(x,t) = \frac{\partial}{\partial x} \lim_{\varepsilon \to 0} \int^x u(s,t,\varepsilon)ds.$$

The process described so far corresponds to a leading order, or better to a slow scale, analysis. The next step is to capture the behavior in the fast scale. Indeed, we use the solution of the basic integral equation to recover information on the solution to the discrete equations (1) and (2) in the case of ε small but nonzero. This in turn allows us to discern the formation of "shock regions" in the x and t plane in which the solution becomes oscillatory and to describe the basic features of these oscillations.

Our procedure is an extension of the method used by Lax and Levermore [6] in studying the zero dispersion limit of the kdV equation.

2. The Korteweg de Vries Equation

In the case of the kdV equation, the associated operator, first derived by Miura, Gardner, and Kruskal [9] is:

(5)
$$\mathcal{L}_\varepsilon(t) = \varepsilon^2 \frac{d^2}{ds^2} - u(s,t,\varepsilon).$$

$\mathcal{L}_\varepsilon(t)$, parametrized by t, is a self adjoint operator acting on square integrable functions $f(s)$ in the real line. The relevance of this operator comes from the unitary equivalence:

(6) $$\mathcal{L}_\varepsilon(t) \sim \mathcal{L}_\varepsilon(0)$$

which is a necessary and sufficient condition for $u(x,t,\varepsilon)$ to satisfy equation (3). This was proven by Peter Lax [5]. The essence of the method of the spectral transformation is to represent the solution $u(x,t,\varepsilon)$ of (3) in terms of some spectral data of the operator (6) and to specify the way in which these data flow in time. In the following we give the relevant spectral data and we state without proof some well known facts about them. More details are given in [12]. We assume that our initial data $v(x)$ is a p-periodic function. The spectral transformation has been obtained in this case by Dubrovinn, Mateev and Novikov [2] and by McKean, van Moerbeke and Trubowitz [7,8]:

i. the periodic/ antiperiodic eigenvalues

(7) $$\xi_0 > \xi_1 > \xi_2 > \xi_3 > \ldots$$

are defined to be the <u>simple</u> eigenvalues of the problem:

(8) $$\mathcal{L}_\varepsilon(t)\, \psi_\varepsilon(s,t) = \xi \psi_\varepsilon(s,t)$$

with boundary conditions

(9) $$\psi_\varepsilon(x,t) = \alpha \psi_\varepsilon(x+p, t)$$

$$\frac{\partial \psi_\varepsilon}{\partial s}(x,t) = \alpha \frac{\partial}{\partial s} \psi_\varepsilon(x+p,t)$$

where $\alpha = \pm 1$.

ii. The Dirichlet eigenvalues:

(10) $$\nu_1 > \nu_2 > \nu_3 > \nu_4 > \ldots$$

are defined by:

(11a) $$\mathcal{L}_\varepsilon(t)\, \psi_\varepsilon(s,t) = \nu \psi_\varepsilon(s,t)$$

(11b) with $\psi_\varepsilon(x,t) = \psi_\varepsilon(x+p,t) = 0$.

The ξ_i's are independent of x and t while the dependence of v_i on (x,t) is expressed as an autonomous system of ordinary differential equations which give $\frac{\partial v_i}{\partial x}$ and $\frac{\partial v_i}{\partial t}$ in terms of $v_1, v_2,\ldots$. This system can be integrated explicitly to give a system of algebraic equations from which the v_i's are determined.

The spectrum $\sigma(\mathcal{L}_\epsilon)$ of $\mathcal{L}_\epsilon$ as an operator in $L^2(-\infty, \infty)$ consists of the union of intervals referred to as "spectral bands":

(12) $$\sigma(\mathcal{L}_\epsilon) = [\xi_1, \xi_0] \cup [\xi_3, \xi_2] \ldots$$

In the generic case this is an infinite union of disjoint intervals.

In the special case of finite gap potentials there is only a finite odd number of simple periodic/antiperiodic eigenvalues. The spectrum is then the union of finitely many intervals and a half line.

We now examine the asymptotic behavior of the spectrum of the operator at time zero:

(13) $$\mathcal{L}_\epsilon(0) = \epsilon^2 \frac{d^2}{ds^2} + v(s)$$

in the limit $\epsilon \to 0$.

It can be shown that as $\epsilon \to 0$:

i) There is no spectrum in the region of the spectral variable $\xi > v_{max} + \rho$ where ρ is arbitrarily small.

ii) The contributions to the solution $u(x,t,\epsilon)$ from the spectral data in the region $\xi < v_{min}$ can be neglected [12].

In the region $v_{min} < \xi < v_{max}$ the spectrum is derived by the W.K.B. method.

Let n_i^2 be the midpoint of the ith spectral band and let $2\delta_i$ be its width i.e.

(14) $$n_i^2 = \frac{1}{2}(\xi_{2i-1} + \xi_{2i-2}), \quad 2\delta_i = \xi_{2i-2} - \xi_{2i-1}.$$

For simplicity we take $v(x)$ to reach a local maximum (and a local minimum) at exactly one point x in each period.

The W.K.B. calculation [12] shows that if $\varepsilon \to 0$ and $i \to \infty$ in such a way that $n_i \to n$ with $v_{min} < n^2 < v_{max}$ then:

(15) $$n_i - n_{i+1} \sim \frac{\pi\varepsilon}{\phi(n)}, \quad \log\delta_i \sim -\gamma(n)/\varepsilon$$

where

(16) $$\phi(n) = \text{Re} \int_x^{x+p} \frac{n}{(v(s)-n^2)^{1/2}} ds,$$

(17) $$\gamma(n) = \text{Re} \int_x^{x+p} (n^2 - v(s))^{1/2} ds.$$

Here, Re denotes the real part and the square root is chosen so that ϕ and γ are positive. Thus in the region $v_{min} < \xi < v_{max}$ the spectral bands are exponentially small in width and have a separation of order ε as $\varepsilon \to 0$. In the following we assume the initial data normalized so that:

(18) $$v_{max} = 1, \quad v_{min} = 0.$$

We now examine the asymptotic behavior of the Dirichlet eigenvalues $v_i(x,t,\varepsilon)$ in the limit $\varepsilon \to 0$. We have:

(19) $$\xi_{2i}(\varepsilon) < v_i(x,t,\varepsilon) < \xi_{2i-1}(\varepsilon)$$

As x and t vary v_i oscillates in the interval $[\xi_{2i}, \xi_{2i-1}]$ which we call the ith gap. The equations which govern this motion are given by (23) in integrated form.

We are now ready to write the specific forms of equations (1) and (2) for the periodic kdV equation. As $\varepsilon \to 0$, equation (1) takes the forms [12]:

(20a) $$u(x,t,\varepsilon) = -\xi_0 + 2\sum_{i=1}^{N} \{v_i(x,t,\varepsilon) - \frac{\xi_{2i-1} + \xi_{2i}}{2}\} + o(1)$$

(20b) $$u(x,t,\varepsilon) = \sum_{i=1}^{N} \varepsilon \frac{d}{dx} \int^{v_i(x,t,\varepsilon)} \frac{\varepsilon P(v)}{R(v)^{1/2}} dv + o(1)$$

Here $N = N(\varepsilon)$ is such that $\xi_{2N} \to 0$ as $\varepsilon \to 0$. The $o(1)$ terms contain the contributions from the spectral region $\xi < 0 = \nu_{min}$ which as said earlier we can neglect.

The functions $R(\nu)$, $P(\nu)$ are given by:

$$(21) \qquad R(\nu) = \prod_{i=0}^{2N} |\nu - \xi_i|$$

$$(22) \qquad P(\nu) = \prod_{i=1}^{N} (\nu - n_j^2).$$

In the same limit equation (2) takes the form [12]:

$$(23) \qquad \sum_{i=1}^{N} \int_{\nu_j(x_0,t_0,\varepsilon)}^{\nu_j(x,t,\varepsilon)} \frac{\varepsilon\, P(\nu)\, d\nu}{(\nu-n_j^2)R(\nu)^{1/2}} = x-x_0 - 4\eta^2(t-t_0).$$

Finally in equations (20b) and (23) the sign of the square root $R(\nu_j)^{1/2}$ is defined by:

$$(24) \qquad \operatorname{sgn} \frac{P(\nu_j)}{R(\nu_j)^{1/2}} = -\operatorname{sgn} \frac{\partial \nu_j}{\partial x} = \operatorname{sgn} \frac{\partial \nu_j}{\partial t}$$

As indicated by (19) $\nu_j(x,t,\varepsilon)$ is constrained to take values in the jth gap. Equations (24) indicate that the sign of $R(\nu_j)^{1/2}$ changes when ν_j reaches the endpoint of the jth gap and turns back, as x and t vary. $R(\nu_j)^{1/2}$ can be made single valued, if it is defined to take positive and negative values respectively on two different copies of the interval $[\xi_{2j}, \xi_{2j-1}]$. The corresponding endpoints of the two copies are then identified and topologically ν_j varies on a circle. This in turn means that the value of the integral in equation (23) corresponding to ν_i depends on how many times ν_i winds around the ith circle as x and t vary. The winding number is clearly independent of the path on the x-t plane connecting points (x_0, t_0) and (x,t).

3. Asymptotic Calculations

Our next step is to calculate the integrals in equation (20) and (23). We omit the calculations in this outline and we present the result:

Theorem 1: Let

(25) $$\sigma_j = \sigma_j(x,t,\epsilon) = \text{sgn} \frac{\partial v_j(x,t,\epsilon)}{\partial x} = \pm 1,$$

(26) $r_j(x,t,\epsilon) =$

$$\frac{1}{2} \sigma\epsilon\log\{|v-n_j| + [(v-n_j)^2 - \delta_j^2]^{1/2}\}^{-1} \Big|_{\substack{v=v_j(x,t,\epsilon)\\\sigma=\sigma_j}} + \Big|_{\substack{v=v_{j-1}(x,t,\epsilon)\\\sigma=\sigma_{j-1}}}$$

where δ_j is defined in (14),

(27) $w_j(x,t,\epsilon) = m_{2j-1}(x,t,\epsilon) - m_{2j-2}(x,t,\epsilon)$ where $m_0 = 0$

and $m_k(x,t,\epsilon)$ is the number of times one of the v_i's (necessarily a unique one) takes the value ξ_k as the space-time variable changes from (x_0,t_0) to (x,t). The integer w_j roughly corresponds to the difference of the winding numbers of the adjacent j^{th} and $j-1^{st}$ gaps.

Then the equations (20b) and (23) become as $\epsilon \to 0$:

(28) $$\int^x u(x',t,\epsilon)dx' = -2 \sum_{i=1}^{N} \epsilon\sigma_i n_i + \text{const.} + o(1)$$

(29) $$\{r_j + \sum_{i \neq j} \epsilon [1/2(1-\sigma_i) - w_i]\log |\frac{n_i - n_j}{n_i + n_j}|\}\Big|^{(x,t)}_{(x_0,t_0)} + w_j(x,t,\epsilon)\gamma(n_j) =$$

$$= n_j(x-x_0) - 4n_j^3 (t-t_0) + o(1)$$

Remark: The r_j appears as the singular contribution to the sum in (23) from the terms having $i = j$ and $i = j - 1$. We prove the following important estimate:

Theorem 2.

For small nonzero ϵ the relations

(30) $$|v_j(x,t) - n_j^2| \leq e^{-|r_j(x,t,\epsilon)|/\epsilon}, \quad \sigma_j = \text{sgn } r_j(x,t,\epsilon)$$

hold simultaneously either when $i = j$ or when $i = j - 1$.

Proof: Of the two indices j and $j-1$ we let i denote the one which makes the expression

$$-\log\{|v_i - n_j^2| + [(v_i - n_j^2)^2 - \delta_j^2]^{1/2}\}$$

larger. By (26) we have:

$\operatorname{sgn} r_j = \sigma_i$ and

$$\frac{|r_j|}{\varepsilon} \leq -\log\{|v_i - n_j^2| + [(v_i - n_j^2)^2 - \delta_j^2]\} .$$

There follows

$$|v_i - n_j^2| + [(v_i - n_j^2)^2 - \delta_j^2]^{1/2} \leq e^{-r_j/\varepsilon}$$

and $|v_i - n_j^2| \leq e^{-r_j/\varepsilon}$.

q.e.d.

4. The Continuum Limit and the Derivation of the Basic Integral Equation

In equations (28-29) we let $\varepsilon \to 0$ and $j \to \infty$ in such a way that $n_j \to n$ $(0,1)$. We define the density function;

(31.a) $\quad \psi(n,x,t) = -w(n,x,t)\phi(n) + \lim_{h \to 0} \lim_{\varepsilon \to 0} \{\frac{\pi \varepsilon}{h}$ [number of i's such that

$$\sigma_i = -1 \text{ and } n^2 \leq v_i(x,t,\varepsilon) \leq (n+h)^2\}$$

(31.b) $\quad$ where $w(n,x,t) = \lim_{\varepsilon \to 0} w_i(x,t,\varepsilon)$ when $n_j \to n$ as $\varepsilon \to 0$.

It is shown in [12] that the limits exist and $\psi(n,x,t)$ is measurable and bounded in the variable n. Equation (31) and the definition of the density function $\phi(n)$ imply:

(32) $\quad -w(n,x,t)\phi(n) \leq \psi(n,x,t) \leq -w(n,x,t)\phi(n) + \phi(n) .$

The right hand side of (32) corresponds to all σ_i's in (31a) being equal to -1 and the left hand side corresponds to all σ_i's being equal to +1. We write equations (28) and (29) in terms of ψ:

(33) $$\int^x u(x',t,\varepsilon)dx' = \frac{4}{\pi} \int_0^1 \eta[\psi(\eta,x,t)\, w(\eta,x,t)\phi(\eta)]d\eta + o(1)$$

(34) $$\frac{1}{\pi} \int_0^1 \log \left|\frac{\eta_j - \mu}{\eta_j + \mu}\right| \psi(\mu,x,t)\, d\mu + r_j(x,t,\varepsilon) + w_j(x,t,\varepsilon)\gamma(\eta_j) =$$

$$= a_j(x,t) + o(1) \qquad j = 1,2,\ldots$$

where

(35) $$a_j(x,t) = \eta_j(x-x_0) - 4\eta_j^3(t-t_0) + r_j(x_0,t_0,\varepsilon) +$$

$$+ \frac{1}{\pi} \int_0^1 \log \left|\frac{\eta_j - \mu}{\eta_j - \mu}\right| \psi(\mu,x_0,t_0)\, d\mu \; .$$

Following the boundedness of the function ψ we have that $\int_0^1 \log \left|\frac{\eta-\mu}{\eta+\mu}\right| \psi(\mu,x,t)\, dt$ is continuous in η. Equations (34-35) imply that the limit

(36) $$\lim_{\substack{\varepsilon \to 0 \\ \eta_j \to \eta}} \{r_j(x,t,\varepsilon) - r_j(x_0,t_0,\varepsilon) + w_j(x,t,\varepsilon)\gamma(\eta)\} = b(\eta,x,t)$$

is uniform in η and $b(\eta,x,t)$ is continuous. A W.K.B. study of the initial data [12] shows that as $\varepsilon \to 0$ and as $\eta_{j(\varepsilon)} \to \eta$ we have:

(37a) $$r_j(x,0,\varepsilon) \to r(\eta,x,0)$$

(37b) $$\text{and} \quad |r_j(x,t,\varepsilon)| \leq \frac{1}{2}\gamma(\eta_j) + o(1) \; .$$

$r(\eta,x,0)$ is derived by an asymptotic calculation of the initial data. (37b) is obtained by showing that if it is violated, the distance $|v_j - \eta_j^2|$ or $|v_{j-1} - \eta_j^2|$ must be exponentially small for all x,t. This is in contradiction to the initial data.

Remark. If the quantity $w_j\gamma(\eta_j) + r_j$ is known, the numbers w_j and γ are uniquely determined by the fact that w_j is an integer and by (37b). Relations (36)-(37) lead to:

(38) $$\begin{aligned} r_j(x,t,\varepsilon) &\to r(\eta,x,t) \\ w_j(x,t,\varepsilon) &\to w(\eta,x,t) \end{aligned} \qquad \text{as } \varepsilon \to 0,\; \eta_j \to \eta \; .$$

where the functions $r(\eta,x,t)$ and $w(\eta,x,t)$ are piecewise continuous in η and the limits are uniform in η when η is bounded away from the points of discontinuity.

Finally we define the integral operator:

$$(39) \qquad L\psi(\eta) = \frac{1}{\pi} \int_0^1 \log \left|\frac{\eta-\mu}{\eta+\mu}\right| \psi(\mu) \, d\mu .$$

Equations (34)-(35) became in the limit $\varepsilon \to 0$, $\eta_j \to \eta$:

$$(40) \qquad L\psi(\eta,x,t) + r(\eta,x,t) + w(\eta,x,t)\gamma(\eta) = a(\eta,x,t)$$

where

$$(41) \qquad a(\eta,x,t) = \eta(x-x_0) - 4\eta^3(t-t_0) + r(\eta,x_0,t_0) + L\psi(\eta,x_0,t_0) .$$

We now derive the basic integral equation:

Theorem 3

Let w denote an integer. Then

$$(w-\tfrac{1}{2})\gamma(\eta) < a(\eta,x,t) - L\psi(\eta,x,t) < w\gamma(\eta) \quad \text{implies} \quad \psi(\eta,x,t) = (1-w)\phi(\eta)$$

$$(42) \qquad a(\eta,x,t) - L\psi(\eta,x,t) = w\gamma(\eta) \quad \text{implies} \quad -w\phi(\eta) < \psi(\eta,x,t) < (1-w)\phi(\eta)$$

$$w\gamma(\eta) < a(\eta,x,t) - L\psi(\eta,x,t) < (w+\tfrac{1}{2})\gamma(\eta) \quad \text{implies} \quad \psi(\eta,x,t) = -w\phi(\eta) .$$

Proof We rewrite (40) as:

$$(43) \qquad a(\eta,x,t) - L\psi(\eta,x,t) = w(\eta,x,t)\gamma(\eta) + r(\eta,x,t).$$

By the remark following equation (37) proving (42) is equivalent to proving:

$$(44) \qquad \begin{aligned} r(\eta,x,t) < 0 \quad &\text{implies} \quad \psi(\eta,x,t) = (1-w)\phi(\eta) \\ r(\eta,x,t) = 0 \quad &\text{implies} \quad -w\phi(\eta) < \psi(\eta,x,t) < (1-w)\phi(\eta) \\ r(\eta,x,t) > 0 \quad &\text{implies} \quad \psi(\eta,x,t) = -w\phi(\eta). \end{aligned}$$

where $w = w(\eta,x,t)$.

If n is a point of continuity of $r(n,x,t)$ at which $r(n,x,t) \neq 0$, there is $\rho > 0$ such that $|r_j(x,t,\epsilon)| > \rho > 0$ when $|n_j - n|$ and ϵ are small independent of each other. This falls from the uniformity of limits (38). By theorem 2, for $i = j$ or for $i = j - 1$ we have:

$$(45) \qquad \sigma_j = \text{sgn } r_j = \text{sgn } r \quad \text{and} \quad |v_i - n_j^2| < e^{-\rho/\epsilon}$$

Each n_j near n has exponentially close to it v_j or v_{j-1} with corresponding $\sigma = \text{sgn } r$. By the definition of ψ and of the density function ϕ:

$$\text{sgn } r = -1 \quad \text{implies} \quad \psi(n,x,t) = -w\phi(n) + \phi(n)$$
$$\text{sgn } r = +1 \quad \text{implies} \quad \psi(n,x,t) = -w\phi(n) \, . \qquad\qquad \text{q.e.d.}$$

We are in a position to state a main result:

Theorem 4

The weak limit of the solution $u(x,t,\epsilon)$ of the initial value problem (3) with the smooth periodic initial data normalized by (18) is given by:

$$(46) \qquad \bar{u}(x,t) = \frac{\partial}{\partial x} \lim_{\epsilon \to 0} \int^x u(x',t,\epsilon)dx' = \frac{4}{\pi} \frac{\partial}{\partial x} \int_0^1 n[\psi(n,x,t) + w(n,x,t)\phi(n)]$$

where $\psi(n,x,t)$ is the solution of the basic integral equation (41). The theorem is a consequence of relations (32) and (41).

Theorem 4 gives precise meaning to the statements in the introduction that the projection of the solution $u(x,t,\epsilon)$:

$$Pu = \bar{u}(x,t) = \frac{\partial}{\partial x} \lim_{\epsilon \to 0} \int^x u(x',t,\epsilon)dx'$$

is given by (46) as a functional of the solution ψ of the basic integral equation (41).

5. The Solution of the Basic Integral Equation

The solution of equation (41) is obtained by its reduction to the Riemann-Hilbert problem solved by Lax and Levermore in their treatment of (3) with single well initial data. We describe the results which are relevant for this outline: For each $t > 0$ the strip of the $x-n^2$ plane $0 \leq n^2 \leq 1$ is divided into two regions, one in which $\psi(n,x,t)$ is an integral multiple of $\phi(n)$ (equivalently $r(n,x,t) \neq 0$) and one in which $\psi(n,x,t)$ is not an integral multiple of $\phi(n)$ (equivalently $r(n,x,t) = 0$). The relevance of these relations is obvious by inspection of (41). For each t, the boundary separating the two regions in the x-n^2 plane is a connected curve G_t, p-periodic in the x variable. The significance of this curve is discussed in section 7. This curve is specified at some t as the graph of a p-periodic, in general multivalued, function of x. We denote its values by:

(47) $\qquad \beta_0^2(x,t) > \beta_1^2(x,t) > \ldots > \beta_{2m}^2(x,t) \qquad$ (see fig. 1)

They lie in the interval [0,1] or in general $[v_{min}, v_{max}]$.

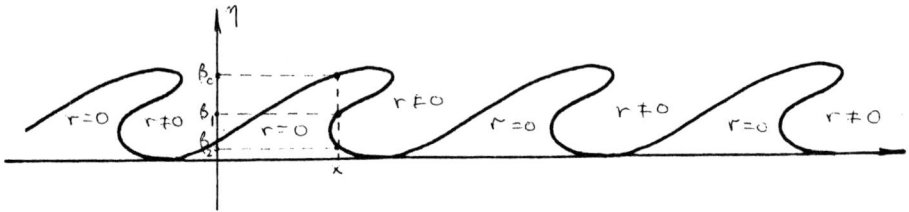

Figure 1

We have

(48)
$r(n,x,t) = 0$ when $n \in [0, \beta_{2m}] \cup [\beta_{2m-1}, \beta_{2m-2}] \cup \ldots \cup [\beta_1, \beta_0]$

$r(n,x,t) \neq 0$ when $n \in [\beta_{2m}, \beta_{2m-1}] \cup \ldots \cup [\beta_2, \beta_1] \cup [\beta_0, 1]$.

The evolution equations for the β_k's are exactly the ones derived by Lax and Levermore [6 eqn 5.27] in their treatment of the problem with single well initial data:

(49)
$$(\beta_k^{2m+1} + \alpha_2 \beta_k^{2m-1} + \ldots \alpha_{m+1} \beta_k) \frac{\partial \beta_k}{\partial t} +$$
$$+ [12\beta_k^{2m+3} - 6(\sum \beta^2)\beta_k^{2m+1} + \gamma_2 \beta_k^{2m-1} + \ldots + \gamma_{m+1}\beta_k] \frac{\partial \beta_k}{\partial x} = 0$$

$k = 0, 1, 2, \ldots 2m$.

The coefficients α_i and γ_j are known functions of the β_i's.

At t=0 we have the result:

(50) $$m(x,0) = 0 \quad \text{and} \quad \beta_0^2(x,0) = v(x)$$

where $v(x) = -u(x,0)$ as in (3).

In this case there is only β_0 and equations (49) become:

(51) $$\frac{\partial}{\partial t}(-\beta^2) - 6(-\beta^2)\frac{\partial}{\partial x}(-\beta^2) = 0 \;.$$

After equation (51) shocks, the curve G_t becomes the graph of a multivalued function of x and a region in the x-t plane emerges having m=1. Equations (49) are in this case identical with Whitman's modulation equations [13]. For higher m's equations (49) coincide with the modulation equations derived by Flaschka, Forest, and McLaughlin [3]. The connection between the β_k's and modulation theory will become apparent in the following section.

The evolution of the β_k's contains the singular phenomenon of shocking and the generation of more β's. Lax and Levermore circumvent this difficulty by parametrizing the curve G_t as $x = x(s,t)$ $n = n(s,t)$. The initial conditions (49) become:

(52) $$x(s,0) = s \qquad n = v(s)^{1/2}$$

and the evolution equations are:

(53) $$n = v(s)^{1/2}$$
$$\frac{\partial x(s,t)}{\partial t} = \frac{12n^{2m+3} - 6(\sum \beta_k^2)n^{2m+1} + \gamma_2 n^{2m-1} + \ldots + \gamma_{m+1} + n}{n^{2m+1} + \alpha_2 n^{2m-1} + \ldots + \alpha_{m+1} n}$$

Once the curve G_t is known, the function $\psi(n,x,t)$ is derived by the Lax Levermore procedure.

6. The Emergence of Oscillations

We study the solution in the vicinity of a fixed point in space and time (x_0, t_0). We make the change of variable:

(54) $$x = x_0 + \varepsilon y, \quad t = t_0 + \varepsilon \tau$$

and we work in the stretched variables y and τ. By abuse of notation we denote $u(x,t,\varepsilon)$ and $v_j(x,t,\varepsilon)$ by $u(y,\tau,\varepsilon)$ and $v_j(y,\tau,\varepsilon)$ respectively. The β_k's are assumed to be constant in the variables y and τ. To keep expressions simple we specialize our main theorem to the case $m(x_0,t_0) = 1$ i.e. where there are only $\beta_0, \beta_1, \beta_2$.

Theorem:

As $\varepsilon \to 0$ $u(y,\tau,\varepsilon)$ has the representation

(55) $$u(y,\tau,\varepsilon) \sim -\beta_0^2 + 2[\mathcal{N}(y,\tau) - \frac{\beta_1^2 + \beta_2^2}{2}]$$

where (y,τ) satisfies the relations:

(56) $$\beta_2^2 < \mathcal{N}(y,\tau) < \beta_1^2$$

(57) $$\frac{\partial}{\partial y} \sim 2[(\beta_0^2 - \mathcal{N})(\beta_1^2 - \mathcal{N})(\beta_2^2 - \mathcal{N})]^{1/2}$$

(58) $$\frac{\partial}{\partial \tau} \sim -2(\beta_0^2 + \beta_1^2 + \beta_3^2)[(\beta_0^2 - \mathcal{N})(\beta_1^2 - \mathcal{N})(\beta_2^2 - \mathcal{N})]^{1/2}$$

Remark 1

Equations (55-58) are the standard representation of the single phase periodic solution $u(y,\tau,\varepsilon)$, known as a cnoidal solution, of the KdV equation:

(59) $$u_\tau - 6uu_y + U_{yyy} = 0$$

In the case $m > 1$ we obtain similarly a representation of the m-phase quasi-periodic solution of (59). The corresponding periodic/multiperiodic eigenvalues are the β_k^2's and the corresponding Dirichlet eigenvalues are the $\mathcal{N}_i$'s.

Remark 2

The solution $u(x,t,\varepsilon)$ is quasiperiodic in the fast variable x/ε, t/ε with the spectral parameters β_k varying slowly according to (49). This establishes the connection between the β_k's and the modulation equations which was stated earlier.

The proof of the theorem rests on a plausible conjecture which is explained in the following sketch of proof.

Sketch of proof:

We first prove (55).

We examine the microstructure of the v_i's as defined in the second section. According to (48)

(60)
$$r(v_j,x,t) = 0 \text{ when } v_j \quad [0,\beta_2^2] \cup [\beta_1^2, \beta_0^2]$$

$$r(v_j,x,t) \neq 0 \text{ when } v_j \quad (\beta_2^2,\beta_1^2) \cup (\beta_0^2,1)$$

Study of region with $r \neq 0$: Theorem 2, equation (6) and the first of relations (38) give:

(61) $\quad v_j \in (\beta_0^2,1)$ implies $v_j = n_j^2$ + error exponentially small in ε.

The contribution of the v_j's in $(\beta_0^2, 1)$ to the sum in (20a) is:

(62)
$$2 \sum_{\{i: v_j \in (\beta_0^2,1)\}} \{v_j - \frac{\xi_{2i-1} + \xi_{2i}}{2}\} \sim 1 - \beta_0^2 \quad \text{as} \quad \varepsilon \to 0.$$

In the region (β_2^2, β_1^2) where again $r \neq 0$ not all v_j's can be determined through theorem 2. The theorem dictates the configuration of fig. 2 in which one v_j denoted by N does not need to be close to some n_j^2.

Fig. 2

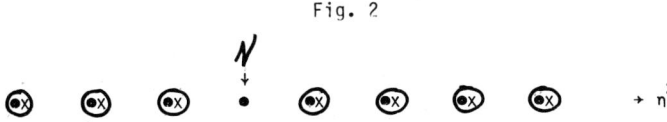

In this figure "x" denotes an n_j^2,

"●" denotes a v_j and a circle around them indicates that their distance decays exponentially as $\varepsilon \to 0$. As y and τ vary, N oscillates in the interval $[\beta_2^2, \beta_1^2]$ effectively passing from one gap to its

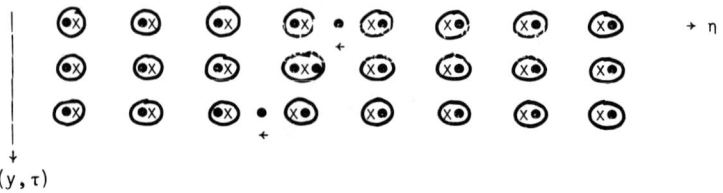

(y, τ)

Fig. 3.

neighboring one as shown in fig. 3. The contribution to (20a) from the region (β_2^2, β_1^2) as $\varepsilon \to 0$ is:

$$(63) \quad 2 \sum_{\{i\,:\,v_i \in (\beta_2^2, \beta_1^2)\}} (v_i - \frac{\xi_{2i-1} + \xi_{2i}}{2}) \sim - (\beta_1^2 - N) + (N - \beta_2^2) =$$

$$= 2N - \beta_1^2 - \beta_2^2$$

Study of region r=0

In this case theorem 2 is of no help. We make the following assumption which we can prove at t=0.

Assumption of molecular chaos:

If $r(\eta, x, (t) = 0$, there is $\varepsilon \ll \ell(\varepsilon) \ll 1$ as $\varepsilon \to 0$ such that:

$$(64) \quad \sum_{\{i\,:\,\eta^2 < v_i < \eta^2 + \ell(\varepsilon)\}} (v_i - \frac{\xi_{2i} + \xi_{2i-1}}{2}) \ll \ell(\varepsilon) \text{ as } \varepsilon \to 0 .$$

In other words, in the regions having $r = 0$ the local average of the deviation of the v_i from the center of its interval is of order higher then ε. The assumption gives immediately:

$$(65) \quad \sum_{\{i\,:\,r(v_i)=0\}} (v_i - \frac{\xi_{2i} + \xi_{2i-1}}{2}) \ll 1 \quad \text{as } \varepsilon \to 0$$

Using (62) (63) and (65) in (20a) where $\xi_0 = 1$ we obtain (55).

The proof of (57)-(58) goes along similar lines. For example to prove (57) we start from the "μ representation" of the solution $u(x,t,\varepsilon)$; our v_i and ξ_k

are the $-\mu_i$ and $-\lambda_k$ respectively of the more traditional notation (e.g. see [3,12]):

$$\text{(66)} \qquad \frac{\partial \nu_i}{\partial y} = \frac{\pm 2\{\prod_{k=0}^{2N} |\xi_k - \nu_i|\}^{1/2}}{\prod_{\substack{k=1 \\ k \neq i}}^{N} (\nu_i - \nu_k)} \qquad i = 1, 2, \ldots N$$

we apply this to $\nu_i = \mathcal{N}$

$$\text{(67)} \qquad \frac{\partial \mathcal{N}}{\partial y} = \frac{2\{\prod_{k=0}^{2N} |\xi_k - \mathcal{N}|\}^{1/2}}{\prod_{\substack{k=1 \\ k \neq 1}}^{N} (\mathcal{N} - \nu_k)} \quad .$$

We perform the following cancellations:

a/ $\{\nu_k: r(\nu_k, x, t) \neq 0\}$: By theorem 2 we define either $\tilde{\nu}_k = \nu_k$ or $\tilde{\nu}_k = \nu_{k-1}$ so that $\tilde{\nu}_k$ satisfies:

$$\text{(68)} \qquad \tilde{\nu}_k = \eta_k^2 + \text{error exponentially small as } \varepsilon \to 0$$

Thus:

$$\text{(69)} \qquad \frac{\{|\xi_{2k-1} - \mathcal{N}||\xi_{2k-2} - \mathcal{N}|\}^{1/2}}{|\mathcal{N} - \tilde{\nu}_k|} = 1 + O(\varepsilon^{-M})$$

for any $M > 0$.

b/ $\{\nu_k: r(\nu_k, x, t) = 0\}$: In this region as $\varepsilon \to 0$

$$\text{(70)} \qquad \frac{|\xi_{2k-1} - \mathcal{N}||\xi_{2k} - \mathcal{N}|^{1/2}}{|\mathcal{N} - \nu_k|} \sim 1 - \frac{\nu_k - \frac{1}{2}(\xi_{2k-1} + \xi_{2k})}{\mathcal{N} - \nu_k} + O(\varepsilon^2) \quad .$$

By the molecular chaos assumption the product of these factors tends to 1 as $\varepsilon \to 0$. The factors which remain after cancelations a/ and b/ have been performed are exactly the factors in the right hand side of (57). q.e.d.

There is still one piece of information which eludes us. We do not have an initial condition for equation (57), therefore we know the solution only up to a phase shift. The calculation of the phase shift (or phase shifts when $m>1$) involves higher order terms and it seems to be beyond the reach of this method.

7. Conclusions

In conclusion I want to incorporate the results into a global picture:

1/ Let (x,t) be any fixed point. As $\varepsilon \to 0$ the solution $u(x + \varepsilon y, t + \varepsilon \tau, \varepsilon)$ as a function of y and τ tends either to a constant, or to a periodic or to a quasiperiodic solution of (59) according to whether respectively $m(x,t) = 0$ or $m(x,t) = 1$ or $m(x,t) > 1$. The pointwise limit:

$$(71) \qquad \lim_{\varepsilon \to 0} u(x + \varepsilon y, t + \varepsilon \tau, \varepsilon)$$

exists in a subsequence $\varepsilon_h \to 0$ and is a constant, periodic or quasiperiodic solution of (59). Changing subsequences only affects the phase shifts.

3/ We define the associated limiting Schrödinger operator in the (y, τ) variables:

$$(72) \qquad \mathcal{L}(x,t;\tau) = \frac{d^2}{dy^2} - \lim_{\varepsilon \to 0} u(x + \varepsilon y, t + \varepsilon \tau, \varepsilon)$$

Its spectrum is independent of the subsequence. It is also independent of τ since (71) satisfies (59). We denote the spectrum of $\mathcal{L}(x,t;\tau)$ by $\sigma(x,t)$.

4/ The curve G_t which evolves according to (53) and gives the β_k's and m acquires a more intuitive meaning. For each t, it is the boundary of the set $\sigma(x,t)$ in the $x - \eta^2$ plane. For example at $t = 0$

$$\mathcal{L}(x,0;\tau) = \frac{d^2}{dy^2} - u(x,0) = \frac{d^2}{dy^2} + v(x)$$

and $\sigma(x,0) = (-\infty, v(x)) = (-\infty, \beta_0^2(x,0))$

5/ Given G_t (71) can be reconstructed up to phase shifts.

6/ The points in the x-t plane where $m(x,t)$ is discontinuous i.e. the boundary points of the shock regions, are the points (x,t) at which the vertical line at x in the $x-\eta^2$ plane is tangent to G_t at some point.

References

1. E. Date and S. Tanaka "Periodic Multisoliton Solutions of the Korteweg de Vries Equation and Toda Lattice", Supplement of the Progress of Theoretical Physics No. 59. 1976 pp. 107-125.

2. B.A. Dubrovnin, V.B. Mateev, S.P. Novikov, "Nonlinear equations of Korteweg de Vries type, finite zoned linear operators and Abelian varieties" Uspekhi Mat. Nauk 33, 1976, pp. 55-136.

3. H. Flaschka, M.G. Forest and D.W. McLaughlin, "Multiphase Averaging and the Inverse Spectral Solution of hte Korteweg-de Vries Equation" Comm. Pure Appl. Math 33, 1980, pp. 739-784.

4. C.S. Gardner, J.M. Greene, M.D. Kruskal, and R.M. Miura, "A method for solving the Korteweg de Vries Equation, Phys. R. Letters, Vol. 19, 1967, pp. 1095-1097.

5. P.D. Lax, "Integrals of Nonlinear Equations of Evolution and Solitary Waves" Comm. Pure Appl. Math. Vol 21, 1968, pp. 467-490.

6. P.D. Lax and C.D. Levermore, "The Small Dispersion Limit of the Korteweg de Vries Equation" I,II,III Comm. Pure Appl. Math 36, 1983, pp. 253-290, 571-594, 809-829.

7. H.P. McKean and P. van Moerbeke, "The Spectrum of Hill's Equation" Inventiones Math. 30, 1975, pp. 217-374.

8. H.P. McKean and E. Trubowitz, "Hill's Operator and Hyperelliptic Function Theory in the PResence of Infinitely Many Branch Points", Comm. Pure Appl. Math 29, 1976, pp. 146-226.

9. R.M. Miura, C.D. Gardner and M.D. Kruskal, Korteweg de Vries Equation and Generalizations. II. Existence of Conservation Laws and Constants of Motion". J. Math. Phys. Vol. 9, 1968, pp. 1204-1209.

10. S. Venakides "The Zero Dispersion Limit of the Korteweg de Vries Equation with Non-Trivial Reflection Coefficient" Comm. Pure Appl. Math Vol. 38, 1985. p. 125-155.

11. S. Venakides, "The Generalization of Modulated Wavetrains in the Solution of the Korteweg de Vries Equation" Comm. Pure Appl. Math., to appear.

12. S. Venakides "The Zero Dispersion Limit of the Korteweg de Vries Equation with Periodic Initial Data, AMS Transaction, to appear.

13. G.B. Whitham, "Linear and Nonlinear Waves" Wiley Intersience, New York, 1974.

STABILITY OF FINITE-DIFFERENCE APPROXIMATIONS FOR HYPERBOLIC INITIAL-BOUNDARY-VALUE PROBLEMS

ROBERT F. WARMING AND RICHARD M. BEAM

NASA Ames Research Center
Moffett Field, CA 94035/USA

Abstract

We consider the stability of finite-difference approximations to hyperbolic initial-boundary-value problems (IBVPs) in one spatial dimension. A complication is the fact that generally more boundary conditions are required for the discrete problem than are specified for the partial differential equation. Consequently, additional "numerical" boundary conditions are required and improper treatment of these additional conditions can lead to instability and/or inaccuracy. For a linear homogeneous IBVP, a finite-difference approximation with requisite numerical boundary conditions can be written in vector-matrix form as $u^{n+1} = Cu^n$ where C is a matrix operator. Lax-Richtmyer stability requires a uniform bound on C^n (i.e., C to the nth power) in some matrix norm for $0 \leq t = n\Delta t \leq T$. One would like to have an algebraic test for Lax-Richtmyer stability. For a matrix C of dimension J (denoted by C_J), a theorem in linear algebra relates $\|C_J^n\|$ to the spectral radius of C_J as $n \to \infty$, with J fixed. We state a conjecture which extends this theorem to difference approximations for IBVPs where the matrix size J increases linearly with n as $n \to \infty$ which corresponds to mesh refinement in both space and time. The asymptotic behavior of $\|C_J^n\|$ is related directly to the eigenvalues from the von Neumann analysis of the Cauchy problem and the eigenvalues from the normal mode analysis of Gustafsson, Kreiss, and Sundström for the left- and right-quarter plane problems. The conjecture is corroborated by examples where the matrix norm of C_J^n is computed numerically at a fixed time as the mesh is refined. An additional conjecture relates the spectral radius of the matrix C_J as $J \to \infty$ to the spectral radius of an auxiliary Dirichlet problem and the eigenvalues from the normal mode analysis of the left- and right-quarter plane problems.

1. Introduction

If one considers a difference approximation to a hyperbolic initial-boundary-value problem (IBVP), then, in general, the interior difference scheme requires more boundary conditions than the analytical boundary conditions needed for the partial differential equation (PDE). These additional boundary conditions for the finite-difference equations are often called "numerical" boundary conditions. In this paper any numerical procedure used to provide a numerical boundary condition is called a "numerical boundary scheme" (NBS). Improper treatment of the NBS can lead to instability even though one starts with an interior scheme which is stable for the pure-initial-value or Cauchy problem.

The homogeneous interior difference scheme together with the requisite homogeneous boundary conditions (analytical plus numerical) can be written in vector-matrix form as $u^{n+1} = Cu^n$. Lax-Richtmyer stability requires a uniform bound on C^n (i.e., C to the nth power) in some matrix norm for $0 \leq n\Delta t \leq T$. Here n is the time index and Δt the time increment. If the matrix operator C is normal, then the spectral radius condition is a necessary and sufficient condition for stability. However, for difference approximations to hyperbolic IBVPs, the matrix C is not normal and we know of no algebraic criteria for Lax-Richtmyer stability, i.e., for proving that $\|C^n\|$ is uniformly bounded.

During the 1960s and early 1970s Godunov and Ryabenkii [1], Kreiss [4,5], Osher [6,7], and Gustafsson, Kreiss, and Sundström [2] developed a stability theory for difference approximations to IBVPs. For the purposes of this paper, we refer to this theory as GKS (Gustafsson, Kreiss, Sundström). An advantage of the GKS theory is that it provides an algebraic test which is necessary and sufficient for stability but at the expense of using a complicated norm. A survey of the essential notions and difficulties arising in the stability of discrete IBVPs can be found in [13]. A discussion of the applicability of theoretical stability analyses to nonlinear gas-dynamics problems is given by Yee [14].

This paper was motivated by a theorem in linear algebra which relates $\|C^n\|$ to the spectral radius of C as $n \to \infty$. This theorem assumes that the matrix C is of fixed dimension J. The purpose of this paper is to present a conjecture which extends this theorem to difference approximations for hyperbolic IBVPs where the matrix size J increases linearly with n as $n \to \infty$. This corresponds to mesh refinement in both space and time for $t = n\Delta t = constant$. The asymptotic behavior of $\|C^n\|$ is related directly to the eigenvalues from the von Neumann analysis of the Cauchy problem and the eigenvalues from the normal mode analysis of Gustafsson, Kreiss, and Sundström for the left- and right-quarter problems. An additional conjecture relates the spectral radius of the matrix C as $J \to \infty$ to the spectral radius of an auxiliary Dirichlet problem and the eigenvalues from the normal mode analysis for the left- and right-quarter plane problems.

This paper is organized as follows. Sections 2 to 8 provide background for the conjectures presented in sections 9 and 11. In section 2 we review the IBVP for a model hyperbolic equation and in section 3 we consider the Lax-Wendroff scheme with an NBS as a prototype discrete IBVP. (Here the term "discrete IBVP" means a finite-difference approximation (including boundary conditions) to an IBVP for a PDE.) Section 4 reviews the Lax-Richtmyer stability theory for a discrete IBVP. In section 5 we review the Cauchy stability of the Lax-Wendroff scheme. In sections 6 and 7 we illustrate the practical difficulty of establishing Lax-Richtmyer stability for discrete IBVPs. Section 8 reviews the GKS-stability theory. A conjecture for an algebraic test for Lax-Richtmyer stability is stated and discussed in section 9. An auxiliary Dirichlet problem is defined in section 10. A conjecture for the spectral radius of C is presented in section 11.

2. Initial-Boundary-Value Problem for a Model Hyperbolic Equation

For simplicity we restrict our attention to the stability of finite-difference approximations to the IBVP for the model hyperbolic equation

$$\frac{\partial u}{\partial t} + c\frac{\partial u}{\partial x} = 0, \quad 0 \leq x \leq L, \quad t \geq 0 \qquad (2.1a)$$

where c is a real constant. Initial data are given at $t = 0$

$$u(x,0) = f(x), \qquad 0 \leq x \leq L \qquad (2.1b)$$

and the problem is well-posed if a boundary condition is prescribed at $x = 0$

$$u(0, t) = g(t) \quad \text{for} \quad c > 0 \tag{2.1c}$$

or at $x = L$

$$u(L, t) = g(t) \quad \text{for} \quad c < 0. \tag{2.1d}$$

In this paper the boundary conditions required for a well-posed PDE problem will be referred to as "analytical boundary conditions."

The stability analysis of a discrete IBVP by the GKS theory involves three auxiliary problems. The first is the pure-initial-value or Cauchy problem defined by

$$\frac{\partial u}{\partial t} + c \frac{\partial u}{\partial x} = 0, \quad -\infty < x < \infty, \quad t \geq 0$$
$$u(x, 0) = f(x). \tag{2.2}$$

The other two are related quarter-plane problems and in defining them we assume (without loss of generality) that $c > 0$. The related left-quarter plane problem is defined by

$$\frac{\partial u}{\partial t} + c \frac{\partial u}{\partial x} = 0, \quad -\infty < x \leq L, \quad t \geq 0$$
$$u(x, 0) = f(x), \tag{2.3}$$

and the related right-quarter plane problem is defined by

$$\frac{\partial u}{\partial t} + c \frac{\partial u}{\partial x} = 0, \quad 0 \leq x < \infty, \quad t \geq 0$$
$$u(x, 0) = f(x), \quad u(0, t) = g(t). \tag{2.4}$$

3. A Prototype Difference Approximation for the Model IBVP

To obtain a difference approximation of the model equation (2.1a) a mesh is introduced in (x, t) space with increments Δx and Δt and indexing defined by $x = j\Delta x$ and $t = n\Delta t$. The spatial domain $0 \leq x \leq L$ is divided into J equally spaced increments, i.e., $J\Delta x = L$. The finite-difference solution will be denoted by u_j^n. As a prototype (explicit) finite-difference approximation for the model equation (2.1a), we consider the Lax-Wendroff scheme

$$u_j^{n+1} = u_j^n - \frac{\nu}{2}(u_{j+1}^n - u_{j-1}^n) + \frac{\nu^2}{2}(u_{j+1}^n - 2u_j^n + u_{j-1}^n), \quad j = 1, 2, \ldots, J-1 \tag{3.1}$$

where $\nu = c\Delta t/\Delta x$ and $|\nu|$ is defined to be the Courant number.

As mentioned in the introduction, most finite-difference approximations for hyperbolic equations require more boundary conditions than those given for the PDE. We illustrate this by applying the Lax-Wendroff scheme on the finite domain $0 \leq x \leq L$. Here we assume $c > 0$ and the analytical boundary condition (2.1c) is given on the left boundary. The computational stencil for the Lax-Wendroff scheme is illustrated by the square symbols of Fig. 3.1. The left boundary is advanced by using the analytical boundary condition (2.1c). It is clear that an additional

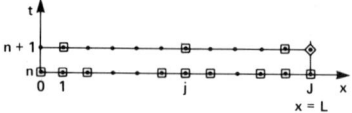

Fig. 3.1. Computational stencil for the Lax-Wendroff scheme.

"numerical boundary scheme" (NBS) is required to advance the solution at u_J^{n+1}, i.e., on the right boundary indicated by the diamond symbol of Fig 3.1.

The NBS should not be imposed arbitrarily but should be determined by using interior information. For example, one could use a one-sided approximation to the spatial derivative u_x. As a prototype NBS for the Lax-Wendroff scheme we consider the following one-sided boundary approximation:

$$u_J^{n+1} = u_J^n - \nu[\alpha u_{J-2}^n - (1+2\alpha)u_{J-1}^n + (1+\alpha)u_J^n] \tag{3.2}$$

where α is a parameter. This NBS is first-order accurate in time and space for any α except $\alpha = 1/2$ in which case it is spatially second-order accurate. If $\alpha = 0$, then (3.2) is simply the one-sided scheme

$$u_J^{n+1} = u_J^n - \nu(u_J^n - u_{J-1}^n). \tag{3.3}$$

The stability of the resulting algorithm, e.g., the interior scheme (3.1) plus the NBS (3.2), is extremely sensitive to the choice of the NBS. We have introduced the parameter α in (3.2) to illustrate this sensitivity. For example, if we choose $\alpha = -1$ in (3.2), then the NBS is

$$u_J^{n+1} = u_J^n - \nu(u_{J-1}^n - u_{J-2}^n). \tag{3.4}$$

We will find (section 8) that the Lax-Wendroff scheme with (3.3) is stable for $0 \leq \nu \leq 1$ while the Lax-Wendroff scheme with (3.4) is unconditionally unstable. Note that (3.3) and (3.4) differ only in the displacement of the approximation of u_x by one grid point.

4. Lax-Richtmyer Stability of a Discrete IBVP (Finite-Spatial Domain)

In considering the Lax-Richtmyer stability of a finite-difference approximation it is appropriate to rewrite the homogeneous interior scheme together with the homogeneous analytical boundary conditions and the NBSs in vector-matrix form as

$$u^{n+1} = Cu^n. \tag{4.1}$$

As an example, we write out the vector-matrix form of the Lax-Wendroff scheme (3.1) with the NBS (3.2). The analytical boundary condition (2.1c) with $g(t) = 0$ is rewritten for the discrete problem as

$$u_0^n = 0. \tag{4.2}$$

For convenience, we rewrite (3.1) as

$$u_j^{n+1} = pu_{j-1}^n + qu_j^n + ru_{j+1}^n, \quad j = 1, 2, \ldots, J-1 \tag{4.3a}$$

where
$$p = \frac{\nu}{2}(1+\nu), \quad q = 1 - \nu^2, \quad r = \frac{\nu}{2}(\nu - 1). \tag{4.3b}$$

Likewise, we rewrite the NBS (3.2) as

$$u_J^{n+1} = zu_{J-2}^n + wu_{J-1}^n + su_J^n \tag{4.4a}$$

where

$$z = -\nu\alpha, \quad w = \nu(1 + 2\alpha), \quad s = 1 - \nu(1 + \alpha). \tag{4.4b}$$

The interior scheme (4.3) together with the NBS (4.4) and the analytical boundary condition (4.2) can be written in vector-matrix form (4.1) where

$$u^n = \begin{bmatrix} u_1^n \\ u_2^n \\ \cdot \\ \cdot \\ \cdot \\ u_{J-1}^n \\ u_J^n \end{bmatrix}, \quad C = \begin{bmatrix} q & r & & & & \\ p & q & r & & O & \\ & \cdot & \cdot & \cdot & & \\ & & \cdot & \cdot & \cdot & \\ & & & \cdot & \cdot & \cdot \\ & O & & p & q & r \\ & & & z & w & s \end{bmatrix} \tag{4.5a,b}$$

Here the matrix size is $J \times J$.

Any difference approximation can be written in the vector-matrix form (4.1) which in turn can be rewritten as

$$u^n = C^n u^0 \tag{4.6}$$

where u^0 is the vector of initial values. Since we need some measure of the magnitude of the solution we use a vector norm $\|\cdot\|$. For example the Euclidean or L_2 norm of a J-dimensional column vector u is defined as

$$\|u\|_2 = \left(\sum_{j=1}^J |u_j|^2\right)^{1/2}. \tag{4.7}$$

The following is a conventional definition of Lax-Richtmyer stability:

Definition 4.1. *A difference approximation represented by (4.6) is said to be Lax-Richtmyer stable if there exists a constant $K > 0$ such that for any initial condition u^0*

$$\|u^n\| \leq K\|u^0\| \tag{4.8}$$

for all $n \geq 0$, $0 \leq n\Delta t \leq T$ with T fixed.

Here a specified relation between Δt and Δx is assumed. Since we are considering only hyperbolic problems we assume $\Delta t/\Delta x = $ constant.

Before continuing, we state several relevant results from linear algebra (see, e.g., [3, chapter 1]). Corresponding to any vector norm $\|\cdot\|$ and a matrix C there exists a natural matrix norm defined by

$$\|C\| = \max_{\|u\| \neq 0} \frac{\|Cu\|}{\|u\|}. \tag{4.9}$$

The spectral radius of a square matrix C is defined by

$$\rho(C) = \max_l |z_l(C)| \tag{4.10}$$

where $z_l(C)$ denotes the lth eigenvalue of C. The Euclidean or L_2 matrix norm of C is given by

$$\|C\|_2 = \sqrt{\rho(C^*C)}. \tag{4.11}$$

The adjoint matrix C^* is defined as the conjugate transpose of C, i.e.,

$$C^* = \overline{C}^T$$

where C^T denotes the transpose of C and the bar indicates that each element of C^T is replaced by its complex conjugate.

One can show that the stability Definition (4.1) is equivalent to requiring a uniform bound on $\|C^n\|$, i.e.,

$$\|C^n\| \le K \tag{4.12}$$

for all $n \ge 0$, $0 \le n\Delta t \le T$. (Uniform boundedness of $\|C^n\|$ is the definition of stability given by Lax and Richtmyer [9, p. 45].)

In general, it is not feasible to compute $\|C^n\|$, so we look for practical algebraic tests that provide necessary and/or sufficient conditions for stability, i.e., uniform boundedness of $\|C^n\|$ defined by (4.12). A necessary condition for stability is the spectral radius condition

$$\rho(C) \le 1 + O(\Delta t) \tag{4.13}$$

and a sufficient condition for stability is

$$\|C\| \le 1 + O(\Delta t). \tag{4.14}$$

There are certain special cases where a simple test is both necessary and sufficient for stability. For example, if C is a normal matrix, i.e., $CC^* = C^*C$, then

$$\rho(C) = \|C\|_2 \tag{4.15}$$

and the spectral radius condition (4.13) is necessary and sufficient for stability in the L_2 norm. An example is the pure-initial-value or Cauchy problem with spatially periodic initial data. In this case C is a circulant matrix and consequently C is normal. The test here is equivalent to the von Neumann analysis. In the next section we consider the initial-value problem (IVP) for the Lax-Wendroff scheme.

5. Stability of an Initial-Value Problem

Although our primary interest is the stability of IBVPs, we briefly review the stability of the IVP for the Lax-Wendroff scheme. The solution is assumed to be spatially periodic with period $L = J\Delta x$, and hence

$$u_j^n = u_{j+J}^n \tag{5.1}$$

for any j. Consequently, the scheme can be written in vector-matrix form as (4.1) where

$$C = \begin{bmatrix} q & r & & & & & p \\ p & q & r & & & & \\ & \cdot & \cdot & \cdot & & O & \\ & & \cdot & \cdot & \cdot & & \\ & & & \cdot & \cdot & \cdot & \\ & O & & & p & q & r \\ r & & & & & p & q \end{bmatrix} \qquad (5.2)$$

and p, q, r are given by (4.3b). The $J \times J$ matrix C is circulant and the eigenvalues z_l are given analytically as

$$z_l = 1 - 2\nu^2 \sin^2 \frac{\pi l}{J} - i\nu \sin \frac{2\pi l}{J}, \quad l = 0, 1, \cdots, J - 1. \qquad (5.3)$$

For $|\nu| \leq 1$, the spectral radius is

$$\rho(C) = 1, \qquad 0 \leq |\nu| \leq 1 \qquad (5.4)$$

independent of the number of spatial points J. For $|\nu| > 1$ and J even

$$\rho(C) = 2\nu^2 - 1, \qquad |\nu| > 1. \qquad (5.5)$$

For $|\nu| > 1$ and J odd, $\rho(C)$ approaches (5.5) from below and for large J

$$\rho(C) \approx (2\nu^2 - 1)\left[1 + O\left(1/J^2\right)\right]. \qquad (5.6)$$

Since the circulant matrix (5.2) is normal, one has from (4.15),

$$\rho(C) = \|C\|_2, \qquad \rho^n(C) = \|C^n\|_2$$

and hence

$$\rho^n(C) = \|C^n\|_2 = 1 \qquad \text{for} \qquad 0 \leq |\nu| \leq 1.$$

But for $|\nu| > 1$, $\|C^n\|_2$ is unbounded with increasing n. This, of course, yields the well-known Lax-Richtmyer stability condition $|\nu| \leq 1$ for the Lax-Wendroff scheme with periodic boundary conditions. The spectral radius $\rho(C)$ is plotted in Fig. 5.1 as a function of ν.

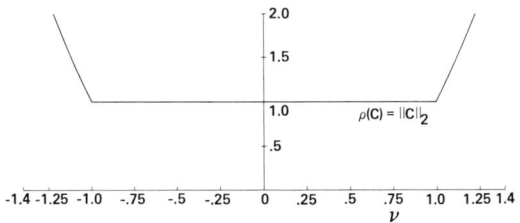

Fig. 5.1. $\rho(C)$ and $\|C\|_2$ of the circulant matrix (5.2).

6. Difficulties in Proving Lax-Richtmyer Stability for IBVPs

In this section we consider the stability of the Lax-Wendroff scheme (3.1) with NBS (3.2). The resulting matrix (4.5b) is not circulant and we illustrate the practical difficulty of establishing Lax-Richtmyer stability for this discrete IBVP. By virtue of the necessary condition (4.13) and the sufficient condition (4.14) for stability we examine $\rho(C)$ and $\|C\|_2$ for two examples.

As a first example, we consider the first-order NBS (3.3), i.e., $\alpha = 0$ in NBS (3.2). A practical problem is that the nonsymmetric matrix (4.5b) has complex eigenvalues (for $|\nu| < 1$) and one cannot (or rather we cannot) obtain a closed-form analytical formula for the eigenvalues of C. We have computed the spectral radius numerically and it is plotted in Fig. 6.1 as a

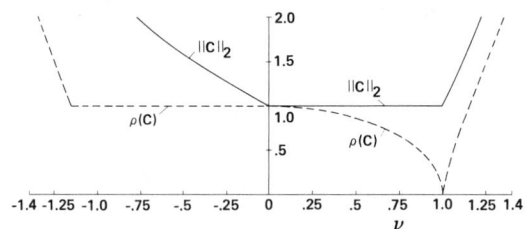

Fig. 6.1. $\rho(C)$ and $\|C\|_2$ of the matrix (4.5b) with $\alpha = 0.0$.

function of ν for $J = 20$. To plotting accuracy, the spectral radius does not change as J increases. One can show analytically for $\nu \geq 1$ that

$$\begin{aligned} \rho(C) < 1, & \quad 1 < \nu < \sqrt{4/3} \\ \rho(C) = 0, & \quad \nu = 1. \end{aligned} \quad (6.1)$$

Hence from (6.1) and Fig. 6.1 one can safely assume

$$\rho(C) \leq 1 \quad \text{for} \quad 0 \leq \nu \leq \sqrt{4/3} \approx 1.1547. \quad (6.2)$$

Negative values of ν correspond to a "reversed problem," i.e., $c < 0$ but with the analytical boundary condition still applied at $x = 0$ and the NBS at $x = L$. For $-\sqrt{4/3} \leq \nu < 0$, the values of $\rho(C)$ which appear to be unity in Fig. 6.1 actually have the behavior $\rho(C) \to 1^+$ as $J \to \infty$. This behavior does not prove instability since one cannot be certain the necessary condition (4.13) is violated. (In fact, the scheme is unstable for $\nu < 0$ (section 8)). It is rather striking to compare the spectral radius of the pure IVP plotted in Fig. 5.1 with the spectral radius of the IBVP plotted in Fig. 6.1.

If $\rho(C) \leq 1$, then the spectral radius condition (4.13) is satisfied, e.g., if we choose $\nu = 1.1$, then $\rho(C) \approx 0.714 < 1$ (see Fig. 6.1). A natural question is whether or not the scheme is stable for $\nu = 1.1$ although we know the Lax-Wendroff scheme with periodic boundary boundary conditions is unstable for $\nu = 1.1$.

In Fig. 6.2 we plot numerically computed values of $\|C^n\|_2$ for several values of J as a function of n. The abscissa is normalized so that the plot is actually a function of time. The normalization is

$$t = n\Delta t = \frac{n\Delta t}{J\Delta x} \quad (6.3)$$

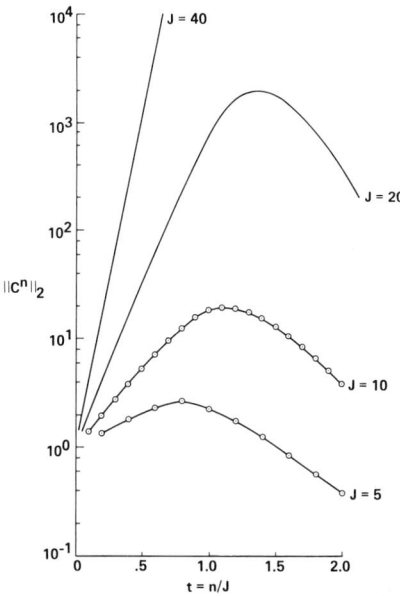

Fig. 6.2. $\|C^n\|_2$ of the matrix (4.5b) as a function of t for several values of J with $\nu = 1.1$ and $\alpha = 0.0$.

where for convenience we have assumed $J\Delta x = 1$ and $\Delta t/\Delta x = 1$. The plotted values correspond to discrete values of time as indicated in the figure by the circle symbols for $J = 5$ and 10. The density of points increases with J and for $J = 20$ and 40 we have simply drawn a continuous curve through the points. If we fix J, then $\|C^n\|_2 \to 0$ as $n \to \infty$ since $\rho(C) < 1$. On the other hand, if we fix t (say, e.g., $t = 1.0$), then we can see from Fig. 6.2 that there is no uniform bound on $\|C^n\|_2$ as J increases and the scheme is unstable for $\nu = 1.1$.

Table 6.1. Asymptotic (large J) numerical values of $\|C\|_2$ for matrix (4.5b) with $\alpha = 0.0$.

ν	$\|C\|_2$
0.0	1.0
.1	1.00345
.2	1.00480
.3	1.00530
.4	1.00530
.5	1.00496
.6	1.00435
.7	1.00352
.8	1.00250
.9	1.00132
1.0	1.0

Although not detectable in Fig. 6.1, $\|C\|_2 > 1$ for $0 < \nu < 1$. Asymptotic (large J) numerical values of $\|C\|_2$ are listed in Table 6.1 as a function of Courant number $0 \leq \nu \leq 1$. The fact that $\|C\|_2 > 1$ does not indicate that the scheme is unstable since condition (4.14) is only a sufficient condition for stability, i.e., Lax-Richtmyer stability requires a uniform bound on $\|C^n\|_2$.

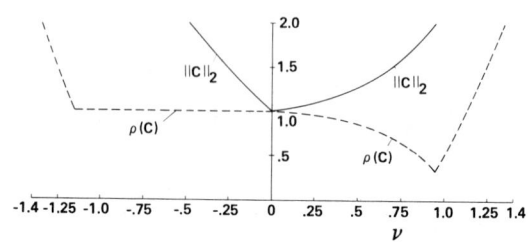

Fig. 6.3. $\rho(C)$ and $\|C\|_2$ of the matrix (4.5b) with $\alpha = 0.5$.

As a second example consider the case $\alpha = 0.5$ in the NBS (3.2). This corresponds to a second-order-accurate approximation to u_x in the NBS. In Fig. 6.3 $\rho(C)$ and $\|C\|_2$ are plotted as a function of ν. For $\nu = 0.5$, $\|C\|_2 = 1.26716$. This is an asymptotic value computed for large J. Again the fact that $\|C\|_2 > 1$ does not indicate that the scheme is unstable. In Table 6.2 we list numerically computed values of $\|C^n\|_2$ for $\nu = 0.5$. Here we see that although $\|C\|_2 > 1$, $\|C^n\|_2$ decreases monotonically with increasing n. The scheme is, in fact, stable for $0 \leq \nu \leq 1$ and unstable otherwise (section 8) although a knowledge of $\|C\|_2$ as plotted in Fig. 6.3 gives no indication of this fact.

Table 6.2. Asymptotic (large J) numerical values of $\|C^n\|_2$ as a function of n for matrix (4.5b) with $\nu = 0.5$ and $\alpha = 0.5$.

n	$\|C^n\|_2$
1	1.26716
2	1.17100
3	1.03425
4	1.01296
5	1.00831
6	1.00130
7	0.99999

For these particular examples it is clear that even if one could analytically determine $\rho(C)$ and $\|C\|_2$ nothing definitive could be concluded about the stability of the scheme. Of course, in principle, one could establish a sufficient condition for Lax-Richtmyer stability by using the energy method to provide an estimate of the solution in terms of the initial data (see inequality (4.8)). However, in practice, this is very difficult to do.

7. Normal-Mode Analysis (Finite-Domain Problem)

The homogeneous vector-matrix equation (4.1) is a first-order difference equation and to solve this equation one looks for a solution of the form

$$u^n = z^n \phi \tag{7.1}$$

where z is a complex constant and ϕ is a J-component vector. Insertion of (7.1) into (4.1) yields (for $z \neq 0$)

$$C\phi = z\phi \tag{7.2}$$

which is the eigenvalue problem for the matrix C where ϕ is the eigenvector and z is the eigenvalue.

The determination of the eigensolutions of a finite-difference operator is sometimes called the "normal mode" analysis. For example, solution (7.1) is called an eigensolution since z is an eigenvalue and ϕ an eigenvector of the matrix operator C. If C has a complete set of eigenvectors, then the eigensolutions are the normal modes and the general solution of (4.1) can be written as

$$u^n = \sum_{l=1}^{J} \alpha_l z_l^n \phi_l \tag{7.3}$$

where z_l and ϕ_l denote the lth eigenvalue and eigenvector of the matrix C and the α_ls are complex constants.

A natural and direct approach for finding the eigensolutions of a matrix operator C is to start with the original partial difference equation plus boundary conditions and look for solutions of the form

$$u_j^n = z^n \phi_j. \tag{7.4}$$

Of course (7.4) is just the component form of (7.1). The resulting equations are called the resolvent equations. Solution of the resolvent equations yields the eigensolutions.

A practical problem is that, in general, for discrete IBVPs such as Lax-Wendroff with an NBS, the eigensolutions cannot be obtained in closed form for the finite-domain problem. Even if the eigenvalues were known, the spectral radius condition (4.13) provides only a necessary condition for Lax-Richtmyer stability. It is rather perplexing that the finite-domain stability analysis can be intractable even for simple scalar one-dimensional schemes.

8. Normal-Mode Analysis (Quarter-Plane Problems)

In the 1960s and early 1970s a stability theory for discrete IBVPs was developed by Godunov and Ryabenkii [1], Kreiss [4,5], Osher [6,7], Gustafsson, Kreiss, and Sundström [2]. For the purposes of this paper, we refer to this theory as the GKS theory. For a brief history of stability theory for difference models of IBVPs see Trefethen [10]. The IBVP on a finite domain for the model hyperbolic equation was defined by (2.1). In addition, three auxiliary problems were defined in section 2: the Cauchy problem and the related left- and right-quarter plane problems. Kreiss [4] and Gustafsson, Kreiss, and Sundström [2] proved the following theorem:

Theorem 8.1. *Consider a difference approximation to an IBVP (e.g., (2.1)) on the finite domain $0 \leq x \leq L, t \geq 0$. Assume that the approximation is stable for the Cauchy problem and stable according to Definition (3.3) of [2] for the related left- and right-quarter plane problems. Then the original problem is also stable.*

For the precise stability definition used by Gustafsson, Kreiss, and Sundström, the reader should refer to their original paper [2]. We should point out that the stability Definition (3.3) of [2] involves a complicated norm. In this paper a difference approximation that satisfies the hypothesis of the above theorem is said to be GKS stable.

For the Lax-Wendroff example considered in this paper the spectral radius condition (4.13) is necessary and sufficient for the Cauchy problem (this is the von Neumann condition). The stability of the left- and right-quarter plane problems are checked by the normal mode analysis [2]. Here we outline the normal mode analysis for the quarter-plane problems.

Stability results obtained by means of the normal mode analysis are based on the resolvent equations. The resolvent equations (for a quarter-plane problem) are obtained by substituting

$$u_j^n = z^n \phi_j \tag{8.1}$$

into the interior scheme and the boundary conditions. In (8.1) z is a complex number and ϕ_j is the eigenfunction. The resolvent equations consist of a difference equation and boundary conditions for the ϕ_j. The general solution of the resolvent equations which is in L_2 has the form

$$\phi_j = \phi_0 \kappa^j, \quad |\kappa| < 1.$$

The approximation is stable for the discrete IBVP if there are no nontrivial solutions of the form

$$|z| > 1, \quad |\kappa| < 1 \tag{8.2}$$

and

$$|z| = 1, \quad |\kappa| = 1 \quad \text{such that if} \quad |z^*| \to 1^+ \quad \text{then} \quad |\kappa^*| \to 1^- \tag{8.3}$$

where z^* and κ^* indicate perturbations off the unit circle. The (necessary) condition that there be no solutions of the form (8.2) is called the Godunov-Ryabenkii condition [1]. A nontrivial solution of the form (8.2) is said to be an eigensolution and the corresponding z an eigenvalue. The much more subtle condition that there be no solutions of the form (8.3) is due to Kreiss [5]. If there are solutions on the unit circle, one must perform a perturbation test. If z approaches the unit circle from outside, one admits only values of $\kappa = \kappa(z)$ which approach the unit circle from inside. A nontrivial solution of the form (8.3) is said to be a generalized eigensolution and the corresponding z a generalized eigenvalue. Henceforth, we refer to an eigenvalue of the form (8.2) or (8.3) as a GKS eigenvalue or GKS generalized eigenvalue.

An excellent account of the GKS-stability theory is given by Trefethen [10]. Trefethen shows that the main result (8.3) of the GKS theory has a simple physical interpretation in terms of group velocity.

We have analyzed the GKS-stability of the Lax-Wendroff scheme (3.1) with the NBS (3.2). The details of the analysis are in a forthcoming NASA report. First one checks the Cauchy stability and finds the well-known stability condition $|\nu| \leq 1$ (section 5). The GKS normal mode analysis

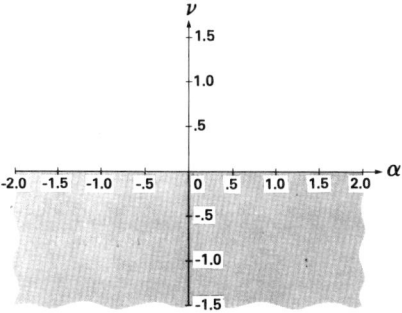

Fig. 8.1. Shaded region indicates a GKS generalized eigenvalue.

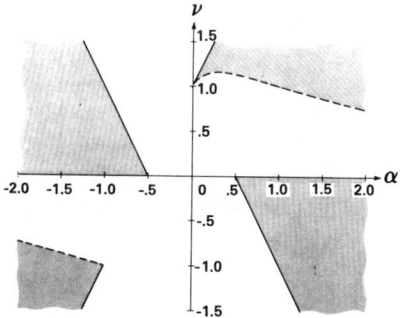

Fig. 8.2. Shaded region indicates a GKS eigenvalue.

was carried out for the quarter-plane problems. The results are shown graphically in the (α, ν) parameter space of Figs. 8.1 and 8.2.

The shaded region of Fig. 8.1 indicates the existence of a GKS generalized eigenvalue for $\nu < 0$ for any value of α. The shaded region of Fig. 8.2 shows values of (α, ν) for which a GKS eigenvalue exists.

The resulting (α, ν) parameter space for which the scheme is GKS stable is shown by the cross-hatched region of Fig. 8.3. Note, for example, that for $\alpha = -1.0$ the resulting scheme is unconditionally GKS unstable.

9. A Conjecture on an Algebraic Test for Lax-Richtmyer Stability

The essential element in both stability and convergence of a finite-difference approximation is the behavior of $\|C^n\|$ for a fixed value of $t = n\Delta t$ as a function of increasing n as the mesh is refined in both space and time. Here the Courant number is assumed to be fixed and hence so is the ratio n/J. As a consequence, the order of the matrix J increases linearly with n as the mesh is refined and this crucial point is emphasized by writing C_J and $\|C_J^n\|$. The asymptotic behavior of $\|C_J^n\|$ for large n (n/J fixed) determines the stability (or instability) of the numerical method

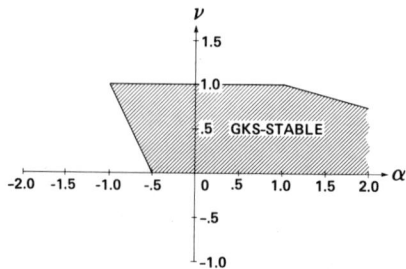

Fig. 8.3. GKS-stability region for Lax-Wendroff scheme (3.1) with NBS (3.2).

since stability requires that $\|C_j^n\|$ be uniformly bounded for all $n \geq 0, 0 \leq n\Delta t = t \leq T$ with T fixed.

We found in section 6 that a direct analytical approach to Lax-Richtmyer stability fails in the sense that one cannot, in general, determine $\rho(C)$ and $\|C\|_2$ analytically let alone $\|C^n\|_2$. On the other hand the GKS theory provides an algebraic check that gives a necessary and sufficient condition for GKS stability. However, the GKS theory does not define stability in terms of the simple Lax-Richtmyer Definition (4.1) (say, in an L_2 norm) but uses a more complicated norm involving integrals over t and solution data given at the boundary and forcing data in the interior. In this section we give a conjecture which relates the two stability theories by generalizations of the following two theorems of linear algebra.

Theorem 9.1. *(Varga [12], p. 65) Let $\|\cdot\|$ be a natural matrix norm and let C be a $J \times J$ matrix with a complete set of eigenvectors and $\rho(C) > 0$. Then*

$$\|C^n\| \approx k\rho^n(C), \qquad n \to \infty \tag{9.1}$$

where k is a positive constant.

The theorem as stated above is a special case of a more general theorem by Varga which is valid for an arbitrary $J \times J$ matrix C. However (9.1) will suffice for our purposes. The second theorem is the following:

Theorem 9.2. *(Varga [12], p. 95) Let $\|\cdot\|$ be a natural matrix norm and let C be an arbitrary $J \times J$ matrix. Then*

$$\lim_{n \to \infty} \|C^n\|^{1/n} = \rho(C). \tag{9.2}$$

In the special case that C has a complete set of eigenvectors, then (9.2) follows directly from (9.1). In addition it follows from (9.1) that

$$\|C^n\|^{1/n} \approx \rho(C)[1 + (\ln k)/n], \qquad n \to \infty.$$

These two theorems play a fundamental role in the rate of convergence of iterative methods for solving linear algebraic systems of fixed dimension J [11, p. 61]. If analogous theorems hold for matrices associated with differences approximations of the form (4.1), then we would have an algebraic check for Lax-Richtmyer stability since one could predict whether or not $\|C_j^n\|$ is

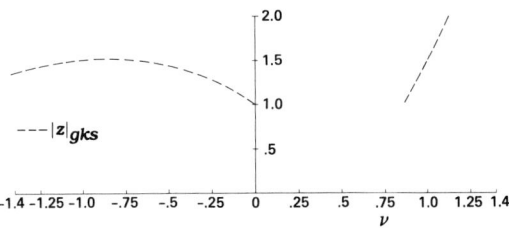

Fig. 9.1. $|z|_{gks}$ for the Lax-Wendroff scheme (3.1) with NBS(3.2) for $\alpha = 1.5$.

uniformly bounded for large n. The above theorems do not apply directly since the dimension of the matrix C is fixed which is not the case for the finite-difference matrix operator C.

Before we give our conjecture we prove a lemma which yields a condition equivalent to the uniform boundedness condition (4.12) of the Lax-Richtmyer stability definition.

Lemma 9.1. *A necessary and sufficient condition for a discrete IVP or IBVP to be Lax-Richtmyer stable is that there exists a positive constant M such that*

$$\|C_J^n\|^{1/n} \leq 1 + M/n, \tag{9.3}$$

for all $n > 0$, $0 < n\Delta t = t \leq T$ with a fixed ratio $\Delta t/\Delta x$.

PROOF.: Necessity: Without loss of generality we assume that $K > 1$ in (4.12). Hence

$$\|C_J^n\|^{1/n} \leq K^{1/n} \leq (1 + K/n),$$

for all $n > 0$. Sufficiency: From (9.3) one has

$$\|C_J^n\| \leq (1 + M/n)^n \leq e^M$$

and hence $\|C_J^n\|$ is uniformly bounded.

Lemma 9.1 conveys no additional useful information unless there is a practical algebraic test for inequality (9.3). Below we state a conjecture which provides an asymptotic estimate of $\|C_J^n\|^{1/n}$ in terms of eigenvalues from the von Neumann analysis of the Cauchy problem and GKS eigenvalues and generalized eigenvalues from the normal-mode analysis of the left- and right-quarter plane problems. As a preliminary we introduce some notation. Consider the von Neumann analysis of the Cauchy problem for a particular scheme. Let $\widehat{C}_J$ denote the circulant matrix associated with the Cauchy problem and denote by $|z|_{von}$ the spectral radius of $\widehat{C}_J$ as $J \to \infty$:

$$|z|_{von} = \lim_{J \to \infty} \rho(\widehat{C}_J) \tag{9.4}$$

where $|z|_{von} \geq 1$. As an example, $|z|_{von}$ for the Lax-Wendroff scheme (3.1) is given by (5.4) and (5.5) and is plotted in Fig. 5.1. In addition, for the discrete IBVP we apply the GKS normal mode analysis to the related left- and right-quarter plane problems. If there exist one or more GKS eigenvalues or generalized eigenvalues, let $|z|_{gks}$ denote the modulus of the maximum modulus eigenvalue or generalized eigenvalue. Obviously $|z|_{gks} \geq 1$ with inequality if there is a GKS eigenvalue.

For example, Fig. 9.1 is a graph of $|z|_{gks}$ as a function of ν for the Lax-Wendroff scheme (3.1) with NBS (3.2) for $\alpha = 1.5$.

We now make the following conjecture:

Conjecture 9.1a. *Consider a discrete IBVP with matrix operator C_J on a finite domain ($J\Delta x = L$). If $t = n\Delta t \leq T$ and $\Delta t/\Delta x$ are fixed so there is a linear relation between n and J, then in the L_2 norm*

$$\lim_{n\to\infty} \|C_J^n\|_2^{1/n} = \max(|z|_{von}, |z|_{gks}). \tag{9.5}$$

If there is no GKS eigenvalue or generalized eigenvalue, then we write

$$\max(|z|_{von}, |z|_{gks}) = |z|_{von}. \tag{9.6}$$

As an example, if we overlay Figs. 5.1 and 9.1 and take $\max(|z|_{von}, |z|_{gks})$, we obtain the result plotted in Fig. 9.2.

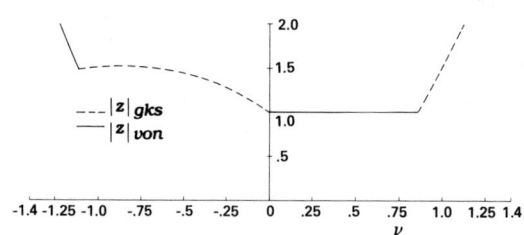

Fig. 9.2. $\max(|z|_{von}, |z|_{gks})$ for the Lax-Wendroff scheme (3.1) with NBS(3.2) for $\alpha = 1.5$.

Conjecture 9.1b. *The asymptotic form of $\|C_J^n\|_2^{1/n}$ is delineated as follows:*
(a) *If $\max(|z|_{von}, |z|_{gks}) = |z|_{gks} > 1$, then*

$$\|C_J^n\|_2^{1/n} \approx |z|_{gks}(1 + (\ln \psi)/n), \qquad n \to \infty \tag{9.7a}$$

where $\psi > 1$ is a positive constant.
(b) *If $\max(|z|_{von}, |z|_{gks}) = |z|_{von} > 1$, then*

$$\|C_J^n\|_2^{1/n} \approx |z|_{von}, \qquad n \to \infty. \tag{9.7b}$$

(The asymptotic form here is the same as (9.7a) except $\psi = 1$.)
(c) *If $\max(|z|_{von}, |z|_{gks}) = |z|_{gks} = 1$, then*

$$\|C_J^n\|_2^{1/n} \approx 1 + \beta(\ln \gamma n)/n, \qquad n \to \infty \tag{9.7c}$$

where β and γ are positive constants.
(d) *If $\max(|z|_{von}, |z|_{gks}) = |z|_{von} = 1$, then*

$$\|C_J^n\|_2^{1/n} \leq 1 + \chi/n, \qquad n \to \infty \tag{9.7d}$$

where χ is a positive constant.

By Lemma 9.1 the first three cases above are Lax-Richtmyer unstable and the last case is Lax-Richtmyer stable. Excluded from our Conjecture (9.1b) are a few delicate borderline cases which we examine in detail in a subsequent paper. For example, case (c) corresponds to a GKS generalized eigenvalue and in (8.3) we assume that the perturbation in z is of the same order as the perturbation in κ. We define the rare cases where the perturbation in z is of higher order than the perturbation in κ to be "borderline" GKS generalized eigenvalues and we treat them separately.

An implication of Conjecture 9.1 is that GKS stability implies Lax-Richtmyer stability. This is consistent with a statement by Trefethen [11, p. 350] that GKS stability may imply L_2 stability. Furthermore, growth rates for the unstable cases a, b, c above are consistent with estimates given by Trefethen [11]. A detailed comparison of our results with those of Trefethen will appear in a subsequent paper where we consider borderline cases.

We illustrate Conjecture 9.1a and the four cases of Conjecture 9.1b by numerical examples.

Case a: In this case $|z|_{gks}$ corresponds to a GKS eigenvalue. From (9.7a) it follows that

$$\|C_J^n\|_2 \approx \psi |z|_{gks}^n, \qquad n \to \infty. \tag{9.8}$$

As a numerical example, consider the point $\alpha = 1.5$, $\nu = 0.9$ in the (α, ν) parameter space shown graphically in Fig. 8.2. For this pair of parameter values there is a GKS eigenvalue and $|z|_{gks} = 1.1315$. In Table 9.1a we tabulate numerically computed values of $\|C_J^n\|_2$ and $\|C_J^n\|_2^{1/n}$ for $n = J$ as a function of mesh refinement, i.e., for increasing values of n. The asymptotic approach of $\|C_J^n\|_2^{1/n}$ to $|z|_{gks}$ is apparent. The constant $\psi = 4.953$ in (9.8) was determined numerically.

Case b: This case may or may not have a GKS eigenvalue. But if there is a GKS eigenvalue, then $|z|_{von} > |z|_{gks} > 1$. From (9.7b) it follows that

$$\|C_J^n\|_2 \approx |z|_{von}^n, \qquad n \to \infty. \tag{9.9}$$

As a numerical example consider the point $\alpha = 0.5$ and $\nu = 1.1$ of Fig. 8.2. Here $|z|_{von} = 1.420$ and there is no GKS eigenvalue. In Table 9.1b we list numerically computed values of $\|C_J^n\|_2$ and $\|C_J^n\|_2^{1/n}$ for $n = J$. The asymptotic approach of $\|C_J^n\|_2^{1/n}$ to $|z|_{von}$ is readily apparent.

If $\max(|z|_{von}, |z|_{gks}) = 1$, then one cannot tell from this limiting value of unity whether the scheme is stable or unstable. We first consider the case where there is a GKS generalized eigenvalue.

Case c: It follows from (9.7c) that

$$\|C_J^n\|_2 \approx (\gamma n)^\beta, \qquad n \to \infty. \tag{9.10}$$

Consider the point $\alpha = 0.5, \nu = -0.5$ of Fig. 8.2. Table 9.1c shows computed values of $\|C_J^n\|_2$ and $\|C_J^n\|_2^{1/n}$ for $n = J$. The constants $\beta = 3/2$ and $\gamma = 5/3$ were determined numerically. The growth rate of $\|C_J^n\|_2$ given by (9.10) is algebraic with increasing n in contrast with the exponential growth of cases a and b given by (9.8) and (9.9). (Compare $\|C_J^n\|_2$ in Tables 9.1a and 9.1c).

If we compare (9.7c) and (9.7d) we see that the weak instability resulting from a GKS generalized eigenvalue manifests itself by the appearance of the weakly growing term $\ln \gamma n$ in the asymptotic expansion (9.7c).

Case d: In this last case there is no GKS eigenvalue or generalized eigenvalue and the scheme is Lax-Richtmyer stable. As an example, consider the point $\alpha = 0.5$ and $\nu = 0.5$ of Fig. 8.2. In

Table 9.1a. Numerical values of $\|C_J^n\|_2$ and $\|C_J^n\|_2^{1/n}$ for matrix (4.5b) with $\alpha = 1.5, \nu = 0.9$.

$J = n$	$\|C_J^n\|_2$	$\|C_J^n\|_2^{1/n}$
10	$.1675 \times 10^2$	1.32555
50	$.2344 \times 10^4$	1.16789
100	$.1129 \times 10^7$	1.14955
150	$.5438 \times 10^9$	1.14350
200	$.2619 \times 10^{12}$	1.14049
250	$.1262 \times 10^{15}$	1.13869
300	$.6077 \times 10^{17}$	1.13748
350	$.2927 \times 10^{20}$	1.13663
400	$.1410 \times 10^{23}$	1.13599

Table 9.1b. Numerical values of $\|C_J^n\|_2$ and $\|C_J^n\|_2^{1/n}$ for matrix (4.5b) with $\alpha = 0.5, \nu = 1.1$.

$J = n$	$\|C_J^n\|_2$	$\|C_J^n\|_2^{1/n}$
10	$.5138 \times 10^2$	1.48279
50	$.5140 \times 10^8$	1.42633
100	$.1914 \times 10^{16}$	1.42174
150	$.7494 \times 10^{23}$	1.42069
200	$.2994 \times 10^{31}$	1.42030
250	$.1209 \times 10^{39}$	1.42013
300	$.4912 \times 10^{46}$	1.42005
350	$.2004 \times 10^{54}$	1.42001
400	$.8200 \times 10^{61}$	1.42000

Table 9.1c. Numerical values of $\|C_J^n\|_2$ and $\|C_J^n\|_2^{1/n}$ for matrix (4.5b) with $\alpha = 0.5, \nu = -0.5$.

$J = n$	$\|C_J^n\|_2$	$\|C_J^n\|_2^{1/n}$
10	$.4325 \times 10^2$	1.45748
50	$.6839 \times 10^3$	1.13946
100	$.2045 \times 10^4$	1.07921
150	$.3827 \times 10^4$	1.05654
200	$.5946 \times 10^4$	1.04441
250	$.8356 \times 10^4$	1.03678
300	$.1102 \times 10^5$	1.03151
350	$.1393 \times 10^5$	1.02764
400	$.1705 \times 10^5$	1.02466

Table 9.1d are numerical values of $\|C_J^n\|_2$ and $\|C_J^n\|_2^{1/n}$. The numbers are rather uninteresting and

$$\|C_J^n\|_2^{1/n} \to 1^-, \qquad n \to \infty.$$

Table 9.1d. Numerical values of $\|C_J^n\|_2$ and $\|C_J^n\|_2^{1/n}$ for matrix (4.5b) with $\alpha = 0.5, \nu = 0.5$.

$J = n$	$\|C_J^n\|_2$	$\|C_J^n\|_2^{1/n}$
10	.97985	.99797
50	.99915	.99998
100	.99987	1.00000
150	.99996	1.00000
200	.99998	1.00000
250	.99999	1.00000
300	.99999	1.00000
350	1.00000	1.00000
400	1.00000	1.00000

10. An Auxiliary Dirichlet Problem

The GKS stability analysis involves three auxiliary problems: the Cauchy problem and the left- and right-quarter plane problems. In this section we consider a fourth auxiliary problem which we call the Dirichlet problem on a finite domain. In the following section, the spectral radius of the Dirichlet problem will be used in a conjecture concerning the spectral radius of the matrix operator C_J associated with a discrete IBVP on a finite domain.

By the auxiliary Dirichlet problem we mean replacing all requisite NBSs by "over-specified" homogeneous boundary conditions. In other words, any grid function value u_j^n required by the interior scheme which falls outside the computational domain $0 \le x \le L$ is replaced by zero. We illustrate this by an example for the Lax-Wendroff scheme (3.1) with $c > 0$. If the Lax-Wendroff scheme is applied at $j = J$, then u_{J+1}^n is to the right of $x = L$. This is, of course, why an NBS is required at $j = J$. In the auxiliary Dirichlet problem one sets

$$u_{J+1}^n = 0 \tag{10.1}$$

which is an over-specified condition. The resulting discrete IBVP can be written in the vector-matrix form (4.1) where the matrix operator C is

$$\widetilde{C}_J = \begin{bmatrix} q & r & & & & \\ p & q & r & & O & \\ & \cdot & \cdot & \cdot & & \\ & & \cdot & \cdot & \cdot & \\ & & & \cdot & \cdot & \cdot \\ & O & & p & q & r \\ & & & & p & q \end{bmatrix} \tag{10.2}$$

where p, q, r are given by (4.3b) and the matrix is of order J. The above matrix is tridiagonal with constant entries and the eigenvalues can be determined analytically. For (10.2), one obtains

$$z_l = (1 - \nu^2) + i\nu\sqrt{1 - \nu^2}\cos[l\pi/(J+1)], \qquad l = 1, 2, \cdots, J.$$

The spectral radius $\rho(\widetilde{C}_J)$ depends weakly on J. For example, for $|\nu| \le 1$ and large J, one finds

$$\rho(\widetilde{C}_J) \approx \sqrt{1 - \nu^2}\,[1 + O(1/J^2)]. \tag{10.3}$$

For the finite-domain problem with L fixed, the asymptotic value $(J \to \infty)$ of the spectral radius is

$$\lim_{J \to \infty} \rho(\tilde{C}_J) = \sqrt{1 - \nu^2}, \qquad |\nu| \leq 1$$
$$= |\nu|\sqrt{\nu^2 - 1} + \nu^2 - 1, \quad |\nu| > 1. \qquad (10.4)$$

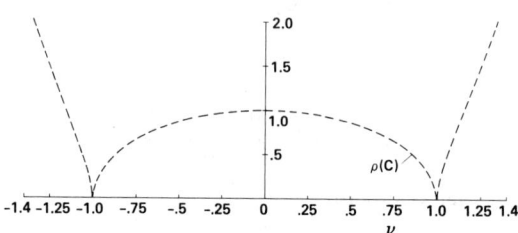

Fig. 10.1. $\rho(C)$ of the matrix (10.2).

This spectral radius is plotted in Fig. 10.1 as a function of ν. It follows directly from (10.4) that

$$\lim_{J \to \infty} \rho(\tilde{C}_J) \leq 1 \qquad for \qquad 0 \leq |\nu| \leq \sqrt{4/3}. \qquad (10.5)$$

The above problem for Lax-Wendroff was first considered by Parter [8]. He used the terminology over-determined rather than over-specified as used here. The spectral radius inequality (10.5) was given in Parter's paper. An over-specified boundary condition is a "proper" NBS if it is consistent with the PDE solution. For example, if an interior wave or disturbance has not reached the boundary and the analytical solution at $x = L$ is $u(L, t) = 0$, then (10.1) is a consistent NBS. Of course, in general, one would not know the PDE solution. In any case, our only interest here is to define an auxiliary problem used in the mathematical analysis.

11. Finite Domain vs. Quarter Plane: A Conjecture on the Eigenvalue Connection

In section 7 we considered the normal mode analysis for the finite-domain problem $0 \leq x \leq L$ with J spatial increments where $J\Delta x = L$. If we keep L fixed and refine the mesh by increasing J, then Δx must necessarily decrease. On the other hand, if we fix Δx and let $J \to \infty$, then $L \to \infty$ and we obtain the related right-quarter plane problem of the GKS theory where the spatial domain is $[0 \leq x < \infty)$. Likewise, if we fix Δx and the right boundary point $x = L$ and increase the number of increments to the left of L, then as $J \to \infty$ we obtain the related left-quarter plane problem of the GKS theory where the spatial domain is $(-\infty < x \leq L]$.

An interesting question is whether or not there is any connection between the eigenvalues of the finite domain $0 \leq x \leq L$ problem as $J \to \infty$ (or $\Delta x \to 0$) and the related right- and left-quarter plane problems. In particular, we are interested in the behavior of the modulus of the eigenvalue of maximum modulus, i.e., the spectral radius as $J \to \infty$. For example, the spectral radii for the Lax-Wendroff scheme for the NBS (3.2) with $\alpha = 0$ and $1/2$ are plotted in Figs. 6.1 and 6.3. In section 7 we noted that, in general, the spectral radius of the finite-domain problem cannot be determined analytically.

In this section we state a conjecture which relates the spectral radius of the finite-domain problem ($0 \leq x \leq L$) as $J \to \infty$ to the eigenvalues of the related left- and right-quarter plane problems and the spectral radius of the auxiliary Dirichlet problem defined in the previous section. In certain special cases such as the Lax-Wendroff scheme, the Dirichlet and quarter-plane eigenvalues can be obtained analytically by the normal mode analysis. In general, one must solve numerically for the roots of polynomial equations that arise in the normal mode analysis.

As a preliminary to stating our conjecture we introduce some notation. Consider first the auxiliary Dirichlet problem on a finite domain. Denote by $|z|_{dir}$ the spectral radius of the matrix operator $\widetilde{C}_J$ as $J \to \infty$:

$$|z|_{dir} = \lim_{J \to \infty} \rho(\widetilde{C}_J). \tag{11.1}$$

For the Lax-Wendroff scheme $|z|_{dir}$ is given analytically by (10.4) and is plotted in Fig. 10.1. In addition, for the discrete IBVP we apply the GKS normal mode analysis to the related left- and right-quarter plane problems. If there exists one or more GKS eigenvalues or generalized eigenvalues, let $|z|_{gks}$ denote the modulus of the maximum modulus eigenvalue or generalized eigenvalue. (This definition of $|z|_{gks}$ was given previously in section 9.) In addition, if there exists a solution $|\kappa| < 1, |z| < 1$, let the corresponding modulus of z be denoted by $|z|_{dec}$. This mode decays temporally and is usually not considered in the GKS analysis. Finally, let $|\hat{z}|_{gks}$ be defined as

$$|\hat{z}|_{gks} = \max(|z|_{gks}, |z|_{dec}). \tag{11.2}$$

If a GKS eigenvalue or generalized eigenvalue exists, then obviously $|z|_{gks} > |z|_{dec}$. As an example, for the Lax-Wendroff scheme with NBS (3.2) we plot $|\hat{z}|_{gks}$ as a function of ν for $\alpha = 1.5$ in Fig. 11.1.

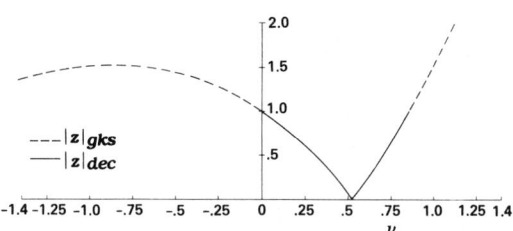

Fig. 11.1. $|\hat{z}|_{gks}$ for the Lax-Wendroff scheme (3.1) with NBS (3.2) for $\alpha = 1.5$.

The solid part of the curve corresponds to the decaying mode $|z|_{dec}$.

For the finite-domain discrete IBVP we make the following conjecture for the asymptotic value of the spectral radius as $J \to \infty$:

Conjecture 11.1. *Consider a discrete IBVP with matrix operator C_J on a finite domain ($J \Delta x = L$). The limit of $\rho(C_J)$ as $J \to \infty$ is*

$$\lim_{J \to \infty} \rho(C_J) = \max(|z|_{dir}, |\hat{z}|_{gks}). \tag{11.3}$$

If $|\hat{z}|_{gks}$ does not exist, then we write

$$\max(|z|_{dir}, |\hat{z}|_{gks}) = |z|_{dir}.$$

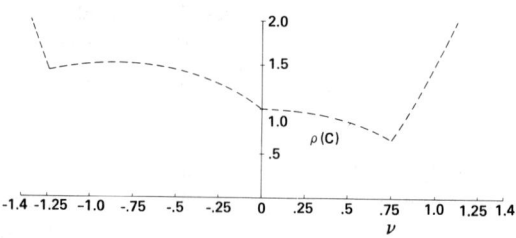

Fig. 11.2. Asymptotic spectral radius of the matrix (4.5b) for $\alpha = 1.5$.

As an example, if we overlay Figs. 11.1 and 10.1 and apply (11.3), we obtain the spectral radius shown in Fig. 11.2. To plotting accuracy, the curve shown in Fig. 11.2 agrees with numerically computed values of $\rho(C_J)$ for $J = 30$.

An interesting computational result is the following. If a GKS eigenvalue or generalized eigenvalue exists and

$$\max(|z|_{dir}, |\hat{z}|_{gks}) = |z|_{gks},$$

then there is a very rapid convergence (with increasing J) of the spectral radius $\rho(C_J)$ of the finite-domain problem to $|z|_{gks}$. We illustrate this with two numerical examples. For $\alpha = 1.5$ and $\nu = -0.5$ there is a GKS eigenvalue and $|z|_{gks} = 1.437500$. In Table 11.1 we list numerically computed values of $\rho(C_J)$ for the finite-domain problem as a function of J. The matrix C_J is given by (4.5b). For $\alpha = 0.0$ and $\nu = -0.5$ there is a GKS generalized eigenvalue and hence $|z|_{gks} = 1.0$. In Table 11.2 we tabulate computed values of $\rho(C_J)$ for the finite-domain problem as a function of J.

Table 11.1 Spectral radius as a function of J for the matrix (4.5b) with $\alpha = 1.5$ and $\nu = -0.5$.

J	$\rho(C_J)$
4	1.428343
6	1.437442
8	1.437500
10	1.437500

Table 11.2. Spectral radius as a function of J for the matrix (4.5b) with $\alpha = 0.0$ and $\nu = -0.5$.

J	$\rho(C_J)$
4	1.048752
6	1.013880
8	1.002809
10	1.000431
12	1.000059
14	1.000008
16	1.000001
18	1.000000

12. Summary

In this paper we have attempted to establish an algebraic test for Lax-Richtmyer stability of finite-difference approximations to IBVPs. Lemma 9.1 and Conjecture 9.1 relate the uniform boundedness condition of the Lax-Richtmyer stability definition to algebraic tests of the von Neumann analysis and the GKS normal mode analysis. In addition, Conjecture 11.1 relates the spectral radius of the matrix operator C_J (for the finite-domain problem) as $J \to \infty$ to the spectral radius of an auxiliary Dirichlet problem and the eigenvalues of the normal mode analysis for the left- and right-quarter plane problems. The conjectures have been corroborated by numerical examples.

References

[1] S. K. Godunov and V. S. Ryabenkii, "Special stability criteria of boundary value problems for non-selfadjoint difference equations," Russ. Math. Surv. 18, 1-12 (1963).

[2] B. Gustafsson, H.-O. Kreiss and A. Sundström, "Stability theory of difference approximations for mixed initial boundary value problems. II," Mathematics of Computation 26, 649-686 (1972).

[3] E. Isaacson, H. B. Keller, "Analysis of Numerical Methods," John Wiley and Sons, New York, 1966.

[4] H.-O. Kreiss, "Difference approximations for the initial-boundary value problem for hyperbolic differential equations," "Numerical Solutions of Nonlinear Differential Equations, Proceedings Adv. Symposium," Madison, Wis., John Wiley and Sons, New York, 141-166 (1966).

[5] H.-O. Kreiss, "Stability theory for difference approximations of mixed initial boundary value problems. I," Mathematics of Computations 22, 703-714 (1968).

[6] S. Osher, "Stability of difference approximations of dissipative type for mixed initial-boundary value problems. I," Mathematics of Computations 23, 335-340 (1969).

[7] S. Osher, "Systems of difference equations with general homogeneous boundary conditions," Trans. Am. Math. Soc. 137, 177-201 (1969).

[8] S. V. Parter, "Stability, convergence, and pseudo-stability of finite-difference equations for an over-determined problem," Numerische Mathematik 4, 277-292 (1962).

[9] R. Richtmyer and K. Morton, "Difference Methods for Initial Value Problems," Interscience Publishers, New York, 1967.

[10] L. N. Trefethen, "Wave Propagation and Stability for Finite Difference Schemes," Thesis, Stanford U., Stanford, Calif., 1982.

[11] L. N. Trefethen, "Instability of Difference Models for Hyperbolic Initial Boundary Value Problems," Comm. Pure Appl. Math. 37, 329-367 (1984).

[12] R. S. Varga, "Matrix Iterative Analysis," Prentice-Hall, Inc., Englewood Cliffs, 1962.

[13] R. F. Warming, R. M. Beam and H. C. Yee, "Stability of difference approximations for initial-boundary-value problems," Proceedings of the Third International Symposium on Numerical Methods for Engineering, Paris, France, March 1983, 1, Pluralis, Paris, 93-117.

[14] H. C. Yee, "Numerical Approximation of Boundary Conditions with Applications to Inviscid Equations of Gas Dynamics," NASA TM-81265, 1981.

CONSTRUCTION OF A CLASS OF SYMMETRIC TVD SCHEMES

H.C. Yee[†]

MS 202A-1, NASA Ames Research Center
Moffett Field, California, 94035

Abstract

A one-parameter family of second-order explicit and implicit total variation diminishing (TVD) schemes is reformulated so that a simpler and wider group of limiters is included. The resulting scheme can be viewed as a symmetrical algorithm with a variety of numerical dissipation terms that are designed for weak solutions of hyperbolic problems. This is a generalization of Roe and Davis's recent works to a wider class of symmetric schemes other than Lax-Wendroff. The main properties of the present class of schemes are that they can be implicit, and, when steady-state calculations are sought, the numerical solution is independent of the time step.

I. Introduction

The notion of total variation diminishing (TVD) schemes was introduced by Harten [1,2]. He derived a set of sufficient conditions which are very useful in checking or constructing second-order TVD schemes. The main mechanism that is currently in use for satisfying TVD sufficient conditions involves some kind of limiting procedure. There are generally two types of limiters: namely slope limiters [3] and flux limiters [4-6]. For a slope limiter one imposes constraints on the gradients of the dependent variables. In contrast, for a flux limiter one imposes constraints on the gradients of the flux functions. For constant coefficients, the two type of limiters are equivalent. The main property of a TVD scheme is that, unlike monotone schemes, it can be second-order accurate and is oscillation-free across discontinuities (when applied to nonlinear scalar hyperbolic conservation laws and constant coefficient hyperbolic systems). Sweby [5] and Roe [6] constructed a class of limiters as a function of the gradient ratio. Most of the current limiters in used are equivalent to members of this class.

Although TVD schemes are designed for transient applications, they also have been applied to steady-state problems [6-10]. It is well known that explicit methods are usually easier to program and often require less storage than implicit methods, but can suffer a loss of efficiency when the time step is restricted by stability rather than accuracy. It is also commonly known that it is not very useful to extend Lax-Wendroff-type schemes to implicit methods, since the resulting schemes are not suitable for steady-state calculations. This is due to the fact that the steady-state solution will depend on the time step. Roe has

---
[†]Research Scientist, Computational Fluid Dynamics Branch

recently proposed a very enlightening generalized formulation of TVD Lax-Wendroff schemes [11]. Roe's result, in turn, is a generalization of Davis's work [12]. It was the investigation of these schemes which prompted the work of this paper. Their formulation has great potential for transient applications, but as it stands is not suitable for an extension to implicit methods.

The aim of this paper is to incorporate the results of Roe [11], with minor modification, to a one-parameter family of explicit and implicit TVD schemes [2,8-9] so that a wider group of limiters can be represented in a general but rather simple form which is at the same time suitable for steady-state applications. The final scheme can be interpreted as a three-point, spatially central difference explicit or implicit scheme which has a whole variety of more rational numerical dissipation terms than the classical way of handling shock-capturing algorithms. In other words, it is a non-upwind TVD scheme or a symmetric TVD scheme. The proposed scheme can be used for time-accurate or steady-state calculations. This note is a summary of a more detailed paper to be published [13]. Only constant coefficient scalar hyperbolic conservation laws are considered here. Extension of the schemes to nonlinear scalar and system of hyperbolic conservation laws, and numerical experiments with two-dimensional airfoil calculations are discussed in the full paper [13].

II. Preliminaries

In this section, a class of explicit and implicit TVD schemes [2] is reviewed. Harten's sufficient conditions for this class of schemes are also stated. This set of conditions is then ultilized in the subsequent sections to construct and reformulate the second-order explicit and implicit TVD schemes of Harten [1,2].

Consider the scalar hyperbolic conservation law

$$\frac{\partial u}{\partial t} + \frac{\partial f(u)}{\partial x} = 0, \tag{2.1}$$

where f is the flux and $a(u) = \partial f/\partial u$ is the characteristic speed. Let u_j^n be the numerical solution of (2.1) at $x = j\Delta x$ and $t = n\Delta t$, with Δx the spatial mesh size and Δt the time step. Consider a one-parameter family of five-point difference schemes in conservation form

$$u_j^{n+1} + \lambda\theta(h_{j+\frac{1}{2}}^{n+1} - h_{j-\frac{1}{2}}^{n+1}) = u_j^n - \lambda(1-\theta)(h_{j+\frac{1}{2}}^n - h_{j-\frac{1}{2}}^n), \tag{2.2}$$

where θ is a nonnegative parameter, $\lambda = \Delta t/\Delta x$, $h_{j\pm\frac{1}{2}}^n = h(u_{j\mp 1}^n, u_j^n, u_{j\pm 1}^n, u_{j\pm 2}^n)$, and $h_{j\pm\frac{1}{2}}^{n+1} = h(u_{j\mp 1}^{n+1}, u_j^{n+1}, u_{j\pm 1}^{n+1}, u_{j\pm 2}^{n+1})$. The function $h_{j+\frac{1}{2}}$ is commonly called a numerical flux function. Let

$$\bar{h}_{j+\frac{1}{2}} = (1-\theta)h_{j+\frac{1}{2}}^n + \theta h_{j+\frac{1}{2}}^{n+1} \tag{2.3}$$

be another numerical flux function. Then (2.2) can be rewritten as

$$u_j^{n+1} = u_j^n - \lambda(\bar{h}_{j+\frac{1}{2}} - \bar{h}_{j-\frac{1}{2}}). \tag{2.4}$$

Here $\bar{h}_{j+\frac{1}{2}} = \bar{h}(u_{j-1}^n, u_j^n, u_{j+1}^n, u_{j+2}^n, u_{j-1}^{n+1}, u_j^{n+1}, u_{j+1}^{n+1}, u_{j+2}^{n+1})$ and is consistent with the conservation law (2.1) in the following sense

$$\bar{h}(u, u, u, u, u, u, u, u) = f(u). \tag{2.5}$$

This one-parameter family of schemes contains implicit as well as explicit schemes. When $\theta = 0$, (2.2) is an explicit method. When $\theta \neq 0$, (2.2) is an implicit scheme. For example, if $\theta = 1/2$, the time-differencing is the trapezoidal formula, and if $\theta = 1$, the time-differencing is the backward Euler method. To simplify the notation, rewrite equation (2.2) as

$$L \cdot u^{n+1} = R \cdot u^n, \tag{2.6}$$

where L and R are the following finite-difference operators:

$$(L \cdot u)_j = u_j + \lambda\theta(h_{j+\frac{1}{2}} - h_{j-\frac{1}{2}}) \tag{2.7a}$$

$$(R \cdot u)_j = u_j - \lambda(1 - \theta)(h_{j+\frac{1}{2}} - h_{j-\frac{1}{2}}). \tag{2.7b}$$

The total variation of a mesh function u^n is defined to be

$$TV(u^n) = \sum_{j=-\infty}^{\infty} |u_{j+1}^n - u_j^n| = \sum_{j=-\infty}^{\infty} |\Delta_{j+\frac{1}{2}} u^n|, \tag{2.8}$$

where $\Delta_{j+\frac{1}{2}} u^n = u_{j+1} - u_j$. Here the general notation convention

$$\Delta_{j+\frac{1}{2}} z = z_{j+1} - z_j \tag{2.9}$$

for any mesh function z is used. The numerical scheme (2.2) for an initial-value problem of (2.1) is said to be TVD if

$$TV(u^{n+1}) \leq TV(u^n). \tag{2.10}$$

The following sufficient conditions for (2.2) to be a TVD scheme are due to Harten [2]:

$$TV(R \cdot u^n) \leq TV(u^n) \tag{2.11a}$$

and

$$TV(L \cdot u^{n+1}) \geq TV(u^{n+1}). \tag{2.11b}$$

Assume the numerical flux h in (2.2) is Lipschitz continuous and (2.2) can be written as

$$u_j^{n+1} - \lambda\theta\left(\tilde{C}^-_{j+\frac{1}{2}}\Delta_{j+\frac{1}{2}}u - \tilde{C}^+_{j-\frac{1}{2}}\Delta_{j-\frac{1}{2}}u\right)^{n+1} = u_j^n + \lambda(1-\theta)\left(\tilde{C}^-_{j+\frac{1}{2}}\Delta_{j+\frac{1}{2}}u - \tilde{C}^+_{j-\frac{1}{2}}\Delta_{j-\frac{1}{2}}u\right)^n, \quad (2.12)$$

where $\tilde{C}^{\mp}_{j\pm\frac{1}{2}} = \tilde{C}^{\mp}(u_j, u_{j\pm 1}, u_{j\pm 2})$ or possibly $\tilde{C}^{\mp}_{j\pm\frac{1}{2}} = \tilde{C}^{\mp}(u_{j\mp 1}, u_j, u_{j\pm 1}, u_{j\pm 2})$ are some bounded functions. Then Harten further showed that sufficient conditions for (2.11) are

(a) if for all j

$$C^\pm_{j+\frac{1}{2}} = \lambda(1-\theta)\tilde{C}^\pm_{j+\frac{1}{2}} \geq 0 \tag{2.13a}$$

$$C^+_{j+\frac{1}{2}} + C^-_{j+\frac{1}{2}} = \lambda(1-\theta)(\tilde{C}^+_{j+\frac{1}{2}} + \tilde{C}^-_{j+\frac{1}{2}}) \leq 1, \tag{2.13b}$$

and

(b) if for all j

$$-\infty < C \leq -\lambda\theta\tilde{C}^\pm_{j+\frac{1}{2}} \leq 0 \tag{2.14}$$

for some finite C. Conditions (2.13) and (2.14) are very useful in guiding the construction of second-order accurate TVD schemes which do not exhibit the spurious oscillation associated with the more classical second-order schemes.

Harten [1-2], Yee et al. and Yee [7-10] investigated a particular form of $C^\pm$. They have shown in a variety of numerical tests that the scheme is quite useful for gas-dynamic calculations. The recent work of Roe [11] suggests a wider class of flux limiters for the Lax-Wendroff-type of TVD schemes which with a minor modification is found to have an immediate application to scheme (2.2). The details will be discussed in the next two sections.

III. A Generalized Formulation of a Class of Symmetric Schemes

In this section, Roe's formulation is reviewed. Then, with a minor modification, his numerical flux is shown to be applicable to a larger class of symmetric schemes. Sufficient conditions for this new class of schemes to be TVD are derived for both the constant coefficient and nonlinear scalar hyperbolic equations.

3.1 Roe's TVD Lax-Wendroff Schemes

Roe [11] has recently developed a generalized formulation of TVD Lax-Wendroff schemes. The form of the schemes is the usual Lax-Wendroff plus a general conservative dissipation term designed in such a way that the final scheme is TVD. For $\partial f/\partial u = a = $ constant, his scheme is written as

$$u_j^{n+1} = u_j^n - \frac{1}{2}\nu(1+\nu)\Delta_{j-\frac{1}{2}}u - \frac{1}{2}\nu(1-\nu)\Delta_{j+\frac{1}{2}}u$$
$$- \frac{1}{2}|\nu|(1-|\nu|)(1-Q_{j-\frac{1}{2}})\Delta_{j-\frac{1}{2}}u$$
$$+ \frac{1}{2}|\nu|(1-|\nu|)(1-Q_{j+\frac{1}{2}})\Delta_{j+\frac{1}{2}}u. \tag{3.1}$$

Here $\nu = a\lambda = a\Delta t/\Delta x$. The first two terms represent the usual Lax-Wendroff scheme and the other two terms represent an additional conservative dissipation. The function $Q_{j+\frac{1}{2}}$ depends on three consecutive gradients $\Delta_{j-\frac{1}{2}}u$, $\Delta_{j+\frac{1}{2}}u$, $\Delta_{j+\frac{3}{2}}u$ and is of the form

$$Q_{j+\frac{1}{2}} = Q(r^-_{j+\frac{1}{2}}, r^+_{j+\frac{1}{2}}), \tag{3.2a}$$

where

$$r^-_{j+\frac{1}{2}} = \frac{\Delta_{j-\frac{1}{2}}u}{\Delta_{j+\frac{1}{2}}u}, \quad r^+_{j+\frac{1}{2}} = \frac{\Delta_{j+\frac{3}{2}}u}{\Delta_{j+\frac{1}{2}}u}. \tag{3.2b}$$

Here $r^\pm$ are not defined if $\Delta_{j+\frac{1}{2}}u = 0$. To avoid the use of extra logic in computer implementation, an equivalent representation will be discussed in reference [13]. If one assumes both Q and Q/r are always positive, then a set of sufficient conditions for (3.1) to be TVD is

$$Q_{j+\frac{1}{2}} < \frac{2}{1-|\nu|} \tag{3.3a}$$

$$\left(Q_{j+\frac{1}{2}}/r^-_{j+\frac{1}{2}}\right) < \frac{2}{|\nu|} \tag{3.3b}$$

$$\left(Q_{j+\frac{1}{2}}/r^+_{j+\frac{1}{2}}\right) < \frac{2}{|\nu|}. \tag{3.3c}$$

Two examples for the function Q are

$$Q(r^-, r^+) = \text{minmod}(1, r^-) + \text{minmod}(1, r^+) - 1, \tag{3.4}$$

and

$$Q(r^-, r^+) = \text{minmod}(1, r^-, r^+). \tag{3.5}$$

Normally the "minmod" function of two arguments is defined as

$$\text{minmod}(x, y) = \text{sgn}(x) \cdot \max\{0, \min[|x|, y \cdot \text{sgn}(x)]\}$$

but within this context

$$\text{minmod}(1, r^{\pm}) = \begin{cases} \min(1, r^{\pm}) & r^{\pm} > 0 \\ 0 & r^{\pm} \leq 0. \end{cases} \qquad (3.6)$$

Other equivalent forms of $Q(r^-, r^+)$ are discussed in Sweby [5] and Roe [6].

Scheme (3.1) is a reformulation of Davis's work [12] in a way which is easier to analyze and includes a class of TVD schemes not observed by Davis. The numerical flux denoted by $h_{j+\frac{1}{2}}^{LW}$ for (3.1) is

$$h_{j+\frac{1}{2}}^{LW} = \frac{1}{2}\left\{a(u_{j+1} + u_j) - [\lambda a^2 Q_{j+\frac{1}{2}} + |a|(1 - Q_{j+\frac{1}{2}})]\Delta_{j+\frac{1}{2}}u\right\}. \qquad (3.7)$$

Scheme (3.1) is second-order accurate in time and space. Observe that by setting $\theta = 0$ in (2.2) and by using (3.7) as the numerical flux, the resulting scheme is (3.1).

3.2 Schemes for Linear Scalar Hyperbolic Equations

If one is to use (3.7) as the numerical flux for (2.2) with $\theta \neq 0$, then the resulting scheme is only useful for transient calculations. For steady-state applications, either one has to restrict the time step in a manner similar to the explicit method or the steady-state solution will depend on the time step. It is emphasized here that the dependence on the time step in steady-state solutions occurs even though the value of Δt is similar to an explicit method. In this case Δt is most often of the same order as Δx; thus the dependence on Δt is less severe. The term that causes this undesirable property is the one with coefficient λ in equation (3.7). Therefore, besides considering the use of (3.7) as the numerical flux for (2.2) when $\theta = 0$, the numerical flux (3.7) with $\lambda a^2 Q_{j+\frac{1}{2}} = 0$ is also considered; i.e., the numerical flux is of the form

$$h_{j+\frac{1}{2}} = \frac{1}{2}\left[a(u_{j+1} + u_j) - |a|(1 - Q_{j+\frac{1}{2}})\Delta_{j+\frac{1}{2}}u\right]. \qquad (3.8)$$

Now the question is, will the new numerical flux (3.8) satisfy the sufficient conditions (2.11)? The answer is yes. It turns out that some of the Q functions that are suitable for the generalized TVD Lax-Wendroff scheme are also suitable for (3.8). The implication is that if one chose the proper Q function, the resulting scheme (2.2) together with (3.8) can be viewed as a symmetrical algorithm with a wide variety of numerical dissipation terms that satisfy the TVD property.

Now with the choice of (3.8), the corresponding $\widetilde{C}^{\pm}$ of equation (2.12) are

$$\widetilde{C}_{j-\frac{1}{2}}^{+} = a\left[1 - \frac{1}{2}Q_{j-\frac{1}{2}} + \frac{1}{2}\left(Q_{j+\frac{1}{2}} / r_{j+\frac{1}{2}}^{-}\right)\right], \qquad a > 0 \qquad (3.9a)$$

$$\widetilde{C}_{j+\frac{1}{2}}^{-} = |a|\left[1 - \frac{1}{2}Q_{j+\frac{1}{2}} + \frac{1}{2}\left(Q_{j-\frac{1}{2}} / r_{j-\frac{1}{2}}^{+}\right)\right], \qquad a < 0. \qquad (3.9b)$$

Therefore, sufficient conditions for this specific numerical flux function (3.8) to be TVD are

$$0 < \lambda(1-\theta)a\left[1 - \frac{1}{2}Q_{j-\frac{1}{2}} + \frac{1}{2}\left(Q_{j+\frac{1}{2}}/r_{j+\frac{1}{2}}^-\right)\right] < 1 \qquad a > 0 \qquad (3.10a)$$

$$0 < \lambda(1-\theta)|a|\left[1 - \frac{1}{2}Q_{j+\frac{1}{2}} + \frac{1}{2}\left(Q_{j-\frac{1}{2}}/r_{j-\frac{1}{2}}^+\right)\right] < 1 \qquad a < 0 \qquad (3.10b)$$

and

$$-\infty < -\lambda\theta\widetilde{C}_{j+\frac{1}{2}}^\pm \leq 0. \qquad (3.11)$$

For $0 \leq \theta \leq 1$ and $\nu \neq 0$, condition (3.10a) is satisfied if

$$Q_{j-\frac{1}{2}} - \left(Q_{j+\frac{1}{2}}/r_{j+\frac{1}{2}}^-\right) < 2 \qquad (3.12a)$$

$$\left(Q_{j+\frac{1}{2}}/r_{j+\frac{1}{2}}^-\right) - Q_{j-\frac{1}{2}} < \frac{2}{\lambda(1-\theta)a} - 2 \qquad (3.12b)$$

$$\lambda a < \frac{1}{1-\theta}, \qquad (3.12c)$$

and condition (3.10b) is satisfied if

$$Q_{j+\frac{1}{2}} - \left(Q_{j-\frac{1}{2}}/r_{j-\frac{1}{2}}^+\right) < 2, \qquad (3.12d)$$

$$\left(Q_{j-\frac{1}{2}}/r_{j-\frac{1}{2}}^+\right) - Q_{j+\frac{1}{2}} < \frac{2}{\lambda(1-\theta)|a|} - 2 \qquad (3.12e)$$

$$\lambda|a| < \frac{1}{1-\theta}. \qquad (3.12f)$$

Since $\lambda|a| \leq \frac{1}{1-\theta}$, the term $\frac{2}{\lambda(1-\theta)|a|} - 2$ is always positive. Therefore the same assumption as Roe can be made; i.e., assume both Q and Q/r are always positive. Then all one has to do is devise a function Q such that

$$Q_{j+\frac{1}{2}} < 2 \qquad (3.13a)$$

$$\left(Q_{j+\frac{1}{2}}/r_{j+\frac{1}{2}}^-\right) < \frac{2}{\lambda(1-\theta)|a|} - 2 \qquad (3.13b)$$

$$\left(Q_{j+\frac{1}{2}}/r_{j+\frac{1}{2}}^+\right) < \frac{2}{\lambda(1-\theta)|a|} - 2 \qquad (3.13c)$$

$$\lambda|a| < \frac{1}{1-\theta}. \qquad (3.13d)$$

With the above choice of Q, the last sufficient condition (3.11) is immediately satisfied. For instance, the two examples given in equations (3.4) and (3.5) satisfy condition (3.13).

For $\theta = 1/2$, scheme (2.2) together with (3.8) and (3.13) is second-order accurate in both space and time. The CFL-like restriction for (2.2) to be TVD in this case is 2. When $\theta = 1$, scheme (2.2) together with (3.8) and (3.13) is unconditionally TVD, but the resulting scheme is first-order in time and second-order in space. When $\theta = 0$, the scheme is explicit, and unlike Roe's schemes, is only first-order in time but second-order in space.

As noted before, the value $r^-_{j+\frac{1}{2}}$ (or $r^+_{j+\frac{1}{2}}$) is not defined if $\Delta_{j-\frac{1}{2}}u$ (or $\Delta_{j+\frac{3}{2}}u$) is finite and $\Delta_{j+\frac{1}{2}}u = 0$. For computer implementation purposes, it might be more convenient to define $Q_{j+\frac{1}{2}}\Delta_{j+\frac{1}{2}}u = \hat{Q}_{j+\frac{1}{2}}$, where $\hat{Q}_{j+\frac{1}{2}}$ is a function of $\Delta_{j-\frac{1}{2}}u$, $\Delta_{j+\frac{1}{2}}u$, and $\Delta_{j+\frac{3}{2}}u$, but not a ratio of those gradients. For this formulation and its extension to the nonlinear case, see reference [13].

3.3 Linearized Version of the Proposed Scheme for Constant Coefficient Equations

For $\theta \neq 0$, scheme (2.2) is implicit. Moreover, this is a genuinely nonlinear scheme in the sense that the final algorithm is nonlinear even for the constant coefficient case. The value of u^{n+1} is obtained as the solution of a system of nonlinear algebraic equations. To solve this set of nonlinear equations noniteratively, a linearized version of (2.2) together with (3.8) is considered. Substituting (3.8) in (2.2), one obtains

$$u_j^{n+1} + \frac{\lambda\theta}{2}\left[au_{j+1} - |a|(1 - Q_{j+\frac{1}{2}})\Delta_{j+\frac{1}{2}}u\right]^{n+1}$$
$$- \frac{\lambda\theta}{2}\left[au_{j-1} - |a|(1 - Q_{j-\frac{1}{2}})\Delta_{j-\frac{1}{2}}u\right]^{n+1} = \text{RHS of (2.2)}. \quad (3.14)$$

Here "RHS of (2.2)" means the right hand side of equation (2.2) with $h_{j+\frac{1}{2}}$ defined in (3.8). Locally linearizing the coefficients of $(\Delta_{j\pm\frac{1}{2}}u)^{n+1}$ in (3.14) by dropping the time index from $(n+1)$ to n, one gets

$$u_j^{n+1} + \frac{\lambda\theta}{2}\left[au_{j+1}^{n+1} - au_{j-1}^{n+1} - |a|(1 - Q^n_{j+\frac{1}{2}})\Delta_{j+\frac{1}{2}}u^{n+1}\right.$$
$$\left. + |a|(1 - Q^n_{j-\frac{1}{2}})\Delta_{j-\frac{1}{2}}u^{n+1}\right] = \text{RHS of (2.2)}. \quad (3.15)$$

Let $d_j = u_j^{n+1} - u_j^n$; i.e., the "delta" notation, equation (3.15) can be written as

$$e_1 d_{j-1} + e_2 d_j + e_3 d_{j+1} = -\lambda(h^n_{j+\frac{1}{2}} - h^n_{j-\frac{1}{2}}) \quad (3.16a)$$

where

$$e_1 = \frac{\lambda\theta}{2}\left[-a - |a|(1 - Q_{j-\frac{1}{2}})\right]^n \quad (3.16b)$$

$$\epsilon_2 = 1 + \frac{\lambda\theta}{2}\left[|a|(1-Q_{j-\frac{1}{2}}) + |a|(1-Q_{j+\frac{1}{2}})\right]^n \qquad (3.16c)$$

$$\epsilon_3 = \frac{\lambda\theta}{2}\left[a - |a|(1-Q_{j+\frac{1}{2}})\right]^n. \qquad (3.16d)$$

The linearized form (3.16) is a spatially five-point scheme and yet it is a tridiagonal system of linear equations. This is because at the $(n+1)$th time level, only three points are involved; i.e., $u_{j-1}^{n+1}, u_j^{n+1}, u_{j+1}^{n+1}$. Although the coefficients ϵ_i involve five points, they are at the nth time level.

It was found in reference [7] that when time-accurate TVD schemes are used as a relaxation method for steady-state calculations, the convergence rate is degraded if limiters are present on the implicit operator. For steady-state applications, one can obtain another TVD linearized form by setting $Q_{j\pm\frac{1}{2}} = 0$ in (3.16); i.e., by redefining (3.16) by

$$\epsilon_1 = \frac{\lambda\theta}{2}(-a - |a|) \qquad (3.17a)$$

$$\epsilon_2 = 1 + \lambda\theta(|a|) \qquad (3.17b)$$

$$\epsilon_3 = \frac{\lambda\theta}{2}(a - |a|). \qquad (3.17c)$$

Scheme (3.16a) together with (3.17) is spatially first-order accurate for the implicit operator and spatially second-order accurate for the explicit operator. Equation (3.17) is considered because no limiter is present for the implicit operator.

3.4 Scheme for Nonlinear Scalar Hyperbolic Conservation Laws

To extend the scheme to nonlinear scalar problems, one simply defines a local characteristic speed

$$a_{j+\frac{1}{2}} = \begin{cases} \Delta_{j+\frac{1}{2}}f/\Delta_{j+\frac{1}{2}}u & \Delta_{j+\frac{1}{2}}u \neq 0 \\ (\partial f/\partial u)|_{u_j} & \Delta_{j+\frac{1}{2}}u = 0 \end{cases} \qquad (3.18)$$

and redefines the $r_{j+\frac{1}{2}}^{\pm}$ in (3.2b) as

$$r_{j+\frac{1}{2}}^- = \frac{|a_{j-\frac{1}{2}}|\Delta_{j-\frac{1}{2}}u}{|a_{j+\frac{1}{2}}|\Delta_{j+\frac{1}{2}}u}, \qquad r_{j+\frac{1}{2}}^+ = \frac{|a_{j+\frac{3}{2}}|\Delta_{j+\frac{3}{2}}u}{|a_{j+\frac{1}{2}}|\Delta_{j+\frac{1}{2}}u}. \qquad (3.19)$$

Unlike the constant coefficient case, $a_{j+\frac{1}{2}}$ and $a_{j-\frac{1}{2}}$ are not always of the same sign. After considering all the possible combinations of the signs of the $a_{j+\frac{1}{2}}$ and $a_{j-\frac{1}{2}}$, a set of sufficient conditions on Q still can be of similar form to (3.13) and is

$$Q_{j+\frac{1}{2}} < 2 \qquad (3.20a)$$

$$\left(Q_{j+\frac{1}{2}}/r_{j+\frac{1}{2}}^-\right) < \frac{2}{\lambda(1-\theta)|a_{j-\frac{1}{2}}|} - 2 \qquad (3.20b)$$

$$\left(Q_{j+\frac{1}{2}}\big/r^+_{j+\frac{1}{2}}\right) < \frac{2}{\lambda(1-\theta)|a_{j+\frac{3}{2}}|} - 2 \qquad (3.20c)$$

$$\lambda|a_{j+\frac{1}{2}}| < \frac{1}{(1-\theta)}. \qquad (3.20d)$$

The numerical flux for the nonlinear case is

$$h_{j+\frac{1}{2}} = \frac{1}{2}\left[(f_{j+1} + f_j) - |a_{j+\frac{1}{2}}|(1 - Q_{j+\frac{1}{2}})\Delta_{j+\frac{1}{2}}u\right]. \qquad (3.21)$$

Observe that when $a_{j+\frac{1}{2}} = 0$, the scheme has zero dissipation. One way is to approximate $|a_{j+\frac{1}{2}}|$ by a Lipschitz continuous function [2]. For example, instead of using (3.21), one can use

$$h_{j+\frac{1}{2}} = \frac{1}{2}\left[(f_{j+1} + f_j) - \psi(a_{j+\frac{1}{2}})(1 - Q_{j+\frac{1}{2}})\Delta_{j+\frac{1}{2}}u\right]. \qquad (3.22)$$

Here ψ is a function of $a_{j+\frac{1}{2}}$ and is of the form

$$\psi(z) = \begin{cases} |z| & |z| \geq \epsilon \\ (z^2 + \epsilon^2)/2\epsilon & |z| < \epsilon \end{cases} \qquad (3.23)$$

or

$$\psi(z) = \begin{cases} |z| & |z| \geq \epsilon \\ \epsilon & |z| < \epsilon \end{cases}, \qquad (3.24)$$

where ϵ is a positive small number [9].

3.5 Alternate Scheme for the Nonlinear Scalar Hyperbolic Problem

A simpler but less rigorous way of extending the constant coefficient case to the nonlinear case is to define a local characteristic speed $a_{j+\frac{1}{2}}$ and keep the restriction on Q the same as in (3.18) and (3.20), but use the $r^\pm_{j+\frac{1}{2}}$ in (3.2b) instead of (3.19). The alternate form requires less computation than the previous more rigorous approach. The relative advantage and disadvantage between these two forms remain to be shown. However, numerical experiments with two-dimensional Euler equations of gas dynamics [7,9] show that the alternate form gives a better shock resolution than the former one (3.18)-(3.22).

As a side remark, a case of Harten's second-order explicit TVD scheme is contained in the class of limiters of Sweby [5] and Roe [6] and is equivalent to a case of Roe's second-order scheme of reference [6]. See Sweby's original manuscript [14] instead of the published version [5] for details. The numerical experiments of Yee et al. and Yee [7-9] with Harten's second-order TVD scheme indicate that the alternate form is favored over the more rigorous approach (3.18)-(3.22). This indication is further endorsed by Davis's numerical experiments [12] with similar examples. Since different limiters have different effects which are highly problem-dependent on the resolution of the numerical solutions, all these possibilities in extending to nonlinear equations require more extensive numerical testing before a clearer picture can be drawn.

3.6 Linearized Version of the Proposed Implicit Scheme for Nonlinear Equations

For the nonlinear case, the situation is slightly more complicated since the characteristic speed $\partial f/\partial u$ is no longer a constant. Substituting (3.22) in (2.2), one obtains

$$u_j^{n+1} + \frac{\lambda\theta}{2}\left[f_{j+1} - \psi(a_{j+\frac{1}{2}})(1 - Q_{j+\frac{1}{2}})\Delta_{j+\frac{1}{2}}u\right]^{n+1}$$
$$- \frac{\lambda\theta}{2}\left[f_{j-1} - \psi(a_{j-\frac{1}{2}})(1 - Q_{j-\frac{1}{2}})\Delta_{j-\frac{1}{2}}u\right]^{n+1} = \text{RHS of (2.2)}. \tag{3.25}$$

Unlike the constant coefficient case, one also has to linearize $f_{j\pm 1}^{n+1}$, $\psi(a_{j\pm\frac{1}{2}}^{n+1})$, and $Q_{j\pm\frac{1}{2}}^{n+1}$. Following the same procedure as in [9], two linearized versions of (3.25) are considered.

Linearized Nonconservative Implicit Form

Adding and substracting f_j^{n+1} on the left-hand-side of (3.25) and using the relation (3.18), one can rewrite (3.25) as

$$u_j^{n+1} + \frac{\lambda\theta}{2}\left[a_{j+\frac{1}{2}}^{n+1} - \psi(a_{j+\frac{1}{2}}^{n+1})(1 - Q_{j+\frac{1}{2}}^{n+1})\right]\Delta_{j+\frac{1}{2}}u^{n+1}$$
$$- \frac{\lambda\theta}{2}\left[-a_{j-\frac{1}{2}}^{n+1} - \psi(a_{j-\frac{1}{2}}^{n+1})(1 - Q_{j-\frac{1}{2}}^{n+1})\right]\Delta_{j-\frac{1}{2}}u^{n+1} = \text{RHS of (2.2)}. \tag{3.26}$$

By dropping the time index of the coefficients of $\Delta_{j\pm\frac{1}{2}}u^{n+1}$ from $(n+1)$ to n, (3.26) becomes

$$\bar{e}_1 d_{j-1} + \bar{e}_2 d_j + \bar{e}_3 d_{j+1} = -\lambda(h_{j+\frac{1}{2}}^n - h_{j-\frac{1}{2}}^n) \tag{3.27a}$$

where

$$\bar{e}_1 = \lambda\theta B^- \tag{3.27b}$$
$$\bar{e}_2 = 1 - \lambda\theta\left(B^- + B^+\right) \tag{3.27c}$$
$$\bar{e}_3 = \lambda\theta B^+ \tag{3.27d}$$

and

$$B^\pm = \frac{1}{2}\left[\pm a_{j\pm\frac{1}{2}} - \psi(a_{j\pm\frac{1}{2}})(1 - Q_{j\pm\frac{1}{2}})\right]^n \tag{3.27e}$$

Again equation (3.27) is a five-point scheme, and yet the coefficient matrix associated with the d_j's is tridiagonal. With this linearization, the method is no longer conservative. Therefore (3.27) is only applicable for steady-state calculations. Again, a spatially first-order accurate implicit operator similar to (3.17) can be obtained for (3.27) by setting $B^{\pm} = \frac{1}{2}[\pm a_{j\pm\frac{1}{2}} - \psi(\pm a_{j\pm\frac{1}{2}})]^n$. Since the limiter does not appear on the left-hand-side, improvement in efficiency over (3.17) might be possible [7,9]. This reduced form is especially useful for multidimensional, nonlinear, hyperbolic conservation laws.

Linearized Conservative Implicit Form

One can obtain a linearized conservative implicit form by using a local Taylor expansion about u^n and expressing $f^{n+1} - f^n$ in the following form

$$f_j^{n+1} - f_j^n = a_j^n (u_j^{n+1} - u_j^n) + O(\Delta t^2), \tag{3.28}$$

where $a_j^n = (\partial f/\partial u)_j^n$. Applying the first-order approximation of (3.28) and locally linearizing the coefficients of $(\Delta_{j\pm\frac{1}{2}} u)^{n+1}$ in (3.25) by dropping the time index from $(n+1)$ to n, one gets

$$u_j^{n+1} + \frac{\lambda\theta}{2}\left[a_{j+1}^n u_{j+1}^{n+1} - a_{j-1}^n u_{j-1}^{n+1} - \psi(a_{j+\frac{1}{2}}^n)(1 - Q_{j+\frac{1}{2}}^n)\Delta_{j+\frac{1}{2}} u^{n+1}\right.$$
$$\left. + \psi(a_{j-\frac{1}{2}}^n)(1 - Q_{j-\frac{1}{2}}^n)\Delta_{j-\frac{1}{2}} u^{n+1}\right] = \text{RHS of (2.2)}. \tag{3.29}$$

Letting $d_j = u_j^{n+1} - u_j^n$, equation (3.29) can be written as

$$e_1 d_{j-1} + e_2 d_j + e_3 d_{j+1} = -\lambda(h_{j+\frac{1}{2}}^n - h_{j-\frac{1}{2}}^n), \tag{3.30a}$$

where

$$e_1 = \frac{\lambda\theta}{2}\left[-a_{j-1} - \psi(a_{j-\frac{1}{2}})(1 - Q_{j-\frac{1}{2}})\right]^n \tag{3.30b}$$

$$e_2 = 1 + \frac{\lambda\theta}{2}\left[\psi(a_{j-\frac{1}{2}})(1 - Q_{j-\frac{1}{2}}) + \psi(a_{j+\frac{1}{2}})(1 - Q_{j+\frac{1}{2}})\right]^n \tag{3.30c}$$

$$e_3 = \frac{\lambda\theta}{2}\left[a_{j+1} - \psi(a_{j+\frac{1}{2}})(1 - Q_{j+\frac{1}{2}})\right]^n. \tag{3.30d}$$

The linearized form (3.30) is conservative and is a spatially five-point scheme with a tridiagonal system of linear equations. Scheme (3.30) is applicable to transient as well as steady-state calculations. But the form of $\tilde{C}^{\pm}_{j\pm\frac{1}{2}}$ for (3.30) is no longer the same as its nonlinear counter part. As of this writing, the conservative linearized form (3.30) has not been proven to be TVD.

For steady-state application, one can use a spatially first-order implicit operator for (3.30) by simply setting all the $Q_{j\pm\frac{1}{2}} = 0$; i.e., redefine (3.30b)-(3.30d) as

$$e_1 = \frac{\lambda\theta}{2}\left[-a_{j-1} - \psi(a_{j-\frac{1}{2}})\right]^n \tag{3.31a}$$

$$e_2 = 1 + \frac{\lambda\theta}{2}\left[\psi(a_{j-\frac{1}{2}}) + \psi(a_{j+\frac{1}{2}})\right]^n \tag{3.31b}$$

$$e_3 = \frac{\lambda\theta}{2}\left[a_{j+1} - \psi(a_{j+\frac{1}{2}})\right]^n. \tag{3.31c}$$

In reference [7,9] this type of linearization proved to be very useful for two-dimensional steady-state airfoil calculations.

Extension of the scalar scheme (3.14), (3.27), or (3.30) to systems of conservation laws can be accomplished by defining at each point a "local" system of characteristic fields, and then applying the scheme to each of the m scalar characteristic equations. Here m is the dimension of the hyperbolic system. Extension of the scalar scheme to higher than one-dimensional systems of conservation laws (for practical calculations) can be accomplished by an alternating direction implicit (ADI) method similar to the one described in Yee et al. and Yee [7,9]. See reference [13] for details.

IV. Concluding Remarks

The present paper was inspired by the work of Roe [11] and Davis [12], and is based on the work of Harten [1-2] and of Harten and the author [7-10]. A one-parameter family of explicit and implicit TVD schemes is reformulated so that a wider group of limiters is included. The current class of schemes as well as Roe and Davis's can be classified as non-upwind TVD schemes or symmetric TVD schemes. The main advantages of the present class of schemes over the ones suggested by Osher and Chakravarthy [15], Roe, or Davis are that: (a) a wider class of time-differencing is included; (b) the implicit scheme allows a natural linearized procedure for a noniterative implicit procedure, and thus might have a greater potential for practical applications, especially for "stiff" problems; (c) when applied to steady-state calculations, the numerical solution is independent of the time step. Furthermore, Roe's formulation can be considered as a member of this family by simply setting $\theta = 0$ and using the numerical fluxes (3.7).

The results of Roe, Davis, and the present formulation provide a more rational way of supplying additional numerical dissipation terms to the commonly known schemes such as the Lax-Wendroff type and some spatially symmetrical explicit and implicit types of schemes. Here the amount of work required to modify existing computer codes with the suggested numerical dissipation terms varies from very minor changes to moderate yet straightforward computer programming. The potential of improving the robustness and accuracy of a wide variety of physical applications is worth the effort of further pursuing the implementation of these ideas into the many existing user-oriented computer codes.

References

[1] A. Harten, A High Resolution Scheme for the Computation of Weak Solutions of Hyperbolic Conservation Laws, NYU Report, Oct., 1981, and J. Comp. Phys., **49**, 357-393 (1983).

[2] A. Harten, On a Class of High Resolution Total-Variation-Stable Finite-Difference Schemes, NYU Report, Oct., 1982; SIAM J. Num. Anal., **21**, 1-23 (1984).

[3] B. van Leer, Towards the Ultimate Conservation Difference Scheme II, Monotonicity and Conservation Combined in a Second Order Scheme, J. Comp. Phys. **14**, 361-370 (1974).

[4] J.P. Boris and D.L. Book, Flux Corrected Transport. I. SHASTA, A Fluid Transport Algorithm That Works, J. Comp. Phys., **11**, 38-69 (1973).

[5] P.K. Sweby, High Resolution Schemes Using Flux Limiters for Hyperbolic Conservation Laws, SIAM J. Num. Analy., **21**, 995-1011 (1984).

[6] P.L. Roe, Some Contributions to the Modelling of Discontinuous Flows , Proceedings of the AMS-SIAM Summer Seminar on Large-Scale Computation in Fluid Mechanics June 27-July 8, 1983, Lectures in Applied Mathematics, **22** (1985).

[7] H.C. Yee, R.F. Warming and A. Harten, Implicit Total Variation Diminishing (TVD) Schemes for Steady-State Calculations, AIAA Paper No. 83-1902, Proc. of the AIAA 6th Computational Fluid Dynamics Conference, Danvers, Mass., July, 1983, also in J. Comp. Phys., **57**, 327-360 (1985).

[8] H.C. Yee, R.F. Warming and A. Harten, Application of TVD Schemes for the Euler Equations of Gas Dynamics, Proceedings of the AMS-SIAM Summer Seminar on Large-Scale Computation in Fluid Mechanics June 27-July 8, 1983, Lectures in Applied Mathematics, **22** (1985).

[9] H.C. Yee, Linearized Form of Implicit TVD Schemes for the Multidimensional Euler and Navier-Stokes Equations, *Advances in Hyperbolic Partial Differential Equations*, a special issue of Intl. J. Comp. Math. Appl., to appear.

[10] H.C. Yee and A. Harten, Implicit TVD Schemes for Hyperbolic Conservation Laws in Curvilinear Coordinates , AIAA Paper No. 85-1513-CP, Proc. of the AIAA 7th Computational Fluid Dynamics Conference, Cinn., Ohio, July 15-17, 1985.

[11] P.L. Roe, Generalized Formulation of TVD Lax-Wendroff Schemes, ICASE Report No. 84-53, October 1984.

[12] S.F. Davis, TVD Finite Difference Schemes and Artificial Viscosity, ICASE Report No. 84-20, June 1984.

[13] H.C. Yee, Construction of Explicit and Implicit Symmetric TVD Schemes and Their Applications, J. Comp. Phys., to appear.

[14] P.K. Sweby, High Resolution Schemes Using Flux Limiters for Hyperbolic Conservation Laws,

U.C.L.A. Report, June 1983, Los Angeles, Calif.

[15] S. Osher and S. Chakravarthy, High Resolution Schemes and the Entropy Condition, SIAM J. Num. Analy., **21**, 955-984 (1984).

Table of Contents from Other Volumes from the Program in Continuum Physics and Partial Differential Equations

Homogenization and effective moduli of materials

October 22 – October 26, 1984

J. L. Ericksen
D. Kinderlehrer
R. Kohn
J.-L. Lions
Conference Committee

M. Bendsoe	Generalized plate models and optimal design
D. Bergman	The effective dielectric coefficient of a composite medium: rigorous bounds from analytic properties
J. Berryman	Variational bounds on Darcy's constant
M. M. Carroll	Micromodeling of void growth and collapse
R. Kohn and G. W. Milton	On bounding the effective conductivity of anisotrpic composites
R. Kohn and M. Vogelius	Thin plates with rapidly varying varying thickness and their relation to structural optimization
G. W. Milton	Modelling the properties of composites by laminates
R. Caflisch, M. Miksis, G. Papinicolaou, and L. Ting	Waves in bubbly liquids
A. C. Pipkin	Some examples of crinkles
P. Sheng	Microstructures and the physical properties of composites
L. Tartar	Remarks on homogenization
J. R. Willis	Variational estimates for the overall response of an inhomogeneous nonlinear dielectric

Theory and applications of liquid crystals

January 21 - January 25, 1985

J. L. Ericksen
D. Kinderlehrer
Conference Committee

Tentative contributors: Berry, G., Brezis, H., Capriz, G., Choi, H. I., Cladis, P., Di Benedetto, E., Gulliver, R., Hardt. R. and Kinderlehrer, D., Leslie, F., Luskin, M., Miranda, M., Ryskin, G., Sethna, J., and Spruck, J.

Amorphous polymers and non-newtonian fluids

March 4 - March 8, 1985

J. L. Ericksen
D. Kinderlehrer
M. Tirrell
S. Prager
Conference Committee

Tentative contributors: Bird, R., Caswell, B., Dafermos, C., Hrusa, W. and Renardy, M., Joseph, D. D., Kearsley, E., Marcus, M. and Mizel, V., Nohel, J. and Renardy, M., Rabin, M., and Tirrell, M., Wool, R. P.

Metastability and incompletely posed problems

May 6 – May 10, 1985

S. Antman
J. L. Ericksen
D. Kinderlehrer
I. Müller
    Conference Committee

Antman, S.	Dissipative mechanisms
Ball, J.	Does rank-one convexity imply quasiconvexity?
Brezis, H.	Metastable harmonic maps
Calderer, M.	Bifurcation of constrained problems in thermoelasticity
Chipot, M. and Luskin, M.	The compressible Reynolds' lubrication equation
Ericksen, J.	Twinning of crystals I
Evans, L. C.	Quasiconvexity and partial regularity in the calculus of variations
Goldenfeld, N.	Introduction to pattern selection in dendritic solidification
Gurtin, M.	Some results and conjectures in the gradient theory of phase transitions
James, R.	The stability and metastability of quartz
Kenig, C.	Continuation theorems for Schrodinger operators
Kinderlehrer, D.	Twinning of crystals II
Lions, J. L.	Asymptotic problems in distributed systems

Metastability and incompletely posed problems

Liu, T.P. Stability of nonlinear waves

Mosco, U. Variational stability and relaxed Dirichlet problems

Müller, I. Simulation of pseudoelastic behaviour in a system of rubber balloons

Pitteri, M. A contribution to the description of natural states for elastic crystalline solids

Rogers, R. Nonlocal problems in electromagnetism

Salsa, S. The Nash-Moser technique for an inverse problem in potential theory related to geodesy

Vazquez, J. Hyperbolic aspects in the theory of the porous medium equation

Vergara-Caffarelli, G. Green's formulas for linearized problems with live loads

Wright, T. Some aspects of adiabatic shear bands

(tentative contents)

Dynamical problems in continuum physics

June 3 - June 7, 1985

J. Bona
C. Dafermos
J. L. Ericksen
D. Kinderlehrer
     Conference Committee

Amick, C.   Solitary water-waves in the presence of surface tension

Beals, M.   Presence and absence of weak singularities in nonlinear waves

Beatty, M.   Some dynamical problems in continuum physics

Beirao da Veiga, H.   Existence and asymptotic behavior for strong solutions of the Navier Stokes equations in the whole space

Bell, J.   A confluence of experiment and theory for waves of finite strain in the solid continuum

Bona, J.   Shallow water waves and sediment transport

Chen, P.   Classical piezoelectricity: is the theory complete?

Keller, J.   Acoustoelasticity

McCarthy, M.   One dimensional finite amplitude pulse propagation in electroelastic semiconductors

Müller, I.   Extended thermodynamics of ideal gases

Pego, R.   Phase transitions in one dimensional nonlinear viscoelasticity: admissibility and stability

Dynamical problems in continuum physics

Shatah, J.   Recent advances in nonlinear wave equations

Slemrod, M.   Dynamic phase transitions and compensated compactness

Spagnolo, S.   Some existence, uniqueness, and non-uniqueness results
                for weakly hyperbolic equations in Gevrey classes

Strauss, W.   On the dynamics of a collisionless plasma

(tentative contents)

RAYMOND H. FOGLER LIBRARY

DATE DUE

**BOOKS ARE SUBJECT TO
RECALL AFTER TWO WEEKS**

~~MAY 2 8 1987~~